Mining Impact on Soil and Water Resources

The subject matter of this book is divided into two sections detailing Soil (focussing on geochemistry, contamination, and remediation) and Water (focussing on hydrogeochemistry, crisis, desertification, and modelling) including case studies, review studies, and essential soil remediation and water. It also explores management practices to explain soil–water interaction, acid mine drainage problems, and contamination levels in water and soil resources. The main topics discussed include soil–water interaction, mining impact on water and soil geochemistry, mining impact on water and soil quality, martials impact, groundwater level depletion, contamination evaluation, health risk assessment, water treatment, soil remediation, remote sensing and geographical information system (GIS), contaminant transport modelling, and water/soil resources management. Emphasis is also given to the new approach to sustainable water and soil resources management.

Features:

- Integrates research in soil and environmental resources management in mining.
- Describes soil resources management in mining regions.
- Covers water geochemistry and contaminant transport modelling.
- Provides solutions for acid mine drainage problems.
- Includes the role of remote sensing and GIS.

This book is aimed at researchers and graduate students in soil resources management, mining, and environment science.

Mining Impact on Soil and Water Resources

Edited by

Ashwani Kumar Tiwari, T.N. Singh,
and Abhay Kumar Singh

CRC Press
Taylor & Francis Group
Boca Raton London New York

CRC Press is an imprint of the
Taylor & Francis Group, an **informa** business

Designed cover image: © Ashwani Kumar Tiwari

First edition published 2025
by CRC Press
2385 NW Executive Center Drive, Suite 320, Boca Raton FL 33431

and by CRC Press
4 Park Square, Milton Park, Abingdon, Oxon, OX14 4RN

CRC Press is an imprint of Taylor & Francis Group, LLC

ISBN: 9781032530154 (hbk)
ISBN: 9781032580975 (pbk)
ISBN: 9781003442554 (ebk)

DOI: 10.1201/9781003442554

Typeset in Times
by codeMantra

To Family Members and Friends

Editors

Ashwani Kumar Tiwari

T.N. Singh

Abhay Kumar Singh

Contents

SECTION II Water

Preface

The Earth is the planet we live on and is the only known planet in the universe to support life. Our Earth is filled with invaluable minerals, ores, fuel, and oil resources for the survival of human life on the Earth. For extracting, we mine the Earth's surface, which directly or indirectly impacts natural resources (i.e., air, water, soil, and biodiversity).

With human interference, depletion and contamination of water and soil are untoward incidents that are leading to tremendous stress and are the most severe environmental concerns for society. Groundwater depletion and soil contamination are severe concerns to the people and policymakers of the world.

Particularly in mining regions, due to mining and related activities, the quality and quantity of water and soil resources are threatened in several parts of the globe. Mine water discharge usually contains unacceptable levels of major ions, metals, and other contaminants that adversely affect the region's water resources. Surface mine expansion and huge overburden piling adversely affect soil health. Moreover, after mining, the mining closure plan/practice is not as per SOP by the concerned authorities in many regions. Due to that, soil, water, and other environmental components are under the solemn threat of contamination.

As we know, soil–water interaction is a natural and common phenomenon; if mining activities contaminate an area's soil directly/indirectly, we contaminate the water resources of that area. Hence, significant ions and metals are common environmental impurities, and their toxicity is problematic for the mining society. The concentrations of some major ions and metals beyond permissible limits in water and soil can harm ecosystems, plants, and animals. Several mining regions across the globe need to have more information/databases on soil–water interaction, acid mine drainage, and water and soil contamination.

Hence, the book, *Mining Impact on Soil and Water Resources*, aims to deliberate on a multilateral approach and address important soil and water-related issues of mining regions across the globe. This book presents the understanding of mining's impact on soil and water, so that most of the points, such as scientific, economic, societal, and environmental, have been covered. This book explains the macro- to micro-level principles to understand two energy and nutrients providing materials, soil and water, and assess the impacts of mining on these entities. In mining regions, other activities such as agriculture, domestic and power-producing industrial activities, and crucial accountable units are simultaneously involved. Still, beyond this, it becomes essential to understand the overall effects being put on the surrounding water and soil environment.

The book is divided into two critical sections: Soil and Water. The Soil section has six chapters covering significant topics such as soil geochemistry, mine waste dump impact, and soil contamination and remediation. The water section has six chapters covering important issues related to water contamination and advanced water resource management approaches in mining areas. The water section covers hydrogeochemistry, metal impact on human and aquatic life, water scarcity, and desertification and modelling.

The book covers Water and Soil Science, Mining, Environmental Sciences, and Geosciences research projects. The book will provide a platform for researchers from the above subjects to interact, share, and discuss concepts to address and manage the numerous issues related to soil and water in mining regions and the risks associated with human and aquatic life. Both sections of this book are anticipated to encourage and inspire meaningful discourse for helpful utilization of the experiences for sustainable mining development and management of soil and water resources.

The book is designed according to the topics/themes mentioned above and can benefit researchers and policymakers from mining, water and soil science, health risk, remediation, remote sensing and geographical information system (GIS), resource modelling, and management from across the globe.

We received an overwhelming response from researchers from Asia and Europe about our call for book chapters. The number of chapters received was significant and appropriate to the theme of the book.

We are grateful to all the authors of the chapters for their information and significant chapter contributions to the book, to reviewers for suggestions and comments to improve the chapter and the book and to the publisher team for their support. We hope this book's outcome will benefit the researchers, mining authorities, engineers, and decision-makers in soil and water resource management.

Editor Biographies

Ashwani Kumar Tiwari is working as an Assistant Professor in the School of Environmental Sciences at Jawaharlal Nehru University, New Delhi, India. His teaching and research areas are water resources management and GIS, hydrogeochemistry, pollution of water resources by geogenic and anthropogenic activities, groundwater–seawater interaction, water quality and health, and mining. He has been awarded the Indian National Chapter–International Association of Hydrogeologists (INC-IAH) Young Scientist Award 2022 for his research contribution to groundwater investigation and management. He is a member of the Indian National Young Academy of Science (INYAS). He is a Chair of the Research Funding Working Group of the Marie Curie Alumni Association (MCAA) and an Executive Member of the Governing Body of the INC-IAH. Also, he is a member of the Steering Committee of the Early Career Hydrogeologists Network (ECHN), IAH. He was a postdoctoral researcher at the Department of Environment, Land and Infrastructure Engineering, Politecnico di Torino, Turin, Italy, for around 4 years. He has done an MSc and an MPhil in Environmental Science and completed his PhD degree in Environmental Science from the Indian Institute of Technology (Indian School of Mines), Dhanbad, India. He was awarded Erasmus Mundus and Marie Skłodowska-Curie Actions Scholarships. He has published several research articles in various reputed international and national journals, and many worldwide workers very well appreciate his works in these respective fields. He has served as a reviewer for many international journals.

T.N. Singh is the Director of the Indian Institute of Technology Patna (IITP), India, and Ex-Vice-Chancellor of Mahatma Gandhi Kashi Vidyapith University, Varanasi, India. He is an expert in the fields of rock mechanics, mining geology, and clean energy. He received his PhD degree from the Institute of Technology (BHU) Varanasi, in 1991 and subsequently served the institute until 2003. He has received many prestigious awards, such as the National Mineral Award, the first P. N. Bose Mineral Award, the SEAGATE Excellence Award for Geo-Engineering, and the GSI Sesquicentennial Commemorative Award. He has nearly 30 years of experience in research and teaching, with 50 doctoral theses completed under his supervision, and has authored more than 418 publications in various journals and conferences of national and international repute, with total citations of over 17,490. He is on the governing and advisory councils of several national institutes and universities.

Abhay Kumar Singh is currently working as a chief scientist at the CSIR-Central Institute of Mining and Fuel Research, Dhanbad, India. He has obtained BSc (Hons) and MSc degrees in Geology from Banaras Hindu University (BHU), Varanasi, and MPhil and PhD in Environmental Science from Jawaharlal Nehru University (JNU), New Delhi, India. He has made a good contribution towards the understanding of the Earth surface processes and the impacts of mining and industrial activities on natural resources and their management. He has successfully completed a number of Science and Technology (S&T) and industry-sponsored projects related to rainwater chemistry, groundwater chemistry, groundwater management, mine water environment, and environmental impact assessment (EIA)/environmental management plan (EMP) study of coal and non-coal mines. He has published more than 100 research papers in National and International Journals/Seminar/Symposia, with total citations of over 5700. He has travelled to the United Kingdom, Nepal, Singapore, Malaysia, Italy, and Mauritius in academic-related pursuits. He is the recipient of the Indian Science Congress Young Science Award (1999), INSA Visiting Fellowship (2005), Commonwealth Academic Fellowship (2012), and the prestigious National Geoscience Award (2013).

Contributors

Natalya Abrosimova
Trofimuk Institute of Petroleum Geology and Geophysics
Siberian Branch of the Russian Academy of Sciences
Novosibirsk, Russia

Nikolay Abrosimov
Regional Scientific and Educational Mathematical Center
Tomsk State University
Tomsk, Russia
and
Sobolev Institute of Mathematics
Siberian Branch of the Russian Academy of Sciences
Novosibirsk, Russia

Uttam Biswas Antu
Department of Soil Science
Patuakhali Science and Technology University
Dumki, Patuakhali, Bangladesh

Abhishek Pandey Bharat
Academy of Scientific and Innovative Research (AcSIR)
Ghaziabad, India
and
CSIR-Central Institute of Mining and Fuel Research
Dhanbad, India

Svetlana Bortnikova
Trofimuk Institute of Petroleum Geology and Geophysics
Siberian Branch of the Russian Academy of Sciences
Novosibirsk, Russia

Subhash Chandra
Department of Civil Engineering, GITAM School of Technology
GITAM University
Visakhapatnam, India
and
Multidisciplinary Unit of Research on Translational Initiatives (MURTI)
GITAM University
Visakhapatnam, India

Alexey Edelev
Trofimuk Institute of Petroleum Geology and Geophysics
Siberian Branch of the Russian Academy of Sciences
Novosibirsk, Russia

Soma Giri
Department of Environmental Science
Central University of South Bihar
Gaya, India

Md. Kamrul Haque
Institute of Bangabandhu War of Liberation Bangladesh Studies
National University
Dhaka, Bangladesh

Md. Saiful Islam
Department of Soil Science
Patuakhali Science and Technology University
Dumki, Patuakhali, Bangladesh

Abu Reza Md. Towfiqul Islam
Department of Disaster Management
Begum Rokeya University
Rangpur, Bangladesh
and
Department of Development Studies
Daffodil International University
Dhaka, Bangladesh

Ramesh Janipella
Water Technology and Management Division
CSIR-NEERI
Nagpur, India

Tapos Kormoker
Faculty of Environmental Science and Disaster Management, Department of Emergency Management
Patuakhali Science and Technology University
Dumki, Patuakhali, Bangladesh

Prince Kumar
CSIR-Central Institute of Mining and Fuel Research
Dhanbad, India

Sazal Kumar
School of Environmental and Life Sciences
The University of Newcastle (UoN)
Callaghan, NSW, Australia

Suraj Kumar
School of Environmental Sciences
Jawaharlal Nehru University
New Delhi, India

Mukesh Kumar Mahato
Department of Environmental Studies
Lakshmibai College, University of Delhi
New Delhi, India

Swapan Mahato
CSIR-Central Institute of Mining and Fuel Research
Dhanbad, India

Isha Medha
Department of Metallurgical and Materials Engineering
Indian Institute of Technology Kharagpur
Kharagpur, India

Paras Ranjan Pujari
CSIR-National Environmental Engineering Research Institute (NEERI)
Nagpur, India

Anita Punia
Bhuparayan Research Foundation
Pilani, India

Resmi S
Geological Survey of India Northern Region
Lucknow, India

Sanjay Kr. Roy
CSIR-Central Institute of Mining and Fuel Research
Dhanbad, India

Olga Saeva
Trofimuk Institute of Petroleum Geology and Geophysics
Siberian Branch of the Russian Academy of Sciences
Novosibirsk, Russia

Anand Singh
CSIR-Central Institute of Mining and Fuel Research
Dhanbad, India

Md. Abu Bakar Siddique
Institute of National Analytical Research and Service (INARS)
Bangladesh Council of Scientific and Industrial Research (BCSIR)
Dhaka, Bangladesh

Rakesh Kr. Singh
CSIR-Central Institute of Mining and Fuel Research
Dhanbad, India

Abhay Kumar Singh
Water Resource Management Group
CSIR-Central Institute of Mining and Fuel Research
Dhanbad, India

Pramod Kumar Singh
Natural Resources and Environmental Management Group C
SIR-Central Institute of Mining and Fuel Research
Dhanbad, India

Saurabh Kumar Singh
Dayal Singh College, University of Delhi
New Delhi, India

T.N. Singh
Indian Institute of Technology Patna
Patna, India

Chandra Shekhar Singh
Department of Mining Engineering
Indian Institute of Technology (Banaras Hindu University)
Varanasi, India

Abhay Kumar Soni
CSIR-Central Institute of Mining and Fuel Research (CIMFR)
Nagpur, India

Mihaela Timofti
Faculty of Sciences and Environment, Department of Chemistry Physics and Environment
Dunarea de Jos University of Galati
Galati, Romania
and
REXDAN Research Infrastructure
Dunarea de Jos University of Galati
Galati, Romania

Ashwani Kumar Tiwari
School of Environmental Sciences
Jawaharlal Nehru University
New Delhi, India

Minhaz Uddin
Department of Environmental Science
Bangladesh Agricultural University
Mymensingh, Bangladesh

Ravi Kumar Umrao
School of Environmental Sciences
Jawaharlal Nehru University
New Delhi, India

Kartik Varwade
CSIR-Central Institute of Mining and Fuel Research
Dhanbad, India

Amit Kr. Verma
Department of Civil and Environmental Engineering
Indian Institute of Technology Patna
Patna, India

Nataliya Yurkevich
Trofimuk Institute of Petroleum Geology and Geophysics
Siberian Branch of the Russian Academy of Sciences
Novosibirsk, Russia

Section I

Soil

1 Metals Contamination in the Agricultural Soils of Mica Mining Areas of Jharkhand, India

Soma Giri, Mukesh Kumar Mahato, and Abhay Kumar Singh

1.1 INTRODUCTION

The possibility of metal deposition in agricultural soil has raised global concerns due to its hazardous nature and ability to accumulate in living organisms via the food pathway (Ma et al., 2015). According to Maxted et al. (2007), metals are possible contributors to the decline in soil fertility. The enrichment of metals in soil can be related to both weathering processes and human activities (Khan et al., 2015). The presence of heavy metals in soils can be ascribed to human activities, which encompass a range of actions such as industrial and vehicular operations, mining activities, waste disposal practices, and agricultural methods such as the utilisation of wastewater for irrigation and the application of chemical pesticides and fertilisers (Manzoor et al., 2006).

Soil has the potential to serve as both a storage and a receptacle for metals derived from human activities, such as mining, industrial operations, and vehicular emissions (Acosta et al., 2011). The presence of metals bound to soil facilitates their potential entry into the food chain, as they tend to accumulate in plants that thrive in such environments (Allibone et al., 2012). In addition, the occurrence of metals in soil has the potential to impact both groundwater and surface water sources through processes such as leaching and run-off, as discussed by Li et al. (2014). According to Izuora et al. (2011), the contamination of soil with metals can potentially endanger the health of both cattle and people. The presence of metals has been associated with several adverse health effects in humans, including cancer, renal disorders, and dysfunction of the neurological, gastrointestinal, immunological, and endocrine systems (Chen et al., 2015; Ganyaglo et al., 2019).

The mica mining areas in Jharkhand are situated within the Great Mica Belt of India, renowned for their superior-quality mica deposits. The mica belt is characterised by the predominance of minerals such as biotite, muscovite, apatite, and hornblende, which are commonly found in association with metals such as aluminium (Al), iron (Fe), and various others in traces. Hence, the presence of metal

DOI: 10.1201/9781003442554-2

contamination can be apprehended in the soil within the vicinity of the mica mining region. Considering this, the present investigation employed a diverse range of indices to assess the extent of metal contamination in the agricultural soils of the region. The study also aims to identify sources of metals in soil through the application of multivariate statistical analysis.

1.2 MATERIALS AND METHODS

1.2.1 Study Area

The study encompasses the mica mining regions located in Jharkhand, which is situated in India's Great Mica Belt. The study region is situated within the geographical coordinates of 84°50′–86°20′E longitude and 24°15′–24°50′N latitude. The climatic conditions in this region are primarily shaped by the southwest monsoon, which gives rise to three distinct seasons: summer, winter, and monsoon (rainy). According to the Central Ground Water Board (CGWB, 2013), the annual average temperature fluctuates between 5°C and 44°C, accompanied by an annual precipitation of 1,126 mm. The region encompasses a variety of geological formations spanning from the Archean to the Recent period. The main formations are Archeanproterozoic, pemocarboniferous and recent. The area comprises of Dharwanian rocks like quartz and schist containing Pegmatite veins in which good-quality mica is found. Archean granites and gneisses are also found. The primary litho-units observed in this area are muscovite-biotite schist, intrusive granites, phyllite-mica schist, and granite gneiss. In addition, dolerite, quartz pegmatite, and quartzite have been sporadically discovered with alluvium deposits found along river systems (CGWB, 2013).

1.2.2 Sampling and Analysis

In this study, soil samples were collected in triplicate from 25 specific locations which primarily consist of agricultural fields within the mica mining regions of Jharkhand (Figure 1.1). Each sample included approximately 1 kg of soil collected from the top layer of the soil (a depth of 0–15 cm). The samples were transported to the laboratory following appropriate packaging and labelling procedures and then exposed to air drying. Subsequently, a composite sample was carefully made for each location by appropriately mixing the three samples taken from the respective site. Subsequently, a soil sample weighing 100 g was collected with the coning and quartering technique. Following a 24-h oven drying period at a temperature of 110°C, the soil samples underwent grinding using a mortar and pestle and subsequently passed through a sieve with a mesh size of 200.

1.2.3 Laboratory Analysis

In total, 0.25 g of soil sample was acid digested using a microwave according to United States Environmental Protection Agency (USEPA) procedure 3052 (USEPA, 1996). This entire digesting process employs a mixture of acids (HNO_3, HF, HCl, and H_2O_2) described in detail by Giri and Singh (2017). Metal concentrations were

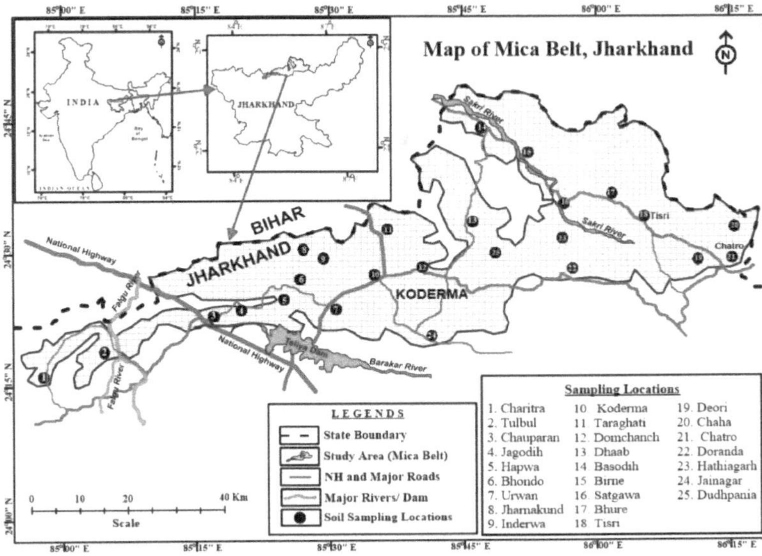

FIGURE 1.1 Sampling sites map of the study area. (Adapted from Giri et al., 2023.)

measured using inductively coupled plasma–mass spectrometry (Perkin Elmer Elan DRC-e ICP-MS).

1.2.4 Principal Component Analysis

Principal component analysis (PCA) is helpful to assess the sources of contaminants in different environmental matrices using a statistical approach. PCA is a method for reducing data variations by transforming the number of studied variables into a lesser number of artificial variables, which are represented by principal components (PCs). The PCs with eigenvalues more than one account for the greatest amount of data variance and are used for analysis since each PC is associated with a source of variables (contaminants) (Kolsi et al., 2013). Varimax rotation is used to reduce the influence of unimportant factors on the retrieved PCs (Closs and Nichol, 1975). A weighting component termed PC score, which measures the correlation between the factor and the variable, is attached to all the variables that were taken into consideration.

1.2.5 Calculation of Multiple Indices for Determining Anthropogenic Influence

1.2.5.1 Enrichment Factor

To determine the degree of anthropogenic influence on the metal content in the soil, the enrichment factor (EF) is calculated for each metal using Eq. (1.1):

$$\text{EF} = (C_X / C_{Fe})_s / (C_X / C_{Fe})_c \tag{1.1}$$

where C_X and C_{Fe} indicate the element X and Fe concentrations in the soil (s) and earth's crust (c), respectively. The metal concentrations were normalised to the textural characteristics of soils using iron as a reference element (Schiff and Weisberg, 1999; Turner and Millward, 2000). This work employs a five-category rating method to represent the degree of anthropogenic pollution (Giri et al., 2013; Sutherland, 2000; Kartal et al., 2006). The categories are as follows:

EF < 2 states deficiency to minimal contamination
EF = 2–5 moderate contamination
EF = 5–20 significant contamination
EF = 20–40 very high contamination
EF > 40 extremely high contamination

1.2.5.2 Geo-Accumulation Index

Anthropogenic influence on the level of metals in soils can also be illustrated by the computations of the Geo-accumulation index (I_{geo}) values proposed by Müller (1979), which can be evaluated using Eq. (1.2). This method compares current and previous concentrations of metals in soils or sediments to estimate pollution levels:

$$I_{geo} = \log_2 \left(C_n / 1.5\, B_n \right) \tag{1.2}$$

where C_n is the concentration of the metal 'n' in the soil of the study area, and B_n is the geochemical background reference value for which the average shale is used in the present study. The constant 1.5 is introduced in the equation considering the natural fluctuations that may have occurred throughout the years in the environment (Christophoridis et al., 2009).

1.2.5.3 Pollution Load Index

The degree of pollution taking into account the cumulative effect of metals in the soil is determined by the pollution load index (PLI) (Tomilson et al., 1980; Giri et al., 2013). PLI is evaluated as the geometric mean of concentration factor (CF) value of n number of studied metals (Eq. 1.3):

$$\text{PLI} = \left(\text{CF}_1 \times \text{CF}_2 \times \text{CF}_3 \times \cdots \times \text{CF}_n \right)^{1/n} \tag{1.3}$$

The index is based on the CF of each metal present in the soil or sediment which is expressed as Eq. 1.4 (Salomons and Forstner, 1984):

$$\text{CF} = \text{Metal content in soil/Metal content in average shale} \tag{1.4}$$

where n denotes the number of metals, and CF denotes the contamination factor.

The PLI is a comparative method for assessing the level of metal contamination and is divided into four categories:

No pollution (PLI < 1)
Moderate pollution (1 < PLI < 2)

Heavy pollution ($2 < PLI < 3$)
Extremely heavy pollution ($3 < PLI$)

1.2.5.4 Nemerow Comprehensive Index

Another indicator, the Nemerow comprehensive index (P_s) (Li et al., 2008), is used to evaluate the environmental quality of agricultural soils surrounding mica mining locations. The index is computed using the following Eqs. (1.5–1.7):

$$C_f^i = \frac{C_s^i}{C_n^i} \tag{1.5}$$

$$mC_d = \left(\sum_{i=1}^{n} C_f^i\right) / n \tag{1.6}$$

$$P_s = \sqrt{\frac{(mC_d)^2 + (C_{f\max}^i)^2}{2}} \tag{1.7}$$

where C^i_s denotes the determined concentration of metal i in the soil sample (in mg/kg). C^i_n is the geochemical background value of metal i (unit: mg/kg) in this investigation. C^i_f is the single pollution index of metal i, $C^i_{f\max}$ is the maximum value of single pollution index of all metals, n is the number of metals measured in the soil sample ($n = 11$ in this study), and mC_d is Hakanson's modified degree of contamination index, which sums all the single pollution indices (C^i_f) to estimate overall metal contamination at a location (Hakanson, 1980).

Brady et al. (2015) classified the mC_d value into seven categories:

$mC_d < 1.5$ as unpolluted;
$1.5 \leq mC_d < 2$ as slightly polluted;
$2 \leq mC_d < 4$ as moderately polluted;
$4 \leq mC_d < 8$ as considerably polluted;
$8 \leq mC_d < 16$ as highly polluted;
$16 \leq mC_d < 32$ as strongly polluted;
$mC_d \geq 32$ as extremely polluted.

The Nemerow index (P_s) is categorised into five groups by Wu et al. (2015): uncontaminated if $P_s < 1$; slightly contaminated if $1 \leq P_s < 2$; moderately contaminated if $2 \leq P_s < 3$; strongly contaminated if $3 \leq P_s < 5$; and seriously contaminated if $P_s \geq 5$ (Ma et al., 2015).

1.3 RESULTS AND DISCUSSION

1.3.1 Metal Distribution in Soil

The metal contents of the 25 agricultural soil samples collected in various locations in Jharkhand's mica mining districts varied greatly. Table 1.1 shows the descriptive statistics for the metals under consideration and depicts the geometric mean

TABLE 1.1
Descriptive Data of Metals (mg/kg) in the Agricultural Soils of the Mica Mining Areas of Jharkhand (*n*=25)

Metals	Range	Geometric Mean	Average Shale	% of Samples Exceeding Average Shale	World Average	% of Samples Exceeding World Average
As	2.44–31.01	13.7	13	56	6.0	96
Al	26,108.2–107,493.3	59,530.4	80,000	12	71,000	32
Ba	107.8–632.0	287.3	580	4	500	20
Cd	0.04–1.00	0.20	0.30	36	0.350	32
Co	0.72–28.8	11.9	19	24	8.0	76
Cr	13.62–299.7	76.6	90	36	70	64
Cu	14.34–278.1	62.6	45	72	30	96
Fe	20,904.5–65,782.2	41,629.1	47,200	32	40,000	68
Mn	79.7–1,294.3	610.3	850	36	1,000	20
Ni	23.7–193.5	52.8	68	28	50	56
Pb	1.52–145.8	28.0	20	76	35	36
Se	0.029–0.24	0.1	0.6	0	0.29	0
Sr	9.34–79.22	36.3	170	0	250	0
V	10.9–97.8	50.0	130	0	90	4
Zn	44.6–2,049.6	106.1	95	40	90	48

of the metals along with their comparison with average shale values (Turekian and Wedepohl, 1961) and global average values (Bowen, 1979). The geometric mean concentration of As, Cu, Pb, and Zn exceeded the average shale metal content. The metals exceeded the average shale values for 56%, 72%, 76%, and 40% of the samples, respectively. In comparison to the world average, the geometric mean concentrations of Co, Ni, and Fe also exceeded the values. The high metal content in the soil can be attributed to mining and vehicular activities.

Significant differences in metal concentrations were observed among various locations. Al, Fe, and Mn concentrations were found to be high in Koderma (Site 10), Domchanch (Site 12), Tisri (Site 18), Deori (Site 19), and Jainagar (Site 24), indicating an impact from mica mining activities in these areas. The highest concentrations of Pb, Zn, Cr, and Ba were found in Chouparan (Site 3), which has a substantial traffic load. The finding implies that the metal concentrations of soil in the current study area are inextricably linked to anthropogenic sources, including mining and vehicular activities.

Pearson's correlation analysis (Table 1.2) is commonly taken up in environmental data analysis to demonstrate the association among various elements. Such associations provide us with some sense of potential origins of the elements. A significant positive correlation between the elements may indicate a common origin of these elements (Ma et al., 2015). However, to confirm this statement, multivariate statistical analyses such as PCA must be conducted in addition to correlation analysis.

TABLE 1.2
Pearson Correlation Matrix among Various Metals in the Soil ($N=25$)

	As	Al	Ba	Cd	Co	Cr	Cu	Fe	Mn	Ni	Pb	Se	Sr	V	Zn
As	1														
Al	−0.006	1													
Ba	0.190	0.235	1												
Cd	0.678**	0.052	−0.391	1											
Co	0.747**	−0.072	0.182	0.502*	1										
Cr	0.008	0.098	0.552**	−0.194	0.316	1									
Cu	0.696**	0.158	−0.061	0.743**	0.716**	0.045	1								
Fe	−0.149	0.596**	0.145	0.015	−0.077	0.309	−0.031	1							
Mn	0.064	0.679**	0.209	0.016	−0.055	0.171	0.037	0.841**	1						
Ni	0.705**	−0.147	−0.192	0.78**	0.672**	−0.066	0.856**	−0.093	−0.121	1					
Pb	−0.07	0.243	0.708**	−0.344	0.178	0.874**	−0.046	0.206	0.101	−0.209	1				
Se	0.391	0.089	−0.051	0.318	0.638**	0.083	0.564**	−0.055	−0.063	0.594**	0.061	1			
Sr	0.556**	0.271	0.059	0.436*	0.587**	−0.102	0.661**	−0.111	−0.052	0.558**	0.003	0.598**	1		
V	0.736**	0.229	0.126	0.644**	0.632**	0.162	0.594**	0.171	0.228	0.635**	0.061	0.382*	0.61**	1	
Zn	−0.116	0.027	0.598**	−0.290	0.116	0.806**	−0.141	0.056	−0.089	−0.205	0.878**	0.023	−0.154	−0.023	1

* Correlation is significant at the 0.05 level (2-tailed).
** Correlation is significant at the 0.01 level (2-tailed).

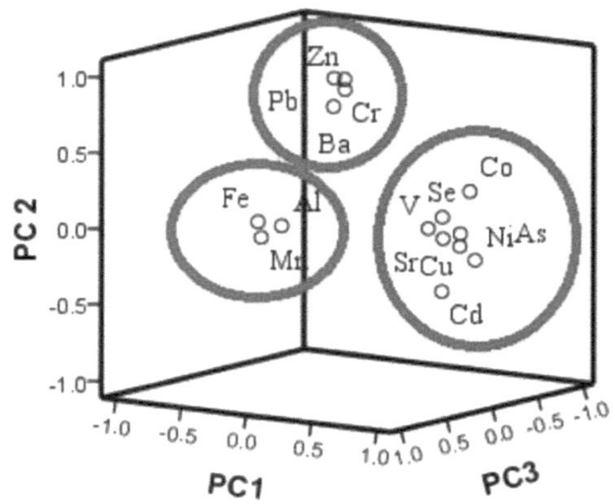

FIGURE 1.2 Principal component loadings (varimax normalised) for metals in mica mining areas.

1.3.2 Principal Component Analysis

PCA provides a better understanding of the interrelationships between the metals in an environmental matrix. It replaces a large data set with a smaller number of components called principal components (PC), which explain most of the dataset's variation. Each of the PCs can be allocated to a source of the components, providing an understanding of their origin (Miller and Miller, 2000). Therefore, PCA was used to determine the source of the metals in the agricultural soils of the Koderma mica mining area.

The findings from PCA are depicted in Figure 1.2, which includes factor loadings, eigenvalues, and explained variance. In this case, 75.86% of the variance was broken down into three variables with eigenvalues greater than 1. The first principal component (PC1), accounting for 35.7% of the variance, revealed that As, Cd, Co, Cu, Ni, Se, Sr, and V had strong loading factors. The component appears to be related to the local lithology. Significant loading for Ba, Cr, Pb, and Zn can be found in the second component, which explains 23.08% of the overall variance in the data. The constituents appear to be associated with human-induced activities such as fossil fuel combustion, waste incineration, and, most importantly, vehicular operations. These combinations of metals are utilised in fuel additives (Ba, Pb, Zn), vehicle tyres (Ba, Pb, Zn), and brake lining fillers (Ba, Cr) (Sharheen, 1975; Hopke et al., 1980; Fishben, 1981; Alloway, 1990; Kennedy and Gadd, 2000; Johnson et al., 2006). Other than that, Cr is utilised in catalytic converters, Zn in lubricants, and Pb in batteries, fuel tanks, and wheel weights (Fishben, 1981; Lohse et al., 2001; Ipeaiyeda and Dawodu, 2008). The third component accounted for 17.04% of the variance in the data and had significant loadings for Al, Fe, and Mn. This factor may be associated with mica mining activity in the research area. The area is noted for its rich mica deposits, and mica is known to be related to Al, Fe, and Mn (Albee, 1965; Greenland et al., 1968; Karunanidhi et al., 2019).

1.3.3 Metal Contamination and Toxicity Assessment in Soil

Metal contamination was found to be moderate to very high in the agricultural soils in the study area, by enrichment factor analysis (Table 1.3). The EF ranged from 0.19–3.9

TABLE 1.3
Enrichment Factor of Metals in Soil with Respect to Different Locations of the Mica Mining Areas of Jharkhand

Location	As	Al	Ba	Cd	Co	Cr	Cu	Fe	Mn	Ni	Pb	Se	Sr	V	Zn
Charitra	1.65	0.77	0.88	0.86	0.98	0.88	1.49	1.00	1.23	0.63	1.65	0.21	0.23	0.37	1.53
Tulbul	1.30	1.13	0.73	0.69	0.86	0.70	2.12	1.00	0.75	0.95	1.37	0.32	0.41	0.56	1.19
Chouparan	1.27	0.70	1.17	0.36	1.37	3.57	1.24	1.00	0.85	0.42	7.81	0.28	0.13	0.49	23.10
Jagodih	1.05	0.91	0.63	0.93	0.95	0.95	1.61	1.00	0.84	1.20	1.29	0.33	0.29	0.59	1.17
Hapwa	0.92	0.89	0.27	1.34	0.68	0.57	1.61	1.00	0.85	0.56	0.88	0.25	0.36	0.56	0.50
Bhondo	2.43	0.82	0.31	3.40	1.55	1.14	6.30	1.00	0.77	2.90	1.13	0.36	0.40	0.77	0.96
Urwan	0.94	0.89	0.89	0.13	0.69	1.39	1.09	1.00	0.70	0.37	3.65	0.27	0.28	0.29	2.22
Jharnakund	2.36	0.96	0.64	1.36	0.61	1.05	1.66	1.00	0.79	1.87	1.69	0.66	0.45	0.51	1.06
Inderwa	3.74	0.81	0.85	2.50	2.72	1.20	4.14	1.00	0.74	2.76	1.63	0.68	0.72	1.15	1.10
Koderma	1.32	1.00	0.76	0.72	0.81	0.76	1.43	1.00	1.26	0.51	1.65	0.23	0.17	0.37	0.59
Taraghati	1.59	0.79	0.90	0.74	0.44	1.36	0.92	1.00	1.07	0.79	3.00	0.05	0.26	0.65	2.05
Domchanch	0.80	0.70	0.36	0.77	0.33	0.72	1.15	1.00	1.09	0.69	0.79	0.20	0.17	0.32	0.89
Dhaab	1.14	0.53	0.30	1.53	0.65	0.25	0.52	1.00	0.15	0.60	0.12	0.15	0.09	0.14	0.88
Basodih	1.27	0.37	0.60	0.69	0.78	0.69	1.37	1.00	0.99	1.30	0.74	0.27	0.09	0.42	0.60
Birne	1.61	0.75	0.43	1.23	1.51	1.10	2.13	1.00	0.66	1.51	1.49	0.41	0.48	0.57	0.75
Satgawan	0.81	0.76	0.45	0.88	0.54	0.99	1.21	1.00	0.84	0.74	2.02	0.27	0.19	0.49	0.74
Bhure	0.19	0.77	0.43	0.17	0.04	0.91	0.75	1.00	0.89	0.36	0.50	0.16	0.09	0.12	0.75
Tisri	0.68	0.94	0.32	0.39	0.37	0.87	0.65	1.00	1.17	0.38	1.85	0.22	0.09	0.30	1.23
Deori	0.76	0.68	0.30	1.09	0.70	0.93	0.92	1.00	0.87	0.81	1.28	0.19	0.14	0.40	0.63
Chahal	3.90	1.74	1.62	2.48	2.58	1.11	7.90	1.00	1.12	2.97	3.27	0.67	0.87	1.10	2.03
Chatro	1.73	1.21	0.47	1.65	0.91	1.31	2.12	1.00	0.88	0.90	1.56	0.31	0.17	0.55	0.98
Doranda	0.97	0.83	0.43	0.72	1.29	1.52	4.07	1.00	1.00	1.37	3.26	0.33	0.37	0.36	1.08
Hathiagarh	0.63	1.04	0.97	0.40	0.40	1.35	0.99	1.00	0.47	0.85	5.31	0.31	0.28	0.33	14.48
Jainagar	1.03	1.24	0.44	1.14	0.35	0.50	1.53	1.00	1.37	0.42	1.91	0.22	0.31	0.40	1.21
Dudhpania	1.29	0.74	0.61	1.17	0.77	1.16	1.50	1.00	0.47	1.26	2.11	0.42	0.29	0.44	0.91

TABLE 1.4
Metal Geo-Accumulation Index (I_{geo}) in Soil with Respect to Different Locations in Jharkhand's Mica Mining Areas

I_{geo} Value	I_{geo} Class	Soil Quality (Muller, 1979)	Metals
>5	6 (>64-fold increase)	Extremely polluted	
4–5	5	Highly polluted to very highly polluted	
3–4	4 (25-fold increase)	Highly polluted	Zn
2–3	3	Moderately polluted to highly polluted	Cu, Pb
1–2	2 (five-fold increase)	Moderately polluted	Cd, Cr, Cu, Pb
0–1	1 (double the background value)	Unpolluted to moderately polluted	As, Co, Cu, Mn, Ni, Pb, Zn
<0	0	Background concentration	As, Al, Ba, Cd, Co, Cr, Cu, Fe, Mn, Ni, Pb, Se, Sr, V, Zn

for As, 0.37–1.74 for Al, 0.27–1.62 for Ba, 0.13–3.4 for Cd, 0.04–2.72 for Co, 0.25–3.57 for Cr, 0.52–7.9 for Cu, 0.15–1.37 for Mn, 0.36–2.97 for Ni, 0.12–7.81 for Pb, 0.05–0.68 for Se, 0.09–0.87 for Sr, 0.12–1.15 for V, and 0.5–23.1 for Zn. The EF indicated significant contamination with respect to Cu, Pb, and Zn in about one-third of the locations. Very high contamination (EF > 20) with respect to Zn was observed in one location, Chouparan, which is subjected to intensive automobile traffic. For the same site, the maximum EF for Pb was also calculated. The EF revealed no pollution in the study area in terms of Al, Ba, Mn, Se, Sr, and V. However, there were a few areas where moderate contaminations of As, Cd, Co, Cr, and Ni were found, which can be linked to geogenic sources and anthropogenic factors such as mining and industries.

Muller (1969) proposed an I_{geo} for quantifying metal accumulation in sediments or soil, which consisted of seven classes (0–6) indicating various degrees of enrichment above background values ranging from unpolluted (class 0) to extremely polluted (class 6) soil quality (Table 1.4). Large fluctuations in the I_{geo} values for the metals in the study area were seen as metals varied from class 0 to class 4, suggesting large spatial variability, representing the background status for certain metals to highly polluted status for others. The majority of the variance was recorded for Zn, where I_{geo} values ranged from −1.68 to 3.85, placing it in the I_{geo} class of 0–4, indicating an unpolluted to highly contaminated state. The I_{geo} values for the metals Cu and Pb at some places indicated moderate to high pollution (class 3). The extent of pollution by different metals, as shown by the CF, indicated Zn and Pb to be the highest contaminants of the soil in the studied area. The CFs for the various elements followed the following sequence for most of the locations: Zn<Pb<Cu<As<Cr<Cd<Fe<Ni<Mn <Al<Co<Ba<V<Se<Sr.

PLI was evaluated to assess the cumulative pollution effect of several metals at various sampling locations, as shown in Figure 1.3. A PLI value of more than 1 calculated for a place supports metal contamination (Cabrera et al., 1999). The average PLI value in the research area was calculated to be 0.70, with a minimum value of 0.23 and a high value of 1.01 (for the locality Doranda). Only two of the study area's locations are classified as moderately polluted in terms of PLI (Table 1.5).

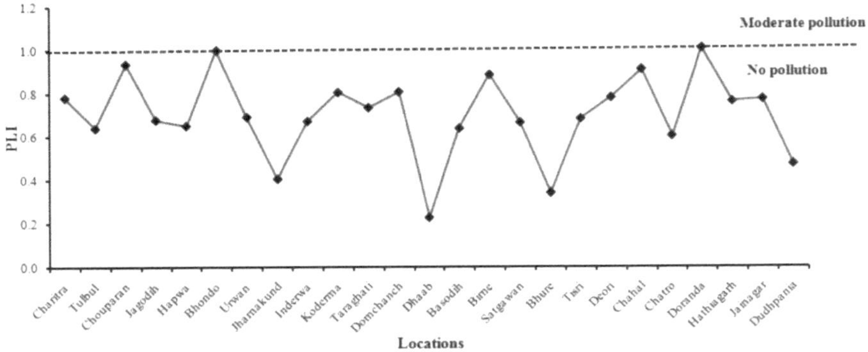

FIGURE 1.3 Pollution load index at different locations of the mica mining areas.

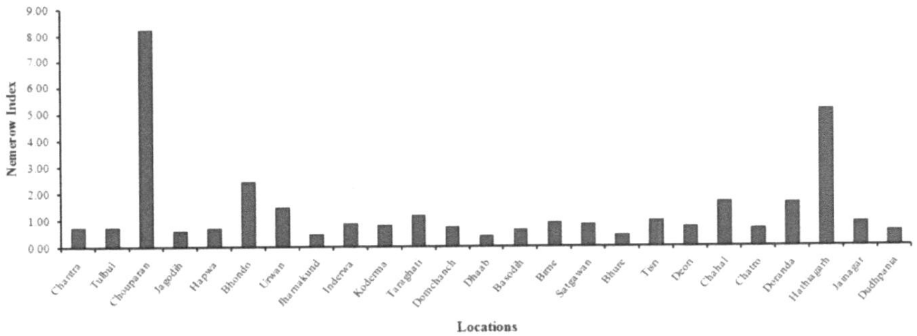

FIGURE 1.4 Nemerow index for soil at different locations of mica mining areas of Jharkhand.

The mC_d values in the study area ranged from 0.32 to 2.58. According to the mC_d rating, the areas ranged from 'unpolluted' ($mC_d < 1.5$) to 'moderately contaminated' ($2 \leq mC_{d<}4$). The average impact of all observed metals is included in the computation of mC_d, reducing the role of one metal. As a result, even if the soils are heavily contaminated with one of the metals, the skewness for the mC_d readings will be small. The present investigation illustrates this, as some of the soils are significantly contaminated with Zn. However, the influence of a single metal is evaluated considerably more carefully in the Nemerow index, which takes into account both the average and maximum CFs as per Eq. (1.7) (Brady et al., 2015). The Nemerow Index values for the various locations in the mica mining area ranged from 0.37 to 8.21 (Figure 1.4), with two of the locations (Hathiagarh and Chouparan) indicating significantly contaminated conditions ($P_s \geq 5$). Thus, the results show that metal contamination in the soils of the research area is mild to significant, with Pb and Zn being the most concerning metals after accounting for EF, I_{geo}, CF, PLI, and Nemerow Index. The areas that were affected by traffic pollution had the highest level of contamination.

TABLE 1.5
Contamination Factor of Metals in Soil along with Pollution Load Index (PLI) with Respect to Different Locations of the Mica Mining Areas of Jharkhand

Code	Location	As	Al	Ba	Cd	Co	Cr	Cu	Fe	Mn	Ni	Pb	Se	Sr	V	Zn	PLI
1	Charitra	1.65	0.77	0.88	0.86	0.98	0.88	1.49	1.00	1.23	0.63	1.65	0.21	0.24	0.37	1.53	0.78
2	Tulbul	1.07	0.93	0.60	0.57	0.71	0.57	1.74	0.82	0.62	0.78	1.13	0.26	0.34	0.46	0.98	0.64
3	Chouparan	1.18	0.66	1.09	0.33	1.28	3.33	1.16	0.93	0.79	0.39	7.29	0.26	0.12	0.46	21.57	0.94
4	Jagodih	0.87	0.76	0.52	0.77	0.79	0.78	1.33	0.83	0.69	0.99	1.06	0.27	0.24	0.49	0.96	0.68
5	Hapwa	0.91	0.88	0.27	1.33	0.67	0.57	1.60	0.99	0.85	0.56	0.87	0.25	0.36	0.56	0.50	0.65
6	Bhondo	2.39	0.81	0.30	3.33	1.52	1.12	6.18	0.98	0.75	2.85	1.11	0.35	0.39	0.75	0.94	1.00
7	Urwan	0.94	0.90	0.90	0.13	0.69	1.40	1.10	1.01	0.71	0.37	3.69	0.27	0.28	0.29	2.24	0.69
8	Jharnakund	1.05	0.43	0.28	0.60	0.27	0.46	0.73	0.44	0.35	0.83	0.75	0.29	0.20	0.23	0.47	0.41
9	Inderwa	1.86	0.40	0.42	1.25	1.36	0.60	2.07	0.50	0.37	1.38	0.82	0.34	0.36	0.57	0.55	0.67
10	Koderma	1.48	1.12	0.86	0.81	0.91	0.85	1.61	1.12	1.41	0.58	1.85	0.25	0.19	0.42	0.66	0.81
11	Taraghati	1.52	0.76	0.87	0.71	0.42	1.30	0.88	0.96	1.03	0.76	2.88	0.05	0.25	0.62	1.97	0.74
12	Domchanch	1.12	0.97	0.50	1.07	0.46	1.00	1.60	1.39	1.52	0.96	1.11	0.28	0.23	0.44	1.23	0.81
13	Dhaab	0.70	0.33	0.19	0.93	0.40	0.15	0.32	0.61	0.09	0.37	0.08	0.09	0.06	0.08	0.54	0.23
14	Basodih	1.31	0.38	0.61	0.71	0.80	0.71	1.41	1.03	1.02	1.34	0.76	0.27	0.09	0.43	0.61	0.64
15	Birne	1.55	0.73	0.41	1.19	1.46	1.06	2.05	0.96	0.64	1.46	1.43	0.40	0.47	0.55	0.72	0.89
16	Satgawan	0.79	0.74	0.44	0.85	0.53	0.96	1.17	0.97	0.81	0.72	1.96	0.27	0.18	0.47	0.72	0.67
17	Bhure	0.19	0.75	0.41	0.17	0.04	0.89	0.73	0.97	0.87	0.35	0.48	0.16	0.09	0.12	0.73	0.34
18	Tisri	0.84	1.16	0.39	0.49	0.45	1.07	0.80	1.23	1.44	0.47	2.28	0.27	0.11	0.37	1.51	0.69
19	Deori	0.97	0.86	0.38	1.38	0.89	1.17	1.16	1.27	1.10	1.02	1.63	0.25	0.17	0.50	0.79	0.78
20	Chahal	2.05	0.91	0.85	1.30	1.35	0.58	4.15	0.53	0.59	1.56	1.72	0.35	0.46	0.58	1.07	0.92
21	Chatro	1.25	0.88	0.34	1.19	0.66	0.95	1.54	0.72	0.64	0.65	1.13	0.23	0.12	0.39	0.71	0.61
22	Doranda	0.97	0.83	0.43	0.72	1.29	1.52	4.07	1.00	1.00	1.37	3.26	0.33	0.37	0.36	1.08	1.01
23	Hathiagarh	0.59	0.97	0.90	0.37	0.37	1.25	0.92	0.93	0.44	0.79	4.92	0.29	0.26	0.31	13.43	0.77
24	Jainagar	1.12	1.34	0.47	1.23	0.38	0.55	1.65	1.08	1.48	0.46	2.07	0.24	0.34	0.43	1.31	0.78
25	Dudhpania	0.79	0.45	0.37	0.71	0.47	0.71	0.91	0.61	0.29	0.77	1.29	0.25	0.18	0.27	0.56	0.47

1.4 CONCLUSION

The geometric mean concentrations of some metals (e.g., As, Cu, Pb, and Zn) in the agricultural soils of the mica mining areas in Jharkhand were found to be greater than the average shale values. The elevated levels of metals found in the soil can be ascribed to a combination of natural and human-induced processes, as evidenced by the results of the multivariate analysis. PCA was conducted, resulting in the identification of three components that collectively accounted for 75.9% of the variability observed in the data. Among these factors, two were found to be probably associated with human activities, while the remaining factor was indicative of geogenic origins. The results of the enrichment factor and I_{geo} analysis revealed significant levels of pollution in the agricultural soils of the study area, particularly in relation to Zn, followed by Pb and Cu. The assessment of the collective impact of various metals on pollution was conducted using the PLI and Nemerow index. The results indicated that two of the sampled areas exhibited an exceedingly high level of metal pollution. However, 72% of the locations fell inside the uncontaminated group according to the Nemerow index. The areas experiencing significant vehicular traffic had elevated levels of metal contamination, with zinc (Zn) being the primary metal of interest among all the metals studied. The study recommends the regular assessment of heavy metal levels in the soil and the use of improved soil management techniques within the region.

ACKNOWLEDGEMENTS

The authors are thankful to the Department of Science and Technology for providing the necessary funding for the study under the DST-Women Scientist Scheme-A (Grant No. SR/WOS-A/EA-28/2018). Also, the authors are grateful to the Director and the research group of Water Resource Management (WRM) of CSIR-Central Institute of Mining and Fuel Research, Dhanbad, for providing the needed laboratory amenities and other logistic support to carry out the study.

REFERENCES

Acosta, J.A., Faz, A., Martínez-Martínez, S., Zornoza, R., Carmona, D.M. and Kabas, S. 2011. Multivariate statistical and GIS-based approach to evaluate heavy metals behavior in mine sites for future reclamation. *Journal of Geochemical Exploration* 109(1–3):8–17.

Albee, A.L. 1965. Distribution of Fe, Mg, and Mn between garnet and biotite in natural mineral assemblages. *Journal of Geology* 73(1):155–64.

Allibone, R., Cronin, S.J., Charley, D.T., Neall, V.E., Stewart, R.B. and Oppenheimer, C. 2012. Dental fluorosis linked to degassing of ambrym volcano, Vanuatu: A novel exposure pathway. *Environmental Geochemistry and Health* 34(2):155–70.

Alloway, B.J. 1990. Soil processes and the behaviour of metals. In: *Heavy Metals in Soils*, ed. B.J. Alloway, 11–37. London: Blackie.

Bowen, H. J. M. 1979. *Environmental Chemistry of Elements*. New York: Academic Press.

Brady, J., Ayoko, G, Martens., W. and Goonetilleke, A. 2015. Development of a hybrid pollution index for heavy metals in marine and estuarine sediments. *Environmental Monitoring and Assessment* 187(5):1–14.

Cabrera, F., Clemente, L., Diaz Barrientos, E., Lopez, R. and Murillo, J.M. 1999. Heavy metal pollution of soils affected by the Guadiamar toxic flood. *Science of the Total Environment* 242:117–129.

Central Ground Water Board. 2013. *Ground water information booklet, Koderma, Jharkhand State*. Patna, India: Central Ground Water Board, Ministry of Water Resources, Government of India.

Chen, H., Teng, Y., Lu, S., Wang, Y. and Wang, J. 2015. Contamination features and health risk of soil heavy metals in China. *Science of the Total Environment* 512–513:143–53.

Christophoridis, C., Dedepsidis, D. and Fytianos, K. 2009. Occurrence and distribution of selected toxic metals in the surface sediments of Thermaikos Gulf, N. Greece. Assessment using pollution indicators. *Journal of Hazardous Materials* 168:1082–1091.

Closs, L.G. and Nichol, I. 1975. The role of factor and regression analysis in the interpretation of geochemical reconnaissance data. *Canadian Journal of Earth Sciences* 12(8):1316–30.

Fishbein, L. 1981. Sources, transport and alterations of metal compounds: An overview. I. Arsenic, beryllium, cadmium, chromium and nickel. *Environmental Health Perspectives* 40:43–64.

Ganyaglo, S.Y., Gibrilla, A., Teye, E.M., Owusu-Ansah, E.D.G.J., Tettey, S., Diabene, P.Y. and Asimah S. 2019. Groundwater fluoride contamination and probabilistic health risk assessment in fluoride endemic areas of the Upper East Region, Ghana. *Chemosphere* 233:862–72.

Giri, S. and Singh, A.K. 2017. Ecological and human health risk assessment of agricultural soils based on heavy metals in mining areas of Singhbhum copper belt, India. *Human and Ecological Risk Assessment: An International Journal* 23(5):1008–1027.

Giri, S. and Singh, A.K. 2023. Fluoride and metals in the agricultural soils of mica mining areas of Jharkhand, India: Assessing the ecological and human health risk. *Soil and Sediment Contamination: An International Journal*. doi:10.1080/15320383.2023.2208675

Giri, S., Singh, A.K. and Tewary, B.K. 2013. Source and distribution of metals in bed sediments of Subarnarekha River, India. *Environmental Earth Sciences* 70(7):3381–3392.

Greenland, L.P., Gottfried, D. and Tilling, R.I. 1968. Distribution of manganese between coexisting biotite and hornblende in plutonic rocks. *Geochimica et Cosmochimica Acta* 32(11):1149–1163.

Hakanson, L. 1980. An ecological risk index for aquatic pollution control. A sedimentological approach; *Water Research* 14(8):975–1001.

Hopke, P.K., Lamb, R.E. and Natusch, D.F.S. 1980. Multielemental characterization of urban roadway dust. *Environmental Science and Technology Letters* 14(2):164–172.

Ipeaiyeda, A.R. and Dawodu, M. 2008. Heavy metals contamination of topsoil and dispersion in the vicinities of reclaimed. *Bulletin of Chemical Society of Ethiopia* 22(3):339–348.

Izuora, K., Twombly, J.G., Whitford, G.M., Demertzis, J., Pacifici, R. and Whyte, M.P. 2011. Skeletal Fluorosis from Brewed Tea. *Journal of Clinical Endocrinology and Metabolism* 96(8):2318–2324.

Johnson, J., Schewel, L. and Graedel, T.E. 2006. The contemporary anthropogenic chromium cycle. *Environmental Science and Technology Letters* 40(22):7060–7069.

Kartal, S., Aydin, Z. and Tokalioglu, S. 2006. Fractionation of metals in street sediment samples by using the BCR sequential extraction procedure and multivariate statistical elucidation of the data. *Journal of Hazardous Materials* 132:80–89.

Karunanidhi, D., Aravinthasamy, P., Subramani, T., Wu, J. and Srinivasamoorthy, K. 2019. Potential health risk assessment for fluoride and nitrate contamination in hard rock aquifers of Shanmuganadhi River basin, South India. *Human and Ecological Risk Assessment* 25(1–2):250–270.

Kennedy, P. and Gadd, J. 2000. *Preliminary Examination of Inorganic Compounds Present in Tyres, Brake Pads and Road Bitumen in New Zealand*. Wellington, New Zealand: Prepared by Kingett Mitchell Ltd for Ministry of Transport, November 2000, Revised October 2003.

Khan, A., Khan, S., Khan, M.A., Qamar, Z. and Waqas, M. 2015. The uptake and bioaccumulation of heavy metals by food plants, their effects on plants nutrients, and associated health risk: A review. *Environmental Science and Pollution Research* 22(18):13772–13799.

Kolsi, S.H., Bouri, S., Hachicha, W. and Dhia, H.B. 2013. Implementation and evaluation of multivariate analysis for groundwater hydrochemistry assessment in arid environments: A case study of HajebElyoun–Jelma, Central Tunisia. *Environmental Earth Sciences* 70(5):2215–2224.

Li, W.X., Zhang, X.X., Wu, B., Sun, S.L., Chen, Y.S., Pan, W.Y., Zhao, D.Y. and Cheng, S.P. 2008. A comparative analysis of environmental quality assessment methods for heavy metal-contaminated soils. *Pedosphere* 18(3):344–352.

Li, Z., Ma, Z., van der Kuijp, T.J., Yuan, Z. and Huang, L. 2014. A review of soil heavy metal pollution from mines in China: Pollution and health risk assessment. *Science of the Total Environment* 468:843–853.

Lohse, J., Sander, K. and Wirts, M. 2001. *Heavy Metals in Motor Vehicles 2. Report Compiles for the Directorate general Environment, Nuclear safety and Civil Protection of the Commission of the European Communities*. Okopol: Okopol-Institut fur Ohkologie und Politik GmbH

Ma, L., Sun, J., Yang, Z. and Wang, L. 2015. Heavy metal contamination of agricultural soils affected by mining activities around the Ganxi River in Chenzhou, southern China. *Environmental Monitoring and Assessment* 187:731.

Manzoor, S., Shah, M.H., Shaheen, N., Khalique, A. and Jaffar, M. 2006. Multivariate analysis of trace metals in textile effluents in relation to soil and groundwater. *Journal of Hazardous Materials Letters* 137(1):31–37.

Maxted, A.P., Black, C.R., West, H.M., Crout, N., Mcgrath, S.P. and Young, S.D. 2007. Phytoextraction of cadmium and zinc from arable soils amended with sewage sludge using *Thlaspicaerulescens*: Development of a predictive model. *Environmental Pollution* 150(3):363–372.

Miller, N.J. and Miller, J.C. 2000. *Statistics and Chemometrics for Analytical Chemistry*, 4th edn. Englewood Cliffs, NJ: Pearson Education.

Muller, G. 1979. Schwermetalle in den Sedimenten des RheinsVeranderungenseit 1971. *Umschau* 79:778

Salomons, W. and Forstner, U. 1984. *Metals in the Hydrocycle*. Berlin: Springer.

Schiff, K.C. and Weisberg, S.B. 1999. Iron as a reference element for determining trace metal enrichment in southern California coastal shelf sediments. *Marine Environmental Research* 48:161–176.

Sharheen, D.G. 1975. *Contributions of Urban Roadway Usage to Water Pollution*. EPA-600/2-75-004. EPA.

Sutherland, R.A. 2000. Bed sediment-associated trace metals in an urban stream, Oahu, Hawaii. *Environmental Geology* 39:611–627.

Tomlison, L., Wilson, L.G., Harris, R. and Jeffrey, D.W. 1980. Problems in the assessments of heavy metal levels in estuaries and formation of pollution index. *Helgolander Meeresuntersuchungen* 33:566–575.

Turekian, K. K. and Wedepohl, K. H. 1961. Distribution of the elements in some major units of the earth's crust. *Geological Society of America Bulletin* 72(2):175–192.

Turner, A. and Millward, G.E. 2000. Particle dynamics and trace metal reactivity in estuarine plumes. *Estuarine Coastal and Shelf Science* 50:761–774.

USEPA. 1996. *EPA Method 3052:Microwave Assisted Acid Digestion of Siliceous and Organically Based Matrices*. Washington, DC: US Environmental Protection Agency, Office of Solid Waste and Emergency Response, US Government Printing Office.

Wu, Q., Leung, J.Y.S., Geng, X., Chen, S., Huang, X., Li, H., Huang, Z., Zhu, L., Chen, J. and Lu, Y. 2015. Heavy metal contamination of soil and water in the vicinity of an abandoned e-waste recycling site: Implications for dissemination of heavy metals. *Science of the Total Environment* 506–507:217–225.

2 Geochemistry of Soils Surrounding the Ursk Dump, Kemerovo Region, Russia

Natalya Abrosimova, Svetlana Bortnikova, Alexey Edelev, Nataliya Yurkevich, Olga Saeva, and Nikolay Abrosimov

2.1 INTRODUCTION

Overburden and host rocks, as well as mine tailings, are not used by most enterprises and are stored in special storage facilities (dumps, tailings, and sludge storages). According to the Federal Service for Supervision of Natural Resources, the total area of waste in Russia is 4 million hectares (www.rpn.gov.ru). Mine tailings usually contain elevated levels of metals and metalloids such as Zn, Cu, Pb, Cd, As, and Sb. The large amount of stored mine tailings is exposed to intensive erosion from wind and rain, causing negative impacts on the adjacent soils. Throughout the world, numerous areas of soils are polluted with Pb, Zn, As, Cd, Co, and Cu due to mining activities (Diatta et al., 2008; Vicente-Beckett et al., 2016; Bisone et al., 2016; Ghosh et al., 2023; Yin et al., 2023). Soil pollution due to metal and metalloid contamination is a serious problem and is widely discussed in the scientific literature (Franchini et al., 2007; Agnieszka et al., 2014; Bisone et al., 2016; Abrosimova et al., 2023). Trace elements such as As, Cd, Cr, Hg, Ni, Cu, Pb, and Zn are extremely toxic and carcinogenic to living organisms (Wang et al., 2020; Buch et al., 2021). In agricultural soils contaminated with metals, their absorption by crops has been associated with several adverse health effects, in both humans (Liu et al., 2019) and local fauna (Picaud and Petit, 2013; Naccarato et al., 2020). Metals such as copper and selenium are essential elements for plants and living organisms, but in high concentrations they become toxic. In Chile, copper tailings have been heavily polluted over the last 90 years, and levels of soil copper are about 1,025 ppm (Ginocchio et al., 2002; Espinoza et al., 2022). Determining potentially toxic element levels in soils has become an efficient method of quantifying the degree of contamination at an area influenced by mine tailings (Ech-Charef et al., 2023).

Studying the spatial distribution and identity of contaminated soils in urban and other geographic areas with a clear identification of the source of pollution is important for planning and decision-making (Zhang et al., 2006; Ouabo et al., 2020).

DOI: 10.1201/9781003442554-3

Analysis of the spatial distribution of heavy metal pollution is the basis of soil pollution assessment and remediation (Zhu and Lin, 2010; Ouabo et al., 2020).

The contamination levels of metals in soils were assessed using the following indices: enrichment factor (EF), geoaccumulation index (I_{geo}), and contamination factor (CF). These indices are widely used to assess the presence and intensity of anthropogenic contaminant deposition on soils (Loska et al., 2004; Wei and Yang, 2010; Krishna et al., 2013; Omotoso and Ojo, 2015; Bech et al., 2016; Fagbenro et al., 2021).

The main goal of our study was to assess the contamination of soils with metals and metalloids in relation to the Ursk tailings dumps, because the dumps are located in the village of Ursk and the soils are used for agricultural purposes. The results presented here will provide a scientific basis for preventing and controlling potential toxic metal and metalloid contamination.

2.2 STUDY AREA AND METHODS

The Ursk tailings are located in Kemerovo Region, near the Ursk settlement (Kemerovo Region, Russia, Figure 2.1). The Ursk tailings are waste dumps of high-sulfide tailings left after the mining and cyanide processing of the Novo-Ursk polymetallic deposit. The waste dumps are 10–12 m high piles of sulfide and oxide tailings (Myagkaya et al., 2019; Saryg-ool et al., 2020; Lazareva et al., 2019).

The primary sulfide/sulfate minerals in the tailings are barite ($BaSO_4$), pyrite (FeS_2), jarosite ($KFe_3(OH)_6(SO_4)_2$), and gypsum ($CaSO_4{\cdot}2H_2O$). Pyrite encloses other minerals such as galena (PbS), chalcopyrite (CuFeS), bornite (Cu_5FeS_4),

FIGURE 2.1 Location of the study area and sampling points (a) Quartz-barite and barite-pyrite Ursk waste piles (b, c) photograph by the author.

arsenopyrite ($FeAsS_2$), sphalerite (ZnS), tennantite (Cu_3AsS_3), altaite (PbTe), and geffroyaite ($(Ag,Cu,Fe)_9(Se,S)_8$) (Myagkaya et al., 2016). One pyrite sample contains an inclusion of metallic gold with Cu and Ag impurities. Barite encloses naumannite (Ag_2Se) (Myagkaya et al., 2016). Some grains of barite are intergrown with anglesite ($PbSO_4$), and some grains have inclusions of galena with impurities of Sb (Yurkevich et al., 2019).

The silicate minerals of the tailings include quartz, muscovite, albite, a minor amount of chlorite, and traces of microcline ($K[AlSi_3O_8]$) (Myagkaya et al., 2016).

During field work, soil samples were taken in areas of subradial profiles 350 m long adjacent to the tailings (Figure 2.1). At each site, representative volumes of soil were taken for sample preparation and analysis. Forty-two sampling points were georeferenced using the Global Positioning System (GPS, Figure 2.1). At each sampling point, a 0.5 kg sample was collected from the four corners and center of each plot at a depth of 0–10 cm, and then the five samples were pooled and extensively homogenized to obtain a representative sample.

The collected soil samples were air-dried and then sifted through a 5 mm sieve to remove plant root debris and ground to a density of −200 mesh. The prepared fraction of the soil samples was stored in paper bags for subsequent analysis. The soil samples were analyzed for geochemical and physicochemical composition.

The chemical elements of the soils were determined using the inductively coupled plasma–mass spectrometry (ICP-MS) Elan DRC-e instrument in the certified laboratory of the analytical center Plasma (Tomsk, Russia). All measurements were performed in triplicate for each element. The relative analysis error was 3%–8%.

The pH was analyzed at a soil-to-water ratio of 1:5 using a portable pH meter, HI 9025 C, Hanna Instruments, Milan, Italy.

The assessment of the contamination levels of metals and metalloids in soils was carried out using the EF, CF, and I_{geo}. In our calculations, the concentrations of chemical elements in felsic rocks were taken as a guideline (Vinogradov, 1962), since the host rocks of the Ursk ore field are dacite porphyries (Roslakov et al., 1995; Bolgov, 1937).

EF was calculated by the following formula:

$$\text{EF} = (C_i / C_r)\text{soil sample} / (C_i / C_r)\text{background}, \tag{2.1}$$

where C_i is the concentration of the considered element, and C_r is the concentration of metal adopted as a reference element.

Strontium (Sr) was chosen as the normalizing element because of the low influence of anthropogenic activities on the Sr content.

The EF levels were categorized as follows:

a. values of $0.5 \leq \text{EF} \leq 1.5$ suggest that the trace metal concentration may come entirely from natural weathering processes (Zhang and Liu, 2002);
b. values of $\text{EF} > 1.5$ indicate that a significant portion of the trace metals was delivered from noncrustal materials (Zhang and Liu, 2002; Yongming et al., 2006; Klerks and Levinton, 1989; Sutherland et al., 2000);

c. values of EF < 2 indicate deficiency to minimal enrichment (Yongming et al., 2006);
d. values of EF > 40 indicate extremely high enrichment (Yongming et al., 2006).

Soil contamination by metals and metalloids was estimated using the CF method for sediments originally proposed by Hakanson (1980). This method is now widely used to assess soil metal contamination (Diatta et al., 2008; Nouri, 2016). CF was calculated using the following formula:

$$\text{CF} = (C_i)\text{sample}/(C_r)\text{ref}, \tag{2.2}$$

where C_i is the concentration of chemical elements in the soil sample,

Cr is the reference value.

CF has been divided into five levels:

1. $\text{CF} \leq 1$ soil is not contaminated;
2. $1 < \text{CF} \leq 2$ slightly and moderately contaminated soil;
3. $2 < \text{CF} \leq 3$ moderately or heavily contaminated soil;
4. $3 < \text{CF} \leq 4$ heavily or extremely polluted soil;
5. $\text{CF} > 4$ extremely polluted soil.

I_{geo} was computed using the following equation:

$$I_{\text{geo}} = \text{Log}_2\left[(C_m)/(1.5 \times B_m)\right], \tag{2.3}$$

where C_m is the measured concentration of the element in soil, B_m is the background value in soil.

According to Muller (1969), the I_{geo} for each metal is calculated and classified as:

1. uncontaminated ($I_{\text{geo}} \leq 0$);
2. slightly contaminated ($I_{\text{geo}} \leq 1$);
3. moderately contaminated ($I_{\text{geo}} \leq 2$);
4. moderately to heavily contaminated ($I_{\text{geo}} \leq 3$);
5. heavily contaminated ($I_{\text{geo}} \leq 4$);
6. heavily to extremely contaminated ($I_{\text{geo}} \leq 5$);
7. extremely contaminated ($I_{\text{geo}} > 5$).

Element distribution maps were constructed using ArcGIS 9.30 software (Hillier, 2011). The nature of the spatial distribution of heavy metal pollution can be generated from discrete point data by spatial interpolation (Zhu and Lin, 2010; Qiao et al., 2018). Inverse Distance Weighting (IDW) was used to predict local features of soil pollution. Inverse distance weighted interpolation models have been widely used (Lu and Wong, 2008; Wollenhaupt et al., 1994; Spezia et al., 2012). IDW interpolation is similar to or even better than other interpolations (Aguilar et al., 2005; Gotway et al.,

1996) and is more suitable for spatial prediction of heavy metal pollution in soils (Qiao et al., 2018).

The R software packages (Venables and Smith, 2009) were used for statistical evaluations. The data normality was checked using the Kolmogorov–Smirnov and Shapiro–Wilk tests. Relationships among the chemical elements in soils were studied using Spearman correlations.

2.3 RESULTS AND DISCUSSION

2.3.1 Geochemistry of Soils

The statistical parameters of concentrations of metals and metalloids in the soils, together with the maximum permissible concentrations (MPCs) for soils (SanRaR 1.2.3685–21), are presented in Table 2.1. We refer to Max (Min) in Table 2.1 as the maximal (minimal) value among measurements of a given parameter. Median is the value separating the higher half from the lower half of a data sample.

TABLE 2.1
Statistical Parameters of Element Concentrations of Soils (n=42); Si–Mo in ppm

	Min	Max	Median	IQR	MPC[a]
Si	31,900	353,000	244,850	108,669	–
Ti	380	4,660	2,960	1,614	–
Al	6,590	77,600	50,710	26,922	–
Fe	27,800	196,000	44,927	20,887	–
Mn	38	5,660	795	942	1,500
Na	890	17,400	8,880	6,675	–
K	1,230	13,000	7,500	4,703	–
Mg	690	13,900	5,824	5,072	–
P	140	2,860	684	623	–
Ba	700	50,500	7,996	15,323	–
Cu	20	470	51	32	66
Zn	36	530	147	65	110
Cd	0.023	1.6	0.2	0.169	1
Be	bdl	2.3	0.94	1.049	–
Co	0.63	27	10.3	9.36	5
Pb	14	2,200	171	584	65
As	bdl	350	53	108	5
Sb	0.98	300	18	96	4.5
Se	bdl	127	16	21	–
Mo	0.81	6.7	2.8	2.98	–
pH	2.68	7.58	5.48	2.47	–

bdl, below detection limit; –, no data.

[a] Maximum permissible concentrations (MPCs), SanRaR 1.2.3685-21 for acid soils (SanRaR 1.2.3685-21).

IQR (interquartile range) is defined as the difference between the upper and lower quartile values in a set of data. To assess the normality of data, the Kolmogorov–Smirnov and Shapiro–Wilk tests were applied. The calculated *p*-values were greater than 0.05 for Si, V, Ti, K, Al, Mg, Na, Be, Li, Rb, and Co and less than 0.05 for Fe, Cu, Se, Mo, Cd, Sb, Ba, Zn, Sr, Ag, Sn, P, Mn, and Cr.

The order of major elements concentrations (from high to low) in the soils was Si>Fe>Al>Na, Mg>Ba>K>Mn>Ti>P.

The concentrations of trace metals and metalloids varied between the sample locations. The Cd, Be, and Mo showed minimal variation from one sampling point to another. The mean concentration of Pb was the highest among the trace metals and metalloids studied, while Cd was the lowest, and the order observed for this study was Pb>Zn>As>Cu>Sb>Co>Mo>Be>Cd.

The pH values had a positive correlation with Li, Be, Mg, Al, K, and Co, and a negative correlation with As and Sb. The correlation analysis revealed statistically significant correlations between the following elements (Figure 2.2).

A positive correlation of elements indicates similar input sources and common geochemical behavior. Fe, As, Sb, Pb, Cu, and Zn are the major elements coming from mine tailings (pyrrhotite, pyrite, chalcopyrite, sphalerite, tennantite, and galena) while Co, Cd, and Se exist as impurities in the main sulfide minerals, representing the minor elements (Myagkaya et al., 2016).

Barium has a positive As-Pb correlation, which indicates its source as barium sulfate, which is often associated with galena (Yurkevich et al., 2019).

Thus, the distribution of elements in soils seems to be linked to mine tailings, the landform of the area, and wind direction. During storage, the tailings were oxidized with the formation of secondary metal sulfates such as anglesite (Saryg-ool et al., 2020), which are unstable, easily destroyed, and can be carried by air currents and water flows.

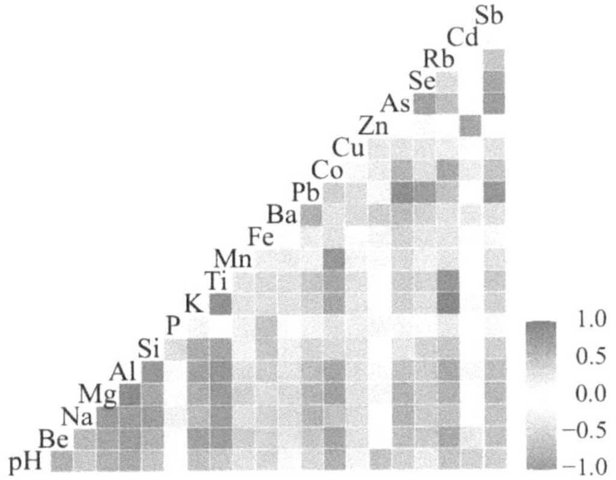

FIGURE 2.2 Spearman rank correlations among all soil properties (physicochemical properties and chemical composition) measured in the 42 study samples. Bold indicate positive and negative correlations, respectively.

For soil contamination assessment, three indices were calculated. The values of the calculated indices for some chemical elements for soil samples from different spots are presented in Figure 2.3.

Variation occurred in EF values (Figure 2.3a): the values of Ba, Pb, Fe, Sb, and As showed extremely high enrichment, while Rb, Be, and Al values indicated deficiency

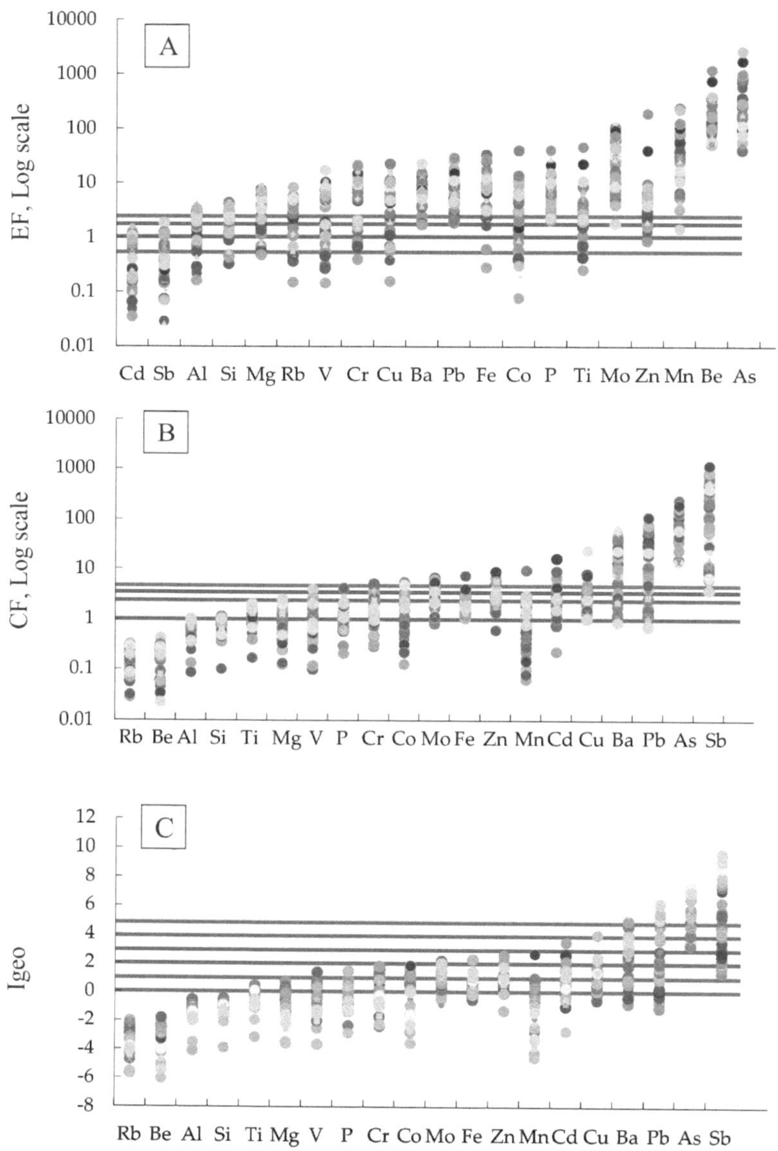

FIGURE 2.3 Assessment of soil pollution with metals and metalloids based on the value of the enrichment factor (a) contamination factor (b) and geoaccumulation index (c).

to minimal enrichment of the soils with these elements. In most of the samples, Rb, Be, and in some samples, Al, Si, Ti, V, Mn, Cr, P, Mg, and Co with EFs values less than 1 indicated no enrichment of the soils. These variations could be from natural and anthropogenic sources, such as natural weathering of the host rocks and mine tailings. However, an EF greater than 1.5 indicates that a significant portion of the trace metals was delivered from noncrustal materials, so these trace metals were likely delivered by mine tailings.

To assess the magnitude of soil contamination with metals and metalloids, CF indices were calculated for each chemical element (Figure 2.3b). Under CF, soils were characterized by different levels of contamination for individual chemical elements: from extreme contamination with Sb and As in all samples, Ba and Pb in most samples, to Mo, Fe, Zn, Cd, and Cu in some samples. The CF value for Rb, Be, Al, Si, Ti, Mg, V, Co, Mn, and P in most samples was less than 1, indicating almost uncontaminated soil. CF values ranging from 1 to 2 for Fe, Zn, and Cu were calculated for most samples, and for Ti, Mg, V, P, Cr, Co, Mn, and Mo for some samples, which indicates low-to-moderate pollution.

I_{geo} for the different metals and metalloids is shown in Figure 2.3c. Moderate pollution ($1 < I_{geo} < 2$) by Mo, Fe, and Zn was observed for most of the soils, while heavy pollution ($3 < I_{geo} < 4$) by Ba and Pb was observed for some soil samples, and extreme pollution by As and Sb for most of the samples. Since Cd, Cu, Ba, Pb, As, and Sb had I_{geo} values >1, the results confirmed that there was considerable contamination for the studied soils.

The spatial distribution of concentrations of chemical elements was constructed, and areas with increased concentrations of chemical elements were identified using ArcGIS software.

The spatial distribution maps of the chemical element concentrations in soil are shown in Figure 2.4. The concentrations of chemical elements in the studied area show considerable variation in spatial distribution. The tailings significantly contribute to chemical element accumulation in soils, especially Fe, Pb, As, Sb, Zn, and Cd. The estimated probability maps also show a trend, with high chemical element concentrations to the northeast, which is caused by prevailing southwest winds in this district, and to the south, which is caused by the direction of rain flow. That is, wind erosion and water flows are the main causes of the pollution of adjacent soils, but in different directions.

2.4 CONCLUSIONS

Metals and metalloids released from the Ursk tailings have affected the neighboring soils. Considerable amounts of As, Sb, Cu, Zn, Cd, Pb, and Fe in soils were associated with mine tailing dumps.

Areoles of contaminated soils with metals and metalloids were identified at a distance from the tailings dump. Water flows pass over and percolate the dumps, incorporating metals and metalloids and subsequently contaminating the soils to the south. Wind erosion also increases the concentration of some elements (Cd, Pb, Zn) in soils to the northeast.

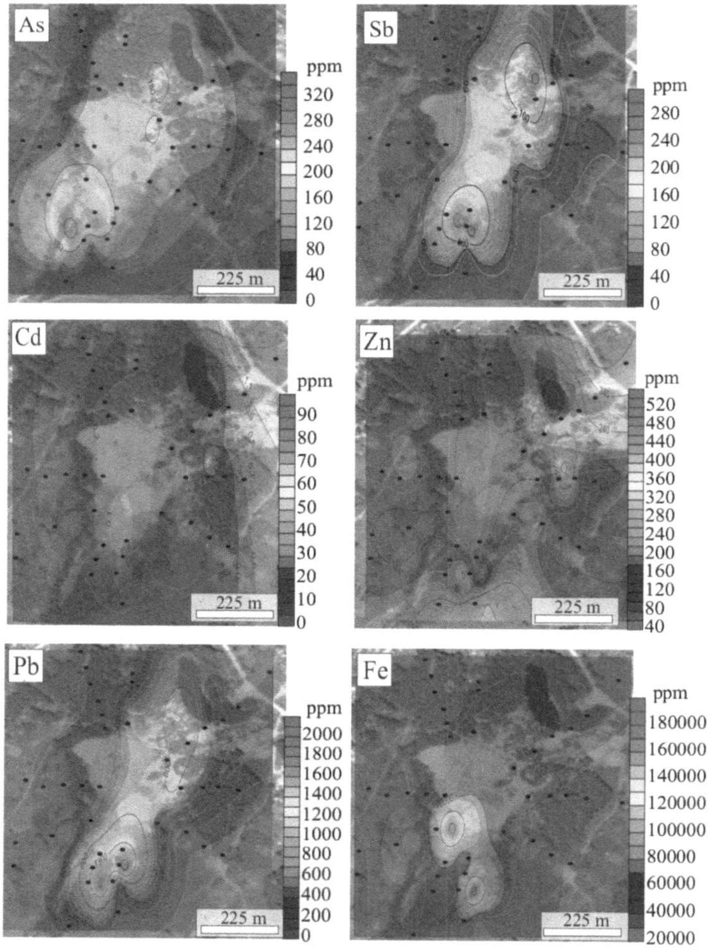

FIGURE 2.4 Estimated probability maps of As, Sb, Cd, Zn, Pb, and Fe concentrations in soils.

Over time, since the storage of tailings began, soils have become secondary pollutants, accumulating toxic elements and passing these elements further along the trophic chains.

The evaluation of total chemical element concentrations in soils surrounding the Ursk tailings provides baseline data for future research and will be a scientific basis for preventing and controlling contamination by heavy metals.

FUNDING

The study was supported by the state contract of the Trofimuk Institute of Petroleum-Gas Geology and Geophysics (project no. FWZZ-2022-0028 and project no. FWZZ-2022-0029).

ACKNOWLEDGMENTS

Our thanks to Jean Kollantai, MSW, Tomsk State University, for assistance with style.

REFERENCES

Abrosimova N., Bortnikova S., Edelev A., Chernukhin V., Reutsky A., Abrosimov N., Gundyrev I. Geochemical and microbiological composition of soils and tailings surrounding the Komsomolsk tailings, Kemerovo Region, Russia. *Bacteria* 2023;2(3):116–28.

Agnieszka B., Tomasz C., Jerzy W. Chemical properties and toxicity of soils contaminated by mining activity. *Ecotoxicology* 2014;23(7):1234–1244.

Aguilar F.J., Aguera F., Aguilar M.A., Carvajal F. Effects of terrain morphology, sampling density, and interpolation methods on grid DEM accuracy. *Photogrammetric Engineering & Remote Sensing* 2005;71:805–816.

Bech J., Roca N., Tume P., Ramos-Miras J., Gil C., Boluda R. Screening for new accumulator plants in potential hazards elements polluted soil surrounding Peruvian mine tailings. *Catena* 2016;136:66–73.

Bisone S., Chatain V., Blanc D., Gautier M., Bayard R., Sanchez F., Gourdon R. Geochemical characterization and modeling of arsenic behavior in a highly contaminated mining soil. *Environmental Earth Sciences* 2016;75:1–9.

Bolgov G.P. Sulfidy Salaira. Urskaya gruppa polimetallicheskikh mestorozhdeniy [Salair sulfides. Ursk group of polymetallic deposits]. *News of the Tomsk Industrial Institute* 1937;58:45–96 (in Russian).

Buch A.C., Niemeyer J.C., Marques E.D., Silva-Filho E.V. Ecological risk assessment of trace metals in soils affected by mine tailings. *Journal of Hazardous Materials* 2021;403:123852.

Diatta J.B., Chudzinska E., Wirth S. Assessment of heavy metal contamination of soils impacted by a zinc smelter activity. *Journal of Elementology* 2008;13:1.

Ech-Charef A., Dekayir A., Jordán G., Rouai M., Chabli A., Qarbous A., El Houfy F.Z. Soil heavy metal contamination in the vicinity of the abandoned Zeïda mine in the Upper Moulouya Basin, Morocco. Implications for airborne dust pollution under semi-arid climatic conditions. *Journal of African Earth Sciences* 2023;198:104812.

Espinoza S.E., Quiroz I.A., Magni C.R., Yáñez M.A., Martínez E.E. Long-term effects of copper mine tailings on surrounding soils and sclerophyllous vegetation in Central Chile. *Water, Air, & Soil Pollution* 2022;233(8):288.

Fagbenro A.A., Yinusa T.S., Ajekiigbe K.M., Oke A.O., Obiajunwa E.I. Assessment of heavy metal pollution in soil samples from a gold mining area in Osun State, Nigeria using proton-induced X-ray emission. *Scientific African* 2021;14:e01047.

Franchini J.C., Crispino C.C., Souza R.A., Torres E., Hungria M. Microbiological parameters as indicators of soil quality under various soil management and crop rotation systems in southern Brazil. *Soil and Tillage Research* 2007;92(1–2):18–29.

Ghosh S., Banerjee S., Prajapati J., Mandal J., Mukherjee A., Bhattacharyya P. Pollution and health risk assessment of mine tailings contaminated soils in India from toxic elements with statistical approaches. *Chemosphere* 2023;324:138267.

Ginocchio R., Toro I., Schnepf D., Macnair N.R. Copper tolerance testing in populations of Mimulus luteus var. variegatus exposed and non-exposed to copper mine pollution. *Geochemistry: Exploration. Environment, Analysis* 2002;2:151–156.

Gotway C.A., Ferguson R.B., Hergert G.W., Peterson T.A. Comparison of kriging and inverse-distance methods for mapping soil parameters. *Soil Science Society of America Journal* 1996;60:1237–1247.

Hakanson L. Ecological risk index for aquatic pollution control, a sedimentological approach. *Water Research Journal* 1980;14:975–1001.

Hillier A. *Manual for Working with ArcGIS 10*. University of Pennsylvania, Pennsylvania, PA, 2011.

Klerks P.L., Levinton J.S. Rapid evolution of metal resistance in a benthic oligochaete inhabiting a metal-polluted site. *Biology Bulletin* 1989;176:135–141.

Krishna A.K., Mohan K.R., Murthy N.N., Periasamy V., Bipinkumar G., Manohar K., Rao S.S. Assessment of heavy metal contamination in soils around chromite mining areas, Nuggihalli, Karnataka, India. *Environmental Earth Sciences* 2013;70(2):699–708.

Lazareva E.V., Myagkaya I.N., Kirichenko I.S., Gustaytis M.A., Zhmodik S.M. In-situ Element Accumulation in natural organic mutter under acid mine drainage influence. *Science of the Total Environment* 2019;660:468–483.

Liu J., Li N., Zhang W., Wei X., Tsang D.C.W., Sun Y., Luo X., Bao Z., Zheng W., Wang J., Xu G., Hou L., Chen Y., Feng Y. Thallium contamination in farmlands and common vegetables in a pyrite mining city and potential health risks. *Environmental Pollution* 2019;248:906–915.

Loska K., Wiechuła D., Korus I. Metal contamination of farming soils affected by industry. *Environment International* 2004;30(2):159–165.

Muller G. Index of geo-accumulation in sediments of the Rhine River. *Geological Journal* 1969;2(3):108–118.

Myagkaya I.N., Gustaytis M.A., Kirichenko I.S., Saryg-ool B.Yu., Lazareva E.V. Acid mine drainage contamination of the ur impoundment environmental geochemistry. *E3S Web of Conferences* 2019;98:09021.

Myagkaya I.N., Lazareva E.V., Gustaitis M.A., Zhmodik S.M. Gold and silver in a system of sulfide tailings. Part 1: Migration in water flow. *Journal of Geochemical Exploration* 2016;160:16–30.

Naccarato A., Tassone A., Cavaliere F., Elliani R., Pirrone N., Sproveieri F., Tagarelli A., Giglio A. Agrochemical treatments as a source of heavy metals and rare earth elements in agricultural soils and bioaccumulation in ground beetles. *Science of The Total Environment* 2020;749:1–41.

Nouri M. Assessment of metals contamination and ecological risk in ait ammar abandoned iron mine soil, Morocco. *Ekológia* 2016;35:32–49.

Official website of the Federal Service for Supervision of Natural Resources. https://www.rpn.gov.ru

Omotoso O.A., Ojo O.J. Assessment of some heavy metals contamination in the soil of river Niger floodplain at Jebba, central Nigeria. *Water Utility Jouranl* 2015;9:71–80.

Ouabo R.E., Sangodoyin A.Y., Ogundiran M.B. Assessment of ordinary Kriging and inverse distance weighting methods for modeling chromium and cadmium soil pollution in E-waste sites in Douala, Cameroon. *Journal of Health and Pollution* 2020;10(26):200605.

Picaud F., Petit D.P. Primary succession of Orthoptera on mine tailings: role of vegetation. *International Journal of Entomology* 2013;4 (1):69–79.

Qiao P., Lei M., Yang S., Yang J., Guo G., Zhou X. Comparing ordinary kriging and inverse distance weighting for soil as pollution in Beijing. *Environmental Science and Pollution Research* 2018;25:15597–608.

Roslakov N.A., Nesterenko G.V., Kalinin Y., Vasiliev I.P., Nevolko A.I., Roslakova N.V., Ivanov O.P. *Gold Contents in Eluvium of the Salair Ridge OIGGM NIC*. Sobolev Institute of Geology and Mineralog, Novosibirsk, Russia, 1995 (in Russian).

SanRaR 1.2.3685-21. Sanitary Rules and Regulations. Hygienic Standards and Requirements for Ensuring the Safety and (or) Harmlessness to Humans of Environmental Factors. APPROVED by the Decree of the Chief State Sanitary Doctor of the Russian Federation Dated January 28, 2021 No 2. Registered with the Ministry of Justice of the Russian

Federation on January 29, 2021, Registration N 62296. With Amendments and Additions from December 30, 2022. https:////base.garant.ru/400274954/ (accessed on 18 May 2023).

Saryg-ool B.O.Y., Myagkaya I.N., Kirichenko I.S., Gustaytis M.A., Shuvaeva O.V., Zhmodik S.M., Lazareva E.V. Redistribution and speciation of elements in gold-bearing sulfide mine tailings interbedded with natural organic matter: case study of Novo-Ursk deposit, Kemerovo Region, Siberia. *Geochemistry: Exploration, Environment, Analysis* 2020;20(3):323–336.

Spezia G.R., Souza E.G., Nóbrega L.H.P., Uribe-Opazo M.A., Milan M., Bazzi C.L. Model to estimate the sampling density for establishment of yield mapping. *Revista Brasileira de Engenharia Agrícola e Ambiental* 2012;16(4):449–457.

Sutherland R.A., Tolosa C.A., Tack F.M.G., Verloo M.G. Characterization of selected element concentrations and enrichment ratios in background and anthropogenically impacted roadside areas. *Archives of Environmental Contamination and Toxicology* 2000;38:428–438.

Venables W.N., Smith D.M. *An Introduction to R*, 2nd edn. Network Theory Limited, London, UK, 2009.

Vicente-Beckett V.A., Taylor McCauley G.J., Duivenvoorden L.J. Metal speciation in sediments and soils associated with acid-mine drainage in Mount Morgan (Queensland, Australia). *Journal of Environmental Science and Health, Part A* 2016;51(2):121–34.

Vinogradov A.P. Average contents of chemical elements in the principal types of igneous rocks of the Earth's crust. *Geochemistry* 1962;7:641–664.

Wang J., Jiang Y., Sun J., She J., Yin M., Fang F., Xiao T., Song G., Liu J. Geochemical transfer of cadmium in river sediments near a lead-zinc smelter. *Ecotoxicology and Environmental Safety* 2020;196:110529.

Wei B., Yang L. A review of heavy metal contaminations in urban soils, urban road dusts and agricultural soils from China. *Microchemical Journal* 2010;94(2):99–107.

Wollenhaupt N.C., Wolkowski R.P., Clayton M.K. Mapping soil test phosphorus and potassium for variable rate fertilizer application. *Journal of Production Agriculture* 1994;7(4):441–448.

Yin Y., Wang X., Hu Y., Li F., Cheng H. Soil bacterial community structure in the habitats with different levels of heavy metal pollution at an abandoned polymetallic mine. *Journal of Hazardous Materials* 2023;442:130063.

Yongming H., Peixuan D., Junji C., Posmentier E.S. Multivariate analysis of heavy metal contamination in urban dusts of Xi'an, Cent, China. *Sci Total Environment* 2006;355:176–186.

Yurkevich N., Bortnikova S., Abrosimova N., Makas A., Olenchenko V., Yurkevich N., Edelev A., Saeva O., Shevko A. Sulfur and nitrogen gases in the vapor streams from ore cyanidation wastes at a sharply continental climate, Western Siberia, Russia. *Water, Air, & Soil Pollution* 2019;230:1–7.

Zhang C. Using multivariate analyses and GIS to identify pollutants and their spatial patterns in urban soils in Galway, Ireland. *Environmental Pollution* 2006;142(3):501–11.

Zhang J., Liu C.L. Riverine composition and estuarine geochemistry of particulate metals in China-weathering features, anthropogenic impact and chemical fluxes. *Estuarine, Coastal and Shelf Science* 54:2002;1051–1070.

Zhu,Q., Lin H.S. Comparing ordinary kriging and regression kriging for soil properties in contrasting landscapes. *Pedosphere* 2010;20(5):594–606.

3 Effects of Revegetation on Abandoned Man-Made Slopes of Chromite Mine in Sukinda Region, India

Anand Singh, Sanjay Kr. Roy, Amit Kr. Verma, Rakesh Kr. Singh, Kartik Varwade, Prince Kumar, Swapan Mahato, and Chandra Shekhar Singh

3.1 INTRODUCTION

In developing nations like India, large infrastructural developments and constructions have shown a stormy growth for the economic and social development of the country. This has led to increase in demands of material and minerals that have to be extracted through the process of mining. There are two types of mining operations prevalent in India, namely underground and opencast. An efficient and economical mode of mining operation is the production of mineral resources through opencast mining. Opencast mines are the most typical reason for degradation of ecosystem (Wu et al., 2019). Exploitation of minerals through opencast mining causes severe environmental disturbance in the area, such as soil erosion (Neshat et al., 2014), ground subsidence (Dong et al., 2015), pollution related to heavy metal (Liu et al., 2017; Ying et al., 2016) with highest impact on degradation of vegetation and soil. Opencast mining changes the topography of the mine area, resulting in the significant change to absorption of soil nutrients and water from the roots of the plant present in the local ecosystem (Cowling et al., 2015; Niu et al., 2019). Disturbances in vegetation of the mine area and damage in the soil quality of the area can lead to imbalances of energy flow and material cycle (Yang et al., 2018). The damage in the soil and ecosystem of the area due to mining process disturbs the structure and function of the prevailing ecosystem in the area (Huang et al., 2015; Zhengfu et al., 2010). Change in structure and function of the ecosystem weakens the ability of resisting interference from external elements and leads to the ecological instability of the mine area (Bian et al., 2018; Sun et al., 2015). Organic matter and nitrogen in the soil play crucial role in growth of vegetation

DOI: 10.1201/9781003442554-4

(Jing et al., 2018), but the contents of organic matter and nitrogen are usually low in soil of mining areas (Liu et al., 2017).

Opencast mining generates huge quantity of overburden materials, which needs to be disposed or relocated at any other place for extraction of valuable minerals. These wastes are dumped in the form of heap-like structures, which undergo several stabilization techniques so that they do not affect the mining activities or harm the livelihood of the nearby population. Stability issue related to waste dump slope needs to be solved for proper functioning of the mining operations in opencast mine. One of the most widely and accepted stabilization techniques used in abandoned man-made waste dump slopes of opencast mine is the process of revegetation (Akers et al., 1974). Restoration of vegetation on abandoned waste dump slopes is important for stability of slopes in long term (Miao & Marrs, 2000; Sever & Makineci, 2009; Bao et al., 2012; Drazic et al., 2012). Root penetration through revegetation in the abandoned mine dump slope helps in improving the geotechnical property of soil and stability of aggregates material. Revegetation is also the process of restoring the ecological balance of the mine area (Jorgensen, 1994). The hydrogeological and mechanical strength of the waste dump slope material changes with the growth of vegetation in mine waste dump slopes (Cherubini & Giasi, 1997). Hydrogeological changes enhance the stability of waste dump as it controls the penetration of rain water, evapotranspiration and reduces the pore pressure in the waste dump soil (Hussain, 1995). Mechanical changes in the waste dump slope material help in enhancing the reinforcement of dump soil material with roots, thereby strengthening the shear property of the dump slope material. Some of the major hydrogeological and mechanical changes in the dump material occur by the effect of root density and its depth (Greenway, 1987; Jha, 1989; Suyama, 1992; Hall et al., 1994). Field experiments and rainfall simulation have resulted in revealing the effect of vegetation on soil erosion and runoff (Calvo-Cases et al., 2003; Hartanto et al., 2003, Biemelt et al., 2005; Cao et al., 2008; Shi et al., 2010; El Kateb et al., 2013). Indigenous plants such as *Shorea robusta* (Sal), *Azadirachta* (Neem), and *Dalbergia sisoo* (Shisham) are known to be having high binding capacity, prevent erosion of soil, and improve the stability of the waste dump slope. The most adopted, structured, and feasible method for soil and water conservation in the mining area is by maintaining a diverse vegetation after the mine is reclaimed (Miao & Marrs, 2000; Josa et al., 2012; Xu, 2012).

This study focuses on the impact of root depth on various physical parameters of the waste dump slope materials deposited in the Chromite mine of Sukinda region, Jajpur, Odisha, India. Three abandoned mine dump slopes of the same mine were selected for the study. Each abandoned mine dump has different types of vegetation prevalent in the dump. The change in physical parameters of the waste dump slope was studied with the change in root depth of vegetation prevailing in each dump. The vegetation with maximum density in the dump slope was selected for studying the physical change in waste dump material. Here, in this case, *Acacia auriculiformis* tree species were found in abundance in all the three dumps of same mine, and hence, this tree is considered for studying the effect of root penetration in the waste dump slope materials.

3.2 STUDY AREA, GEOLOGY, AND HYDROLOGICAL CHARACTERISTICS

Survey of India toposheet nos. 73H/I3, 73G/16, and 73G/12 covers the study area of Sukinda region, Jajpur, Odisha, India. Latitude from 20°45′N to 21°00′N and longitude from 85°45′E to 86°00′E bound the area of study (Behura et al., 2015). Sukinda region shows varied physiographical features such as 20%–25% forest, 5%–10% with Brahmani River basin, and cultivated land of 65%–70%. The highest elevation measured from Mean Sea Level is about 440 m in the Sukinda region, Odisha, India. Chromite ore belt is present in a V-shaped valley structure, which lies between Daitari and Mahagiri hill in the north and south, respectively (Nayak & Kale, 2020). Figure 3.1 shows different geological, geohydrological, geotechnical, geomorphological, and natural potential of the study area.

Area of study lies in Gorumahisani Group of Archaean age, dominant in jasper, oldest grit, ferruginous slate, epidiorite, and conglomerate (Sahoo et al., 2018). Various subgroups such as Charnockite, Khondalite, and Migmatite of Proterozoic

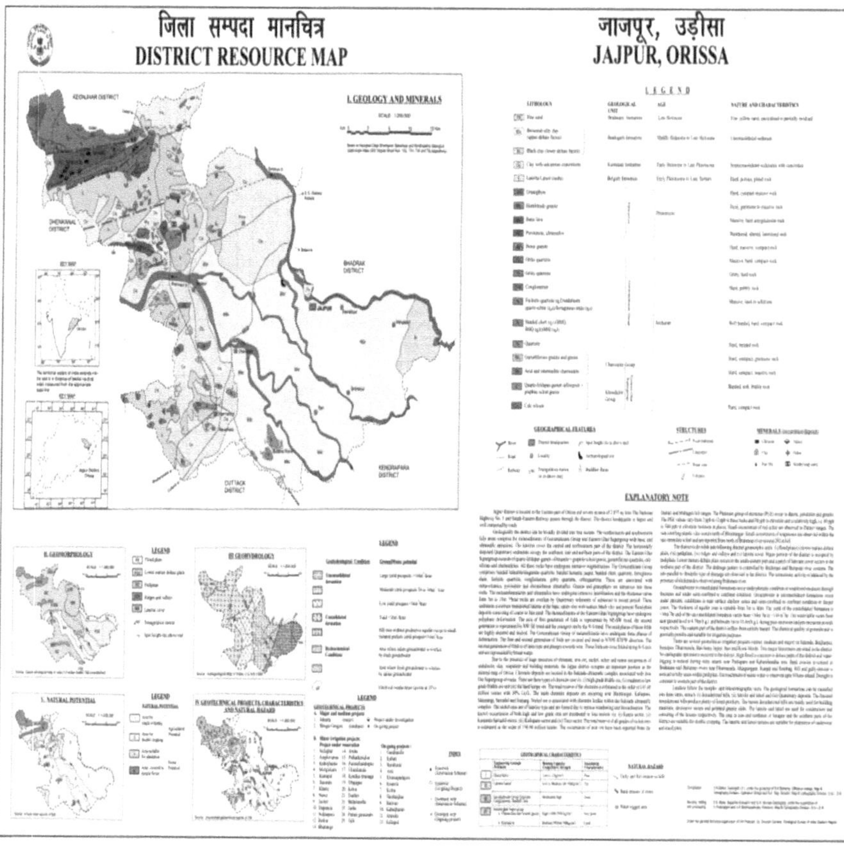

FIGURE 3.1 Jajpur resource map. (Geological Survey of India.)

and Archaean ages are situated above the older rock. Gneiss of super group, Paleo-proterozoic, is represented by the presence of gneiss of tourmaline hornblende, granite, and biotite. The study area consists of majority of sediments from Pleistocene and Holocene ages. The sediments of Pleistocene and Holocene ages are generally lying low, which creates the undulating land, and are labeled as the youngest formation of the study area. The major intrusive, namely Sukinda ultramafics shows an intrusion in the older rock with an average width of 2–5 km and goes for about 20 km in ENE-WNW direction from Kansa in the east to Maruabil and beyond in the west. Chromite bands devoid consists essentially of magnesium and less quantity of pyroxenite in the void of chromite mineral. Peridotite-dunite members found in the area are mostly serpentinized and laterized, but the pyroxenite found in the study area is comparatively fresh.

The study area lies in tropical region where climate is characterized by very hot summers and cool winters. Summer is typically from March to June when daily average maximum temperature ranges from a maximum of 43°C during daytime to a minimum of 16°C at night. Winter is from November to February when daily average maximum temperature during day goes up to 30°C and minimum temperature at night becomes as low as 10°C. The average annual rainfall is about 1,000 mm every year. The Southwest monsoon lasts from mid-June to mid-September, and the area gets more than 75% of the annual rainfall during this period.

3.3 METHODOLOGY

3.3.1 Selection of Plant Specimen

Three inactive dumps of Sukinda region was selected for studying the effect of root penetration on the stability of the inactive dumps in the area. The details of the inactive dumps are provided in Table 3.1. The soil samples and soil with root samples were collected from the dump, after they reached their maximum height, and no further development is planned on the same dump.

Inactive dumps were planted with various vegetation types for stabilizing the dumps. The details of vegetation types found on the inactive dumps are shown in Table 3.2.

Dumps D1, D2, and D3 were physically investigated, and it was observed that the highest number of *Acacia auriculiformis* tree species was present in each of the

TABLE 3.1
Description of Dump Sites in Sukinda

		Location			Dump Angle (°)	
S. No.	Dump Name	Latitude	Longitude	Max. Height of Dump (m)	Bench Angle	Overall Slope Angle
1	D1	21°03′06.5″	85°48′50.8″	19	45°	45°
2	D2	21°02′47.5″	85°48′55.4″	14	29°	29°
3	D3	21°02′29.3″	85°48′57.1″	24	36°	22°

TABLE 3.2
Common Vegetation Found on Overburden Dump Slopes of Sukinda Region

S. No.	Dump Name	Scientific Name of Common Vegetation (Common Name)
1	D1	Mangifera indica (Mango), Psidium guajava (Pjuli), *Acacia auriculiformis (Akasia)*, Melia azedarach Linn. (Mahanimba)
2	D2	*Acacia auriculiformis (Akasia)*, Azadirachta indica (Neem), Albizia lebbeck (Sirisha), Dalbergia sissoo (Sisso)
3	D3	Azadirachta indica (Neem), Dalbergia sissoo (Sisso), *Acacia auriculiformis (Akasia)*, Neolamarckia cadamba (Kadamba), Millettia pinnata (Kranja), Tecoma (Tekama), Azadirachta indica (Neem), Senna hirsuta (Bada Chakunda), Conocarpus lancifolius (Conocarpus), Citrullus Colocynthis (Bara)

three dumps of Sukinda region. Root samples of *Acacia auriculiformis* with soil were collected from all the three dumps for estimating their physical properties. Soil without roots were also collected from the three dumps for estimating the change in the soil physical parameters after the root penetration of *Acacia auriculiformis*. This study deals with the change in physical properties of soil in Sukinda region after the penetration of root in the soil from vegetation.

3.3.2 Collection of Root Specimen and Preparation

Three inactive dumps of Sukinda region were physically investigated, and a suitable plan was made to collect the soil samples with root of *Acacia auriculiformis* plant and without roots. As per the planned method of collection of soil with roots, the surface of the dump was divided into 1*1 m quadrant (Pandey et al., 2014). The quadrant was again subdivided into 10 smaller quadrants for collecting the sample from each smaller quadrant. The samples from each quadrant will ensure the homogeneity in the collection of samples. Soil samples with root were collected after removal of the top soil up to 10 cm on the surface. After the removal of top soil up to 10 cm, soils with root samples were collected after digging the surface up to 30 cm depth. The samples were collected in a cylindrical container as shown in Figure 3.2.

After collecting the soil with root samples up to 30 cm depth, surface was again dig up to the depth of 60 cm from 30 cm depth. The collection of both the samples has been shown in Figure 3.2. The samples of soil with roots were collected in the same way as they are shown in Figure 3.2. Soils without roots were collected after removing the top surface up to 10 cm and where there was no evidence of roots present in the soil.

The roots collected with soil were stored in an air-tight plastic bag so that the soil doesn't lose its moisture content. The collected soil with root and soil without root were immediately carried to the laboratory for testing of the variations in physical parameters of the collected sample.

Figure 3.3 shows the resultant of the geotechnical testing on soil sample prepared in the laboratory.

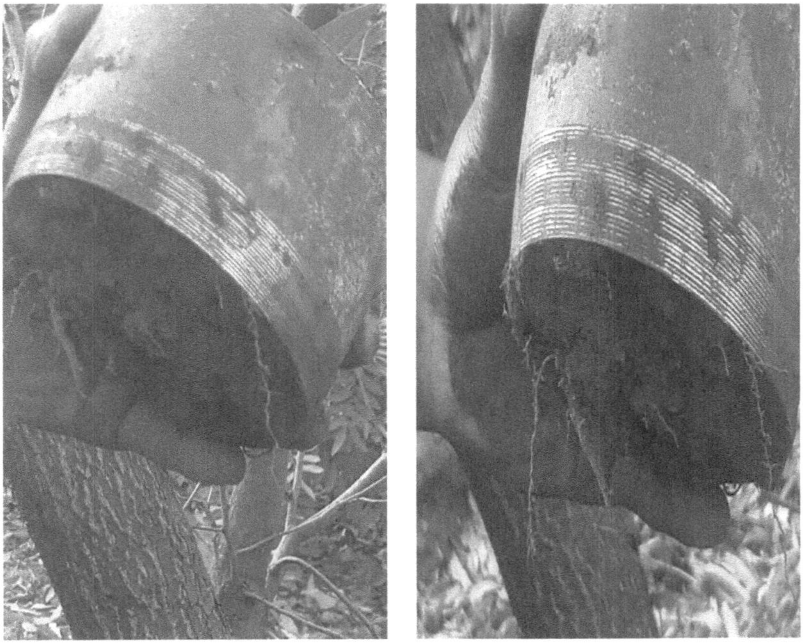

FIGURE 3.2 Soil with root samples collected for laboratory tests from 30 and 60cm.

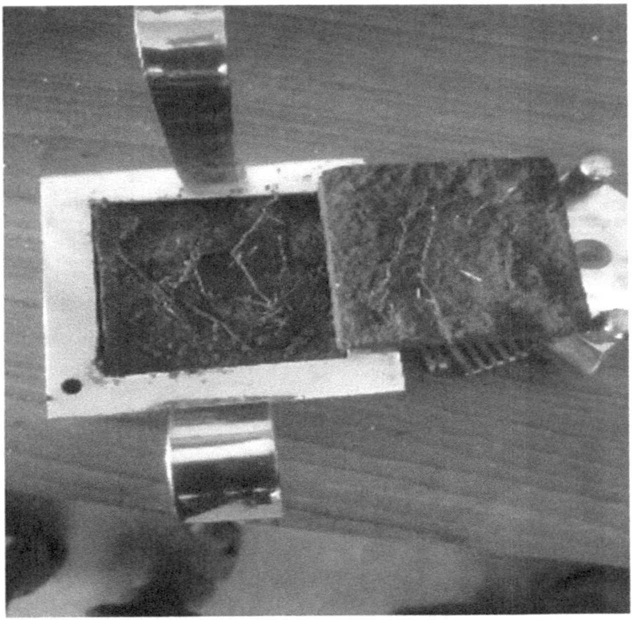

FIGURE 3.3 Preparation of soil with roots sample in the sampler of direct shear strength testing apparatus.

The procedures followed for testing the soil have been mentioned as follows:

1. Soil was air-dried and then sieved by a 2 mm mesh size.
2. The soil was mixed with the water in such a way that the soil moisture content should be in between 12% and 14% (Fan & Su, 2008).
3. Saturated soil was kept in an air-tight container for 24 hr, so that the soil fully soaks the water.
4. Required amount of soil was taken from the sealed container. The soil was kept in sampler and compacted in layers with root.

The aforementioned procedure was followed for preparation of the samples of soil with root. Same procedure for testing of soil without roots was also followed except the introduction of root in the sampler.

3.4 PHYSICAL PROPERTIES OF SOIL OF STUDY AREA

Inactive dumps D1, D2, and D3 were studied for different physical parameters as mentioned in Table 3.3. The changes in the soil properties with and without roots were investigated by following the procedures mentioned in there IS standards code.

It is evident from Table 3.3 that the bulk density of the dump material increases with the depth of root penetration. The density of the dump materials without root is quite low when compared with the density of the dump materials with root. The increase in density of the dump material generates more stress in mine dump slopes, which leads to increase in shear strain and, hence, decreases the safety factor of the mine dump slopes (Nie et al., 2018). In general, bulk density varies in the range of 9.80–13.7 kN/m^3 for undisturbed soil. Plantation is better in the soil with bulk density of sandy material <15.7 kN/m^3, <13.7 kN/m^3 for silty material, and <10.8 kN/m^3 for clayey materials (Muckel & Mausbach, 1997). Abundant amount of lateritic soil having high percentage of oxides of iron generally shows higher bulk density.

TABLE 3.3
Soil Properties of Inactive Dumps with Root and without Roots

Name of Dump	Depth of Root	Bulk Density (kN/m^3)	Moisture Content (%)	Specific Gravity	Void Ratio (e)	Porosity (*n*) (%)	Liquid Limit (%)	Plastic Limit (%)
D1	OB	12.60	17.90	2.47	0.95	48.72	38	32.80
	0–30 cm	16.18	15.35	2.51	0.92	47.90	37.8	32.22
	30–60 cm	16.67	17.80	2.61	0.86	46.23	38.3	27.7
D2	OB	12.70	18.80	2.49	0.89	46.60	36.93	26.51
	0–30 cm	15.79	15.85	2.59	0.79	44.07	34.80	26.20
	30–60 cm	16.77	19.6	2.71	0.69	40.82	37.00	27.3
D3	OB	12.80	19.10	2.50	1.15	59.20	46.02	31.80
	0–30 cm	15.83	15.90	2.64	1.04	50.13	43.29	31.30
	30–60 cm	16.87	19.20	2.73	0.89	36.7	50	35.65

In general, moisture content in soil varies in the range of 10%–45% for the ideal growth of vegetation in lateritic soil. The safety factor of mine slope is maximum when the moisture content in soil is 10%, and the safety factor remains high up to the threshold limit of 30% moisture content after which the factor of safety starts decreasing (Li et al., 2021). Specific gravity is defined as the ratio of material density to the density of water at a fixed temperature. Specific gravity of the dump material without root and dump material with root varies from 2.47 to 2.73, as mentioned in Table 3.3. Void ratio is the ratio of volume of the void to the solid volume, and it ranges from 0.69 to 1.15 as shown in Table 3.3. Porosity is also related to the void ratio of as defined in Eq. (3.1).

$$n = \left(\frac{1}{1+e}\right) * 100 \quad (3.1)$$

Liquid limit of soil is defined as the quantity of water content after which the soil tends to drift and flow. Plastic limit of soil is defined as the amount of water content in the soil after which the fine soil grains cannot be remolded unless they are broken. The physical properties calculated in Table 3.3 directly or indirectly also affect the stability of mine slopes, which can be discussed further after numerical analysis of the mine slopes. This chapter only deals with the change in physical properties of the dump materials when they come in contact with plant roots at different depths. Void ratio and content of moisture in soil were calculated as per the standard procedures mentioned in IS 270 Part 2—1973. Procedures mentioned in IS code 270 Part 5—1985 were followed for calculating the plastic and liquid limit of the dump materials with root and without root.

3.5 RESULT AND DISCUSSION

Physical properties that affect the stability of the mine slopes are generally not considered when the safety factor of the mine slope is calculated by numerical analysis method. Factor of safety of mine slopes is inversely related to the bulk density of the dumping material (Nie et al., 2018). Shear strain increases with the increase in bulk density of the dump material as it increases the shear stress in the mine slope. The penetration of plant roots increases the bulk density of the dump material, but when it is compared with the increase in shear strength properties of the soil, increase in bulk density does not affect much the stability of the mine slopes. Content of moisture in dump materials increases with the depth of the root penetration. Specific gravity shows very little variation up to the root penetration of 60cm from inactive dump surface.

From Figures 3.4–3.6, it is evident that moisture content decreases when root is in the depth of 0–30cm, but it increases gradually when the depth of root is between 30 and 60cm. It can be due to the moisture-holding capacity of roots, which is increased when the depth of root is between 30 and 60cm. Specific gravity of the dump materials increases with increase in the root penetration of plant up to 60cm. Void ratio of the dump material decreases with the increase in depth of the plant root penetration. It results in decreasing the penetration of water in the dump materials. Since porosity

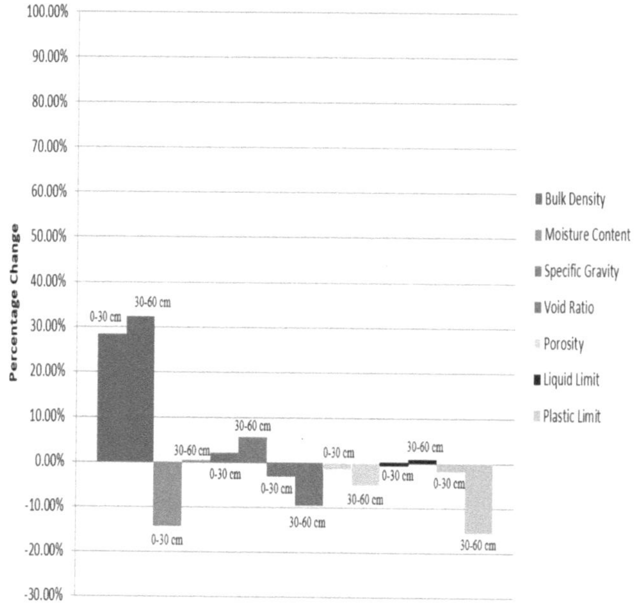

FIGURE 3.4 Change in physical properties of dump material of Dump D1 in presence of root.

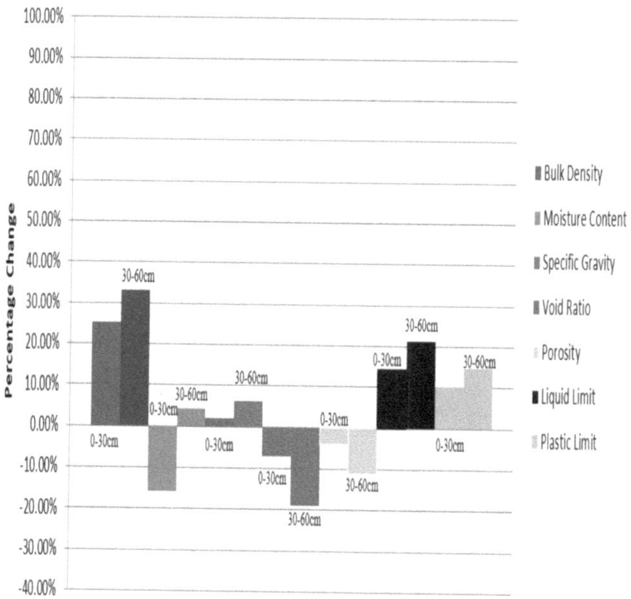

FIGURE 3.5 Change in physical properties of dump material of Dump D2 in presence of root.

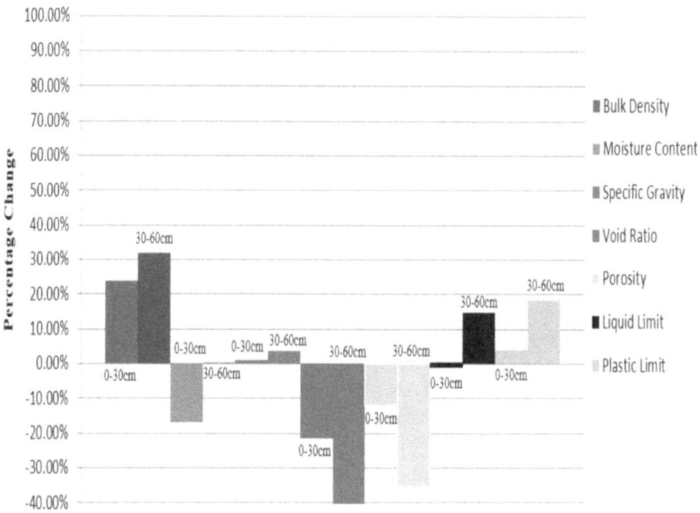

FIGURE 3.6 Change in physical properties of dump material of Dump D3 in presence of root.

is related to the void ration as shown in Eq. (3.1), the porosity of the dump material also decreases with the increase in depth of root penetration up to 60 cm. It is also evident that the liquid and plastic limits of the dump material increase as the root is penetrated inside up to 60 cm (except in the case of D1 for plastic limit). The decrease in plastic limit in the case of D1 needs more study and is considered as an exceptional case for this chapter.

3.6 CONCLUSION

Physical characteristic of dump material affects the shear strength of dump slope in mines. Shear strength property of the dump material plays an integral part of the dump slope when the engineers talk about the stability analysis of the mine slope. From Table 3.3 and Figure 3.4, Figure 3.5, and Figure 3.6, it can be seen that with the penetration of plant roots on inactive mine dump slopes, the physical properties of the dump material change for the better stability of the dump materials. Plant root reduces the moisture content up to 30 cm, decreases the void ratio of the dump materials, reduces the porosity of the dump material, and also affects the liquid limit and plastic limit of the dump materials. Although the bulk density of the dump material increases with the penetration of plant root, it gets compensated with the increase in other physical properties of the dump materials. This study has tried to estimate the change in physical properties of the inactive dump slopes of the mining area after revegetation. Directly or indirectly, physical properties of the dump slope materials play an important part in affecting the geotechnical properties of the dump slope materials. Presence of tree root helps in enhancing the physical properties of the slope materials, which is directly related to the stability of the dump slopes.

REFERENCES

Akers, D. J., Coalgate, J. L., & Muter, R. B. (1974). *Gob Pile Stabilization and Reclamation.* Coal Research Bureau, West Virginia University.

Bao, N., Ye, B., & Bai, Z. (2012). Rehabilitation of vegetation mapping of ATB opencast coal-mine based on GIS and RS. *Sensor Letters*, 10(1–2), 387–393.

Behura, S., Kar, M., & Upadhyay, V. P. (2015). Phytosociological Study of herbs species at two reclaimed sites of Sukinda Chromite mining region of Odisha, India. *Geo-Eco-Trop*, 39(2), 329–342.

Bian, Z. F., Lei, S. G., Jin, D., & Wang, L. (2018). Several basic scientific issues related to mined land remediation. *Journal of China Coal Society*, 43(1), 190–197.

Biemelt, D., Schapp, A., Kleeberg, A., & Grünewald, U. (2005). Overland flow, erosion, and related phosphorus and iron fluxes at plot scale: A case study from a non-vegetated lignite mining dump in Lusatia. *Geoderma*, 129(1–2), 4–18.

Calvo, A., Boix, C., & Imeson, A. C. (2003). Runoff generation, sediment movement and soil water behaviour on calcareous (limestone) slopes of some Mediterranean environments in Southeast Spain. *Geomorphology*, 50(1–3), 269–291.

Cao, Y., Ouyang, Z. Y., Zheng, H., Huang, Z. G., Wang, X. K., & Miao, H. (2008). Effects of forest plantations on rainfall redistribution and erosion in the red soil region of southern China. *Land Degradation & Development*, 19(3), 321–330.

Fan, C. C., & Su, C. F. (2008). Role of roots in the shear strength of root-reinforced soils with high moisture content. *Ecological Engineering*, 33(2), 157–166.

Cherubini, C., & Giasi, C. (1997). The influence of vegetation on slope stability. *International Journal of Civil Engineering and Technology,* 11, 124–131.

Cowling, R. M., Potts, A. J., Bradshaw, P. L., Colville, J., Arianoutsou, M., Ferrier, S., ... & Zutta, B. R. (2015). Variation in plant diversity in mediterranean-climate ecosystems: The role of climatic and topographical stability. *Journal of Biogeography*, 42(3), 552–564.

Dong, S., Samsonov, S., Yin, H., Yao, S., & Xu, C. (2015). Spatio-temporal analysis of ground subsidence due to underground coal mining in Huainan coalfield, China. *Environmental Earth Sciences*, 73, 5523–5534.

Dražić, D., Veselinović, M., Cule, N., & Mitrović, S. (2012). New post-exploitation open pit coal mines landscapes–potentials for recreation and energy biomass production: A case study from Serbia. *Moravian Geographical Reports*, 20(2), 2–16.

El Kateb, H., Zhang, H., Zhang, P., & Mosandl, R. (2013). Soil erosion and surface runoff on different vegetation covers and slope gradients: A field experiment in Southern Shaanxi Province, China. *Catena*, 105, 1–10.

Greenway, D. R. (1987). Vegetation and slope stability. In MG Anderson, KS Richards (eds) *Slope Stability*. Wiely, New York, pp. 187–230.

Hall, B. E., Giles, E. L., & Rauch, H. P. (1994). Experiences with the use of trees in slope stabilization. In *International Conference on Soil Mechanics and Foundation Engineering*, New Delhi, India, pp. 1231–1235.

Hartanto, H., Prabhu, R., Widayat, A. S., & Asdak, C. (2003). Factors affecting runoff and soil erosion: Plot-level soil loss monitoring for assessing sustainability of forest management. *Forest Ecology and Management*, 180(1–3), 361–374.

Huang, Y., Tian, F., Wang, Y., Wang, M., & Hu, Z. (2015). Effect of coal mining on vegetation disturbance and associated carbon loss. *Environmental Earth Sciences*, 73, 2329–2342.

Hussain, A. (1995). Fill compaction-erosion study in reclaimed areas. *Indian Mining Engineering Journal*, 34(6), 19–21.

IS 2720-5 (1985): Methods of test for soils, Part 5: Determination of liquid and plastic limit [CED 43: Soil and Foundation Engineering].

IS 2720-2 (1973): Methods of test for soils, Part 2: Determination of water content [CED 43: Soil and Foundation Engineering].

Jha, A. K. (1989). A note on the root development in dry tropical naturally revegetated coalmine spoils. *Vegetatio*, 85, 67–70.

Jing, Z., Wang, J., Zhu, Y., & Feng, Y. (2018). Effects of land subsidence resulted from coal mining on soil nutrient distributions in a loess area of China. *Journal of Cleaner Production*, 177, 350–361.

Jørgensen, S. E. (1994). Models as instruments for combination of ecological theory and environmental practice. *Ecological Modelling*, 75, 5–20.

Josa, R., Jorba, M., & Vallejo, V. R. (2012). Opencast mine restoration in a Mediterranean semi-arid environment: Failure of some common practices. *Ecological Engineering*, 42, 183–191.

Li, J., Wang, X., Jia, H., Liu, Y., Zhao, Y., Shi, C., ... & Wang, K. (2021). Assessing the soil moisture effects of planted vegetation on slope stability in shallow landslide-prone areas. *Journal of Soils and Sediments*, 21(7), 2551–2565.

Liu, D., Quan, Y., Ren, Z., & Wu, G. (2017). Assessment of heavy metal contamination in soil associated with Chinese coal-fired power plants: A case study in Xilingol, Inner Mongolia. *International Journal of sustainable Development & World Ecology*, 24(5), 439–443.

Liu, X., Bai, Z., Zhou, W., Cao, Y., & Zhang, G. (2017). Changes in soil properties in the soil profile after mining and reclamation in an opencast coal mine on the Loess Plateau, China. *Ecological Engineering*, 98, 228–239.

Miao, Z., & Marrs, R. (2000). Ecological restoration and land reclamation in open-cast mines in Shanxi Province, China. *Journal of Environmental Management*, 59(3), 205–215.

Muckel, G. B., & Mausbach, M. J. (1997). Soil quality information sheets. *Methods for Assessing Soil Quality*, 49, 393–400.

Nayak, S., P, B., & Kale, P. (2020). A review of chromite mining in Sukinda Valley of India: Impact and potential remediation measures. *International Journal of Phytoremediation*, 22(8), 804–818.

Neshat, A., Pradhan, B., & Dadras, M. (2014). Groundwater vulnerability assessment using an improved DRASTIC method in GIS. *Resources, Conservation and Recycling*, 86, 74–86.

Nie, M., Mao, X., & Wang, Y. (2018, May). The influence of soil parameters on the stability of loose slope. In *2018 3rd International Conference on Advances in Materials, Mechatronics and Civil Engineering (ICAMMCE 2018).* Atlantis Press, Amsterdam, pp. 188–192.

Niu, Y., Zhou, J., Yang, S., Chu, B., Ma, S., Zhu, H., & Hua, L. (2019). The effects of topographical factors on the distribution of plant communities in a mountain meadow on the Tibetan Plateau as a foundation for target-oriented management. *Ecological Indicators*, 106, 105532.

Pandey, B., Agrawal, M., & Singh, S. (2014). Coal mining activities change plant community structure due to air pollution and soil degradation. *Ecotoxicology*, 23, 1474–1483.

Sahoo, H. B., Gandre, D. K., Das, P. K., Karim, M. A., & Bhuyan, G. C. (2018). Geochemical mapping of heavy metals around Sukinda–Bhuban area in Jajpur and Dhenkanal districts of Odisha, India. *Environmental Earth Sciences*, 77, 1–17.

Sever, H., & Makineci, E. (2009). Soil organic carbon and nitrogen accumulation on coal mine spoils reclaimed with maritime pine (Pinus pinaster Aiton) in Agacli–Istanbul. *Environmental Monitoring and Assessment*, 155, 273–280.

Shi, Z. H., Yan, F. L., Li, L., Li, Z. X., & Cai, C. F. (2010). Interrill erosion from disturbed and undisturbed samples in relation to topsoil aggregate stability in red soils from subtropical China. *Catena*, 81(3), 240–248.

Sun, Q., Bai, Z., Cao, Y., Xie, M., Hu, X., Jiang, Y., & Lu, Y. (2015). Ecological risk assessment of land destruction in large open-pit mine. *Transactions of the Chinese Society of Agricultural Engineering*, 31(17), 278–288.

Suyama, M. (1992). Assessment of biotechnical slope stability effect for urban forest in Japan. In *International Symposium on Landslides*, A.A. Balkema, Christchurch, pp. 831–836.

Wu, Z., Lei, S., Lu, Q., & Bian, Z. (2019). Impacts of large-scale open-pit coal base on the landscape ecological health of semi-arid grasslands. *Remote Sensing*, 11(15), 1820.

Xu, Y. W. (2012). Slope protection with vegetation for waste dump at Guangyue road in Shanghai. *Advanced Materials Research*, 594, 465–471.

Yang, Z., Zhu, Q., Zhan, W., Xu, Y., Zhu, E., Gao, Y., ... & Chen, H. (2018). The linkage between vegetation and soil nutrients and their variation under different grazing intensities in an alpine meadow on the eastern Qinghai-Tibetan Plateau. *Ecological Engineering*, 110, 128–136.

Ying, L., Shaogang, L., & Xiaoyang, C. (2016). Assessment of heavy metal pollution and human health risk in urban soils of a coal mining city in East China. *Human and Ecological Risk Assessment: An International Journal*, 22(6), 1359–1374.

Zhengfu, B. I. A. N., Inyang, H. I., Daniels, J. L., Frank, O. T. T. O., & Struthers, S. (2010). Environmental issues from coal mining and their solutions. *Mining Science and Technology (China)*, 20(2), 215–223.

4 Phytoremediation of Heavy Metals from Contaminated Sites

A Potential Review for Future Perspectives

Md. Saiful Islam, Tapos Kormoker, Abu Reza Md. Towfiqul Islam, Uttam Biswas Antu, Minhaz Uddin, Sazal Kumar, Md. Kamrul Haque, and Md. Abu Bakar Siddique

4.1 INTRODUCTION

Soil contamination by heavy metals has received significant attention because of its persistence, non-biodegradability, long biological half-lives, and toxicity to flora and fauna (Islam et al. 2015a; Proshad et al. 2018). Toxic metals in soil are commonly found as a result of both natural and man-made activities (Islam et al. 2015b, 2019; Peng et al. 2018). Soil contamination by common heavy metals such as mercury (Hg), lead (Pb), cadmium (Cd), copper (Cu), chromium (Cr), manganese (Mn), zinc (Zn), antimony (Sb), arsenic (As), and aluminum (Al) has now become a critical challenge in recent times of expanding urbanization and industrialization (Duruibe et al. 2007; Khalid et al. 2017). Heavy metals such as Cr, Cu, Cd, Pb, Hg, and As have generally been regarded as the most hazardous elements in the environment and have been put on the USEPA's (US Environmental Protection Agency) list of priority contaminants (Lei et al. 2010). Toxic metals are associated with a variety of health risks, including heart disease, chronic anemia, mental retardation (Iqbal 2012), cancers, and kidney problems (Wuana and Okieimen 2011), neurological system, brain, skin, teeth, and bone problems (Järup 2003). Therefore, reducing the health risks associated with exposure to heavy metals is critical, which can be addressed by eliminating toxic metals from the surroundings, and developing effective ways for heavy metals removal from contaminated soils is crucial.

This study looked into the phytoremediation approach, which is the easiest and cheapest way to clear up heavy metals from contaminated environments, particularly, from soils. Plants with higher bioaccumulation capabilities have shown potential in extracting metals from contaminated soils by phytoextraction, increasing or lowering

DOI: 10.1201/9781003442554-5

metal concentration in the soil, and boosting metal transport from soil to root or root to shoot cells (Glick 2010; Ma et al. 2011a; Rajkumar et al. 2012; Khalid et al. 2017; Liu et al. 2018). Since it does not disrupt the ecosystem, takes minimal labor, and is, therefore, less expensive than classic physicochemical procedures, phytoremediation is an environmentally benign alternative for reclaiming contaminated soil. In recent days, significant advances in the use of modern biotechnology, such as phytoextraction and phytodegradation of heavy metals, have been investigated (Rajakaruna et al. 2006; Souza et al. 2014). The production of various metabolites by root cells and microbes in the rhizosphere has been recognized as an essential element of growing plants for enhanced phytoremediation efficiency (Badri et al. 2009; Segura et al. 2009). To cope with metal toxicity, hyperaccumulator plants use a variety of techniques. Tolerance, for example, is assisted by techniques such as metal repossession, absorption in cell components, separation, and inactivation by biopolymer secretions (Choppala et al. 2014).

Metal removal from polluted soils can be accomplished using a variety of biophysical and chemical approaches. Phytoremediation, on the other hand, is a cost-effective method for removing toxic metals from heavily polluted areas. This type of decontamination is especially crucial in soil with metal pollution because contaminated soils have entirely different features than air or water. This could be because the persistence of toxic metals in soil is lengthier than practically any other substance. As plants are large consumers of metals from primary sources, metal remediation using plants (phytoremediation method) appears to be a viable and appealing alternative in the current conditions.

Therefore, in this review, the authors briefly explain certain cradles, and consequences of metals in the soil environment, aspects influencing the biological availability of metals, and contemporary bioadsorbent approaches such as phytomining, phytostabilization, phytovolatilization, phytoextraction, and others, with a focus on their benefits, drawbacks, mechanisms, and future perspectives, as well as how such strategies are beneficial. Additionally, innovative approaches, such as the use of hyperaccumulator plant–metal interactions, are presented in order to fulfill the purpose of phytoremediation involving metal purification, mobilization, immobilization, and transformation less expensively and more safely.

4.2 HEAVY METALS IN THE ENVIRONMENT

Toxic metals are widely proven to be classified into two classes in terms of crop health, with a few metals, such as Zn, Cu, Fe, Mn, and Ni being required by plants as key ingredients used for optimum plant development at relatively low levels (Fageria et al. 2009; Hemavathy et al. 2019; Lei et al. 2019; Li et al. 2019). Alternatively, substantial amounts are poisonous to floras and can hinder plant progress, deteriorate dirt superiority, reduce crop production, and consequently, reduce food value, posing a well-being hazard to individuals and creatures (Seth 2012). Certain metals, such as Cr, As, Pb, Ag, Se, Cd, and Hg, are known to be bioactive in plants (Seth 2012).

4.2.1 Heavy Metal Sources

As a result of both natural and manmade activities, heavy metals were increased in the surroundings. Certain soil receives toxic metals derived from the parent material from which they were formed as a natural source. Igneous rock and metamorphic rock make up around 95% of the overall earth's crust, whereas sedimentary rocks make up the remaining 5% (Thornton 1981). Metals such as Cd, Co, Zn, Cu, and Ni are much more common in basaltic volcanic rocks, whereas Cu, Zn, Mn, Pb, and Cd are abundant in shales made up of fine inorganic and organic silt.

Contrarily, industries and agricultural activities have led to elevated concentrations of heavy metals in water and soil, while excavation and smelting processes are seen as major recognizable emitters of all these metals in the environment. Phosphate nourishments, manure sludge, and secretions from power plants, metallic industries, city traffic, and cement manufacturing are considered the main cradles of toxic metals in dirt from human activities (Wu et al. 2004; Proshad et al. 2018). Polluted sludge and fertilizer, as well as metal-containing substances, contribute to soil metallic fortification (Sarwar et al. 2010; Czarnecki and Düring 2015). Numerous insecticides, fungicides, and herbicides, which are key sources of certain elements including Cu, As, Zn, and Fe, must be treated in vast proportions for integrated pest management (Alengebawy et al. 2021).

Copper is a widely utilized metal in a variety of items, including batteries, computer chips, semiconductors, mobile phones, sewage pipes, fertilizers, insecticides, paper manufacturing, mining, and smelting (Romero-Cano et al. 2017; Gossuin and Vuong 2018; Rani et al. 2018). Cu may enter the environment as a result of the extraction of copper from industries that manufacture or utilize metallic copper or copper alloys.

Zinc's chemical and mechanical properties are responsible for the majority of its manufacturing use, with galvanizing iron and steel being the most common applications. Other phosphorus-based fertilizers (TSP or triple superphosphate and calcium phosphate) include zinc and cadmium in varying amounts depending on the source of the rock phosphate could be used as a foliar application to replenish crucial micronutrients (Rehim et al. 2014; Imran et al. 2015; Imran and Rehim 2016). Similarly, substances including Zn, B, Fe, Mn, and Cu are used to promote optimum crop growth and production.

Furthermore, inorganic composts, carbon-based compounds containing high concentrations of heavy metals, including farmyard manure, mucks, and sludges, are added to the soil to increase soil productivity or rehabilitate damaged sites (Table 4.1). Cadmium is utilized in many different industries, such as batteries, electronics, metal refining, electroplating, petrochemical products, textile products, agrochemicals, metallurgical processes, photography, and dyeing industries (Basu et al. 2017; Jia et al. 2018; Zheng et al. 2018). Increased Cd concentrations in the air, water, and soil can be spotted near industrial sites, particularly nonferrous mining and metal-handling operations. The recurrent practice of crude surplus streams for irrigation could be a substantial home for heavy metals (Islam et al. 2019). Furthermore, emissions from automobiles that use lead-enriched petrol cause considerable Pb deposition in soils.

4.2.2 Effects of Heavy Metals

In the environment, metals can change their structural properties and become exceedingly toxic to plants and animals (El Ati-Hellal et al. 2021; Papa et al. 2010). Toxic metals including Cu, Fe, Zn, and Mn, as well as non-essential metals

TABLE 4.1
Sources, Harmful Effects, and Some Important Hyperaccumulator Plants of Heavy Metals

Heavy Metal	Possible Sources	Harmful Effects	Plant Species for Metal Accumulation	References
Cd	Rock phosphate, plastic stabilizers, paints and pigments, cement industry, power stations, metal industries	Carcinogenic, mutagenic, and teratogenic; endocrine disruptor; interferes with calcium regulation in biological systems; causes renal failure and chronic anemia	*Azolla pinnata, Eleocharis acicularis, Rorippa globose, Solanum photenocarpum, Salix viminalis, Salix fragilis, Populus deltoids*	Awofolu (2005), Salem et al. (2000), and Skuza et al. (2022)
Zn	Application of pesticides and fertilizers for crop production	Dizziness and fatigue in case of overdose	*Eleocharis acicularis, Arabidopsis helleri (syn. Cardaminopsis halleri), Arabisdopsis paniculate, Thlaspi caerulescens, Thlaspi praecox, Armeria maritima* ssp.	Hess and Schmid (2002) and Dinh et al. (2018)
Pb	Urban traffic using leaded fuels, electric batteries, weedicides, and pesticides	Renal failure, cardiovascular disease, reduced intelligence, short-term memory loss, coordination problems, decreased learning ability in children	*Euphorbia cheiradenia, Thlaspi caerulescens, Euphorbia cheiradenia, Chinopodium album L., Noccaeaca erulescen*	Salem et al. (2000), Padmavathiamma and Li (2007), Wuana and Okieimen (2011), Iqbal, (2012), Li et al. (2020), and Dinh et al. (2018)
Cr	Steel industry, leather industry, contaminated biosolids and manures, fly ash	Rapid hair loss	*Pteris vittata, Brassica Juncea L.*	Salem et al. (2000) and Pasricha et al. (2021)
Cu	Rock phosphate and zinc fertilizers	brain and kidney damage, severe anemia, intestinal irritation	*Euphorbia cheiradenia, Eleocharis acicularis, Elsholtzia splendens, Nakai ex Maekawa*	Salem et al. (2000), Wuana and Okieimen, (2011), Asensio et al. (2018), Lange et at (2017)
As	Pesticides, wood preservatives	As (as arsenate) is an analog of phosphate and thus interferes with essential cellular processes such as oxidative phosphorylation and ATP synthesis	*Corrigiola telephiifolia, Corrigiola telephiifolia, Eleocharis acicularis, Pteris biaurita, Pteris cretica, Pteris vittata L., Pteris quadriaurita, Pteris ryukyuensis*	Tripathi et al. (2007) and Pasricha et al. (2021)

(*Continued*)

TABLE 4.1 (*Continued*)
Sources, Harmful Effects, and Some Important Hyperaccumulator Plants of Heavy Metals

Heavy Metal	Possible Sources	Harmful Effects	Plant Species for Metal Accumulation	References
Ni	Metal industry, kitchen appliances, surgical instruments, batteries and steel alloys	Allergic dermatitis known as nickel itch; inhalation can cause cancer of the lungs, nose, and sinuses; cancers of the throat and stomach have also been attributed to its inhalation; hematotoxic, immunotoxic, neurotoxic, genotoxic, reproductive toxic, pulmonary toxic, nephrotoxic, and hepatotoxic; causes hair loss	*Alyssum bertolonii*, *A. caricum*, *A. corsicum*, *A. murale*	Salem et al. (2000), Khan et al. (2007), Das et al. (2008), Duda-Chodak and Baszczyk (2008), Mishra et al. (2010), and Van Der Ent (2018)
Hg	Mining, coal combustion, surgical instruments, medical waste	Anxiety, autoimmune diseases, depression, difficulty with balance, drowsiness, fatigue, hair loss, insomnia, irritability, memory loss, recurrent infections, restlessness, vision disturbances, tremors, temper outbursts, ulcers, and damage to brain, kidney, and lungs	*Achillea millefolium*, *Marrubium vulgare* L. *Eremochloa ciliaris*, *Erato polymnioides*, *Cyrtomium macrophyllum*	Jianxu et al. (2012), Ainza et al. (2010), Gulati et al. (2010), Ranieri et al. (2020), and Pasricha et al. (2021)

such as Cr, As, Ag, Cd, Pb, Hg, and Se, can harm not only plants but also other species (Hocaoğlu-Ozyiğit et al. 2022; Roy and McDonald 2013; Papa et al. 2010; Roy and McDonald 2013). Metals accumulation in topsoil is currently a serious concern considering the potential dangers and detrimental consequences on the soil ecology by reducing microorganism activity (Yu et al. 2012; Yuan et al. 2014). When heavy-metal toxicity increases in the soil, the enzyme-synthesizing ability of microorganisms is inhibited, and their survival rate is also decreased (Yeboah et al. 2021; Yu et al. 2012; Yuan et al. 2014). Heavy metals have a negative impact on a variety of physiological functions in plants that can even stop the normal plant growth rate, and they can even stop them from growing (Popova et al. 2009; Xu et al. 2009). The lower photosynthetic efficiency and chlorophyll concentration could be the cause of this growth slowdown. The lower photosynthetic efficiency and chlorophyll concentration could be the cause of this growth slowdown (Wang et al. 2022). Glutathione and proline are involved in the plant's response to stress. Heavy metal stress increases the proline accumulation rate of plant cells, which maintains cellular osmotic balance under stress (Aslam et al. 2017). Because of the increased rate of reactive oxygen species (ROS) formation in plants under metal stress conditions, toxic metal poisonousness may induce cell tissue disruption and the collapse of biomolecules and cellular components (Jaishankar et al. 2014).

Agricultural soil pollution is regarded as a serious concern for the environment owing to the extensive distribution of heavy metals, and its effects (acute and long-term) on the growth of plants cannot be neglected (Edelstein and Ben-Hur 2018). Since there are more people engaged in metal toxicity, the detrimental consequences are more likely to be noticed in the cells exposed at first (Rizvi and Khan 2018). Toxic metals impair ionic homeostasis as well as enzymatic activity, and their impacts can be seen in physiological functions as diverse as germination of seeds, development and growth, photosynthetic processes, plant water content, and important metabolism. Heavy metal toxicity causes cellular disruption, oxidative stress, and enzyme and nutrient imbalance at the time of seed germination, which reduces the germination rate (Sethy et al. 2013). At the growth stage of the plant, a higher concentration of heavy metal damages the DNA and denatures the protein of plant cells. As a result, the ultimate growth of the plant is reduced (Syed et al. 2021). Chlorosis, senescence, leaf rolling, low biomass output, limited development, long-term mortality, and reduced seed counts are all symptoms of heavy-metal toxicity (Shahid et al. 2017; Shi et al. 2018). Plants exposed to higher amounts of heavy metals produce reactive oxygen and suffer from oxidative stress. Plant development, metabolic processes, and acceptance of oxidative damage were seen in distinct species due to the phytotoxicity of Cd, Pb, and Zn (Brendova et al. 2016; Oliveira and Tibbett 2018).

Heavy metals entering the food chain, such as Cd and Pb, are of special concern because they can affect humans and other animals (Bundschuh et al. 2012; Ji et al. 2013; Rahman et al. 2014; Islam et al. 2015a). These metals are extremely toxic, even in tiny amounts, and can cause considerable injury (Sarwar et al. 2010). According to the World Health Organization (WHO), the maximum permissible value of heavy metals in plants are Cd (0.2 mg/kg), Fe (425.5 mg/kg), Cu (73.3 mg/kg), Pb (0.3 mg/kg), Ni (67.9 mg/kg), and Zn (99.4 mg/kg) (Mensah et al. 2009). In the soil contaminated with Ni, Cr, As, Cu, Pb, and Cd, Islam et al. (2015c) used seven different food items (milk, egg, fish, meat, vegetables, fruits, and cereals) to conduct a risk assessment for Bangladeshi residents based on heavy metal concentrations in the indigestible components of these food products. The levels of Cr, Ni, As, Cd, and Pb in some foods were found to be higher than the WHO and the Food and Agriculture Organization (FAO) permissible limits indicating that daily intake of these heavy metals from fish, vegetables, grains, and fruits was greater than the standard tolerable daily consumption (MTDI), posing a serious risk to the public in the metal-contaminated area of Bangladesh. Heavy metal exposure in the human body has been linked to several health disorders, depending on the metals, quantity, and oxidation state, among many other things. Table 4.2 shows how some poisonous substances simply induce harm.

TABLE 4.2
Some Rhizobium Inoculants for Heavy Metal Treatment

Rhizobium Inoculant	Plant Species	Heavy Metals	References
Bradyrhizobium sp.	*Glycine max*	As	Bianucci et al. (2018)
Sinorhizobium meliloti CCNWSX0020	*Medicago sativa*	Cu	Chen et al. (2018)
Mesorhizobium loti HZ76	*Robinia pseudoacacia*	Cd, Zn, Pb	Fan et al. (2018)
Pseudomonas sp. CPSB21	*Helianthus annuus*	Cr	Gupta et al. (2018)
Bacillus megaterium, Pseudomonas aeruginosa	*Suaeda nudiflora*	Zn, Pb	Jha et al. (2017)

4.3 MECHANISTIC INSIGHT INTO HEAVY METAL PHYTOREMEDIATION

Heavy metals are often transported from soil to plant as a solution that serves as a readily accessible reservoir of harmful metals and nutrients. Soil pH, organic substances, root biomass, and microbial activity all influence heavy metal absorption (Sarwar et al. 2010). Through xylem vessels via symplastic or apoplastic pathways, heavy metals can be accumulated by the roots that can either be deposited in root cells (phyto-immobilization) or transported to apical portions (sequestration) (Figure 4.1). Heavy metals are often accumulated in plant shoot vacuoles, which may be a crucial mitigation strategy in hyperaccumulators for keeping dangerous metals out of essential cellular metabolism. Metal mobilization in root exudates, metal ion absorption by the root system, transportation toward stems and leaves, metal sequestering into cells, and metal tolerances are the five key mechanisms in phytoremediation (Ali et al. 2013). Metal susceptibility in plants is mediated by processes such as metal bonding to cell membranes, active metal ion transport inside vacuoles, metal ion chelation with peptides and proteins, and complex building within plant tissues (Memon and Schr€oder 2009). Figure 4.1 depicts a high-level overview of phytoremediation systems for the removal of metals and metalloids from contaminated areas. Aside from physiological parameters that govern plant tolerance, the annual turnover of plant biomass and gross toxic metals concentration accumulated yearly are critical in assessing plant phytoremediation capabilities.

4.4 TRADITIONAL CONCEPT FOR HEAVY METAL PHYTOREMEDIATION

The process of employing trees and shrubs to decontaminate toxic heavy metals (Ni, Cr, Cu, Se, Zn, Cd, Pb) polluted soil, water, or air is known as phytoremediation. Plants, for example, can gather large quantities of heavy metals from contaminated areas and deposit them in plant tissues, where they have no harmful effects till the plant is consumed. Remediation of heavy metals using a wide range of plant

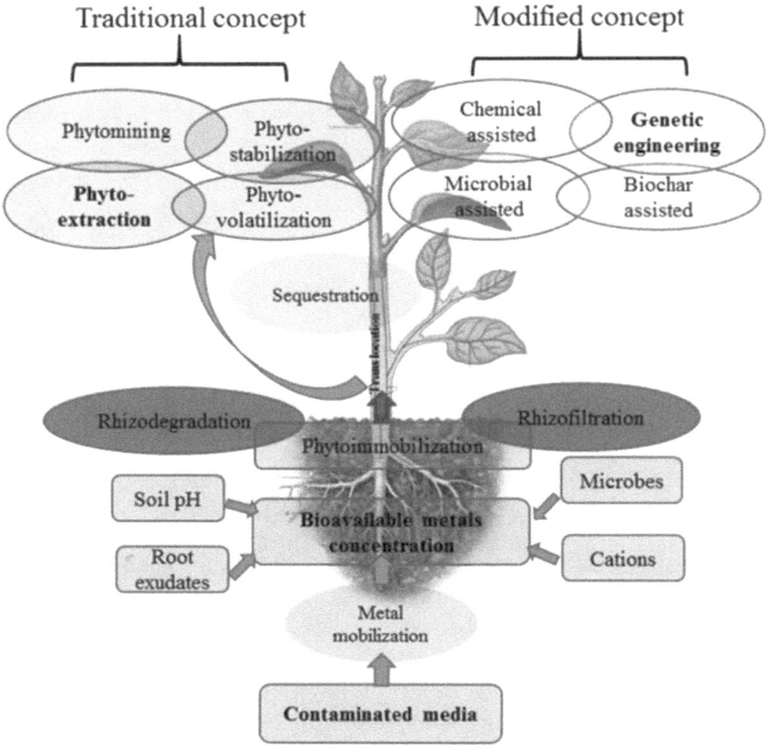

FIGURE 4.1 Mechanisms involved in phytoremediation technology for heavy metals in contaminated media. Metals are mobilized through the activity of plant roots such as changes in soil pH, root exudates production, release of cations, and interaction with microbes. The bioavailable metals are transported to the aerial parts of plant tissue where metals can be phyto-remediate through different mechanisms under traditional (phytomining, phytostabilization, phytoextraction, and phytostabilization) as well as modern concepts (chemically assisted, microbial-assisted, biochar-assisted, and genetic engineering phytoremediation).

species to decompose, extract, contain, or immobilize heavy metals in soil is a modern approach (Glick 2003). Due to its novel, cost-effective approach to more existing treatment procedures employed at heavily polluted locations, this approach has attracted the attention of international scientists. Metal extraction from soil; their concentration in plant cells; heavy metal degradation by numerous abiotic and biotic methods; heavy metal immobilization in the root system; and regulation of run-off and infiltration by plants are some of the techniques used in phytoremediation advancement (Macek et al. 2000).

Several plant and microbial populations are already being studied for their ability to remove toxic metals from polluted soil (Chiang et al. 2006; Glick 2010; Abhilash et al. 2012, 2013; Kcil et al. 2015). However, the potency of any remediation method is determined largely by the three critical variables mentioned here, which are (1) the fundamental nature of plants, such as simple reproduction, phytoproduct

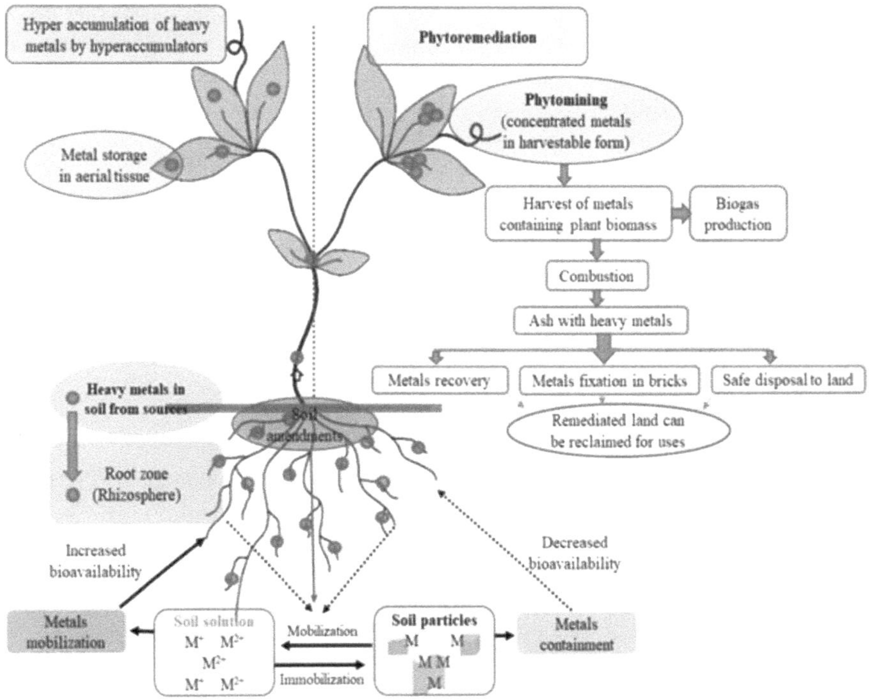

FIGURE 4.2 Phytoextraction overview including different management options of the end product**:** Metals in the bulk soil are taken up by the roots of heavy metal hyperaccumulators, possibly with assistance from rhizosphere interactions. Metals are actively loaded in the xylem and transported to the aerial tissues of the plant. In leaves, the metals are stored, often in vacuoles, at high concentrations. Plants remove contaminating particles making the land safe for use in other sectors. Metals that were dispersed in the soil become concentrated in heavy metal hyperaccumulator's tissues enabling their efficient harvesting for selling (benefits of phytoremediation). The relationship between immobilization, bioavailability, and phytoremediation of toxic heavy metals in soil-plant systems.

production, degradation, and tolerance potential; and (2) microbial population present in the intended soil with plant growth promotion potential to flourish in the polluted environment; and (3) water solubility, long-range transport potential, environmental persistence, and phyto-accumulation of heavy metals are all examples of physicochemical characteristics (Figure 4.2). Remediation using plants technique is classified into four subcategories: (1) phytostabilization, (2) phytovolatilization, (3) phytomining, and (4) phytoextraction (Parmar and Singh 2015; Ramanjaneyulu et al. 2017).

4.4.1 Phytostabilization

Phytostabilization is a plant-based metal degrading method that reduces metal mobility and solubility, therefore decreasing leaching and inclusion in groundwater and the food chain (Ramanjaneyulu et al. 2017; Khalid et al. 2017). Plants excrete excess fluid such as enzymes (amylase, protease, invertase), organic acids (oxalic, acetic, citric, malic, maleic), sugars (sucrose, glucose, fructose, maltose), amino acids, phenolic acids, fatty acids, proteins, lipids, carbohydrates (Liao et al. 2023) inside the root system, which serves as a source of nutrition and a chemoattractant for soil microbes (Girkina et al. 2018), pH control (Yan et al. 1996), minerals and nutrients absorption (Strom et al. 2002), and decontamination of polluted soils (Silva et al. 2004). Heavy metals are physically and chemically immobilized via absorption into the root zone and assimilation with soil additives such as organic materials, sludges, clays, and phosphorus-based agrochemicals (Wuana and Okieimen 2011). Metals are absorbed into plants that can thrive in metal-contaminated areas, thereby limiting metal mobility (Erakhrumen and Agbontalor 2007).

4.4.2 Phytodegradation

Plants digest contaminants by metabolic activities such as dehalogenase and oxygenase that are not dependent on rhizobial microbes (Vishnoi and Srivastava 2008). This process is catalyzed by oxygenases and nitroreductase enzymes, which catalyze and accelerate metal breakdown. Green plants can uptake organic xenobiotics from polluted environments and decontaminate them via metabolic functions. As a result, green plants are known as the "Green livers" of the biosphere. Several studies have reported the use of genetically modified plants (e.g., transgenic poplars) for the phytodegradation of synthetic agrochemicals (Doty et al. 2007).

4.4.3 Phytovolatilization

The process through which plants take pollutants from the soil, transform them into a volatile state, and thereafter release them into the environment by transpiration is known as phytovolatilization (Tangahu et al. 2011). Water molecules from leaf tissue release water-soluble metals into the air. Pollutants are converted to be soluble in water and into nontoxic state during transportation by molecules of water and compartmentation in plant vacuoles, and then they become volatilized (Yan et al. 2020). Metals deposit in the leaf epidermis, from where they might go into the mesophyll. Metal accumulation in mesophyll leads to altered water-soluble metals with an increased evaporation level, which inhibits metal accumulation in mesophyll and has no effect on plants. This technique is useful for detecting organic pollutants as well as heavy metals such as Hg and Se. *Arabidopsis thaliana* and tobacco, for example, may both volatilize Se and Hg (Ojuederie and Babalola 2017; Rayu et al. 2012).

4.4.4 Phytoextraction and Phytomining

Phytoextraction is the most efficient phytoremediation method for extracting metals from contaminated soils, sediments, or water (Asgari Lajayer et al. 2019; Cherian and Oliveira 2005; Milic et al. 2012) (Cherian and Oliveira 2005; Milic et al. 2012). Plants suitable for phytoextraction should have a high rate of growth, aboveground enhanced production of biomass, a highly branched root system, elevated target metal absorption ability, motility of deposited toxic metals from roots to the leaves, decent ability to adapt to existing climatic and environmental conditions, microbial and insect resistance, and simplicity of harvesting and processing (Adesodun et al. 2010; Sakakibara et al. 2011; Shabani and Sayadi 2012). In the natural phytoextraction process, metals are absorbed by root systems and delivered to the plant's aboveground cells (Jadia and Fulekar 2008, Figure 4.1). Plants for phytoextraction should have rapid growth, large biomass, and broad root systems and the capacity to retain large amounts of heavy metals (Sytar et al. 2016). Around the world, hyperaccumulator plants have now been employed in a variety of metalliferous settings (Ghosh and Singh 2005; Usman et al. 2018). Kos et al. (2003) and Guo and Miao (2010) discovered that Arundodonax (a perennial herbaceous plant) accumulated 2.92–4.02 mg/kg and 0.57–1.42 mg/kg of Cd in foliage and roots, respectively, as an example of phytoextraction mechanisms in a Cd polluted soil. Few more studies show the hyperaccumulation of different metals. For example, *Alyssum murale*, a hyperaccumulator of nickel (Ni), has been developed as a commercial crop for the phytoremediation of Ni from soils (Tappero et al. 2007).

Thlaspi caerulescens was used to treat soils that had been heavily contaminated by Zn, Pb, from previous mining operations and Cd-polluted water (El Kheir et al. 2008; Maharajan et al. 2023). Sunflower, or *Helianthus annuus*, is employed in phytoremediation procedures to treat arsenic-contaminated locations because of its capacity to absorb arsenic (Antosiewicz et al. 2008). Besides, Indian mustard (*Brassica juncea*) can accumulate large amounts of chromium (Cr) and other metals, and it is frequently employed in phytoremediation processes (Bluskov et al. 2005). *Xanthium strumarium* L. is also found as a potential species for phytoremediation of Cr (Hasan et al. 2021). *Ipomoea carnea* can accumulate Cr and As from polluted soils and show phytoremediation activities (Sarker et al. 2024).

Biomass of plants carrying metals may be burned to get energy, and the ashes remaining are known as "bio-ore," which can be treated for recovery of metals. Following phytomining, cultivated plant biomass might be utilized to produce biogas, burned up for recovery of metals, put in bricks, or dumped in abandoned sites (Figure 4.2). This method is the most effective for reducing metal contamination concentrations in soil through plant roots and shoots while preserving soil characteristics. Meers et al. (2010) carried out a field trial utilizing maize for energy cultivation in the Campine area of Belgium and the Netherlands and discovered that 30,000–42,000 kWh of renewable energy was created per hectare of land. With the replacement of coal-fired power stations, CO_2 emissions may be reduced by up to 21 tons per hectare annually. Because of the low sulfur concentration of bio-ores, processing them emits less SOx into the atmosphere, making it a more environmentally friendly option than traditional extraction methods. As a result, tight environmental legislation would make bio-based mining for heavy metals phytoremediation a more appealing alternative for scientists (Siddiqui et al. 2009).

4.4.5 Rhizodegradation

The method by which toxic metals and organic contaminants in soil are broken down by microbes in the rhizobial region is known as rhizodegradation (Mukhopadhyay and Maiti 2010). The faster degradation of pollutants in the rhizobial area is most likely related to a rise in the number of microbes and their metabolic activity. Exudates comprising carbohydrates, amino acids, and flavonoids secreted by plants can boost microbial activity in the rhizosphere by 10–100 times. Plant roots exude nutrient-rich secretions that supply carbon and nitrogen to soil microbes and a nutrition-friendly environment for the microbial population. To increase microbial activity, the roots of plants soften the ground and transport moisture to the rhizosphere. Other terms for it include increased rhizosphere microbial degradation, phytostimulation, and plant-regulated bioremediation. Plants also generate enzymes that in collaboration with rhizospheric bacteria degrade organic contaminants in the environment (Yadav et al. 2010 and Verma et al. 2021). Table 4.2 shows some rhizobium inoculants for heavy metal treatment.

4.4.6 Rhizofiltration

It is described as the utilization of plant roots with low contaminant concentrations to uptake, absorb, and discharge toxins from contaminated water sources. The contaminants in the contaminated water were retained or deposited by the root system. Heavy metals are moved into the root zone by adsorption and uptake via precipitation and sequestration (Tangahu et al. 2011; Erakhrumen and Agbontalor 2007). It acts similarly to phytoextraction. These three processes occur in the ground, notably in the roots of plants. The effective outcome was achieved by using plants that were acclimated to the contaminant for rhizofiltration of those pollutants. Rhizofiltration offers numerous benefits, which include the possibility to employ it in-situ or ex-situ as well as the usage of species apart from hyperaccumulators. Plants such as sunflower, Indian mustard, tobacco, rye, spinach, and corn have been studied for their ability to eliminate lead from wastewater, with sunflower producing the best performance. In a wide concentration range (4–500 mg/L), Indian mustard was found to be very effective at removing lead (Raskin et al. 2000). The technique was tested in the field with uranium (U) polluted water at varying concentrations from 21 to 874 g/L; Dushenkov recorded an administered U level of 20 g/L before releasing it (Dushenkov et al. 1997).

A summary of the different techniques of phytoremediation of heavy metals is presented in Table 4.3.

4.5 MODIFIED CONCEPT FOR HEAVY METAL PHYTOREMEDIATION

With numerous research foundations, the use of diverse trees and shrubs to remove metals from polluted soil appears to be an impactful approach to dealing with heavy metal issues. Traditional phytoremediation techniques are still limited in their use on a large scale due to several drawbacks, including the need for a long time, the low phytoextraction ability of hyperaccumulator plants, the fact that merely a minor portion of metals are bioavailable, an absence of information around agronomy, breeding prospective, insect pest and ailment range, and the risk of diet series (Ali et al. 2013). As a result of these inescapable restrictions, researchers are compelled to alter

TABLE 4.3
Summary of the Different Techniques of Phytoremediation of Heavy Metals

Techniques for Phytoremediation	Description of the Mechanism of Phytoremediation
Phytoextraction	Accumulation of heavy metals in harvestable biomass of plants i.e., shoots
Phytofiltration	Sequestration of heavy metals from contaminated waters by hyperaccumulator plants
Phytostabilization	Limiting the mobility and bioavailability of heavy metals in soil by plant roots
Phytovolatilization	Conversion of heavy metals to the volatile form and their subsequent release into the atmosphere
Phytodegradation	Degradation of organic xenobiotics by plant enzymes within plant tissues
Rhizodegradation	Degradation of organic xenobiotics in the rhizosphere by rhizospheric microorganisms
Phytodesalination	Removal of excess salts from saline soils by halophytes

classic phytoremediation concepts in order to reduce these restrictions and safeguard extensive phytoremediation utilizing changed concepts (Figure 4.2).

4.5.1 Chemical-Assisted Phytoremediation

The most important consideration is choosing the right hyperaccumulator plant species for heavy metal phytoremediation. Baker et al. (1994) stated that if the presence of metal and their shoot-to-root ratio are greater than 1 in a plant species, then it indicates that plant shoots can accumulate more heavy metals than their roots. A plant having such traits might be used as a hyperaccumulator species as well as for phytoremediation. Metal concentrations in hyperaccumulator plant aerial parts range from 1,000 to 10,000 mg/kg depending on the concentrations of heavy metals, which can be as low as 100 mg/kg in the particular instance of Cd that is highly detrimental (Kramer 2010). The majority of plants that can be categorized as hyperaccumulators so far match the criteria mentioned above. The downside of employing these plants is that they develop slowly and produce very little aboveground biomass (Saifullah et al. 2009), making them unsuitable for large-scale phytoremediation. Such plants, on the other side, may serve as the basis for developing transgenic plants able to gather sizable quantities of heavy metals while yielding enough biomass. As a result, the phytoremediation capability of non-hyperaccumulator plant species with higher vegetation cover should be studied. Because all these species absorb only a small concentration of metals, chelate-assisted phytoremediation or induced phytoremediation will be more beneficial in improving their absorption capabilities (Saifullah et al. 2009). Synthesized chelators such as diethylene triamine penta-acetic acid (DTPA), ethylene diamine tetra-acetic acid (EDTA), and ethylene glycol tetra-acetic acid (AGTA) are used to boost metal absorption rate (Saifullah et al. 2009; Pereira et al. 2010). In the past few years, some scientists have also focused on a range of methodologies for dealing with metal stress in metal-sensitive trees and shrubs such as woody shrub *Daphne jasmine* (Wiszniewska et al. 2019) and *Arabidopsis thaliana*

(Remy et al. 2016; Saifullah et al. 2009; Sarwar et al. 2010, 2015; Popova et al. 2012). In this regard, an abundant phenolic chemical compound (salicylic acid or SA) is used in relieving stress is an enticing and increasing area of study. Some research on SA in the defense response of plants against metal stress has produced convincing results. According to He et al. (2010), pretreatment of SA in rice seeds (0.1 mM/1 day) enhanced germination rate attributes and growth of seedlings in Cd-containing culture media. According to Popova et al. (2012), administration of SA significantly reduced the negative impacts of Cd on chlorophyll concentration, proline concentration, and foliar moisture contents in maize. As a consequence, SA pretreatment in combination with a chemically aided phytoremediation scheme might well decrease metal toxicity in metal removable plants, resulting in increased biomass production as well as metal extraction capacity. Reduced molecular weight organic acids such as citric acid, acetic acid, malic acid, and oxalic acid can be used as chelating agents to create metal complexes of small-to-moderate consistency. Organic acids seem to be more biodegradable in the ground than synthetic chelators, resulting in less pollution as well as concern for the environment (Souza et al. 2013).

4.5.2 Biochar-Assisted Phytoremediation

Biochar is a highly permeable carbon-containing material created by pyrolyzing organic sources including plant products, green manure, and effluents (Paz-Ferreiro et al. 2014). Biochar has distinct physicochemical properties, including higher pH values, a high surface area for heavy metal adsorption, ash, and carbon contents, and the ability to incapacitate toxic metals, allowing it to be an effective toxic metals remediator. The high surface area of biochar is applicable for absorption to form heavy metal–biochar complexes, whether through metal cation exchange with Na^+, K^+, Ca^{2+}, and Mg^{2+} in the presence of functional groups in biochar (Lu et al. 2012). There are some mechanisms through which biochar can absorb the heavy metals from the soil aqueous solution. Those mechanisms are physical, biological, chemical sorption method, redox precipitation method, batch sorption experiment, precipitation, sorption, complexation, ion exchange, and electrostatic interaction mechanisms (Zhang et al. 2020; Bandara et al. 2020; Gholizadeh et al. 2021; Yuan et al. 2020). The higher pH and alkalinity of biochar can minimize metal absorption and boost precipitation in biochar-supplemented soil. To improve their efficiency against toxic metals, biochar is used in combination with defined phytoremediation strategies (phytomining, phytostabilization, phytovolatilization, and phytoextraction). Biochar is often used to boost the growth of plants and biomass production by roughly 10% (Jeffery et al. 2011; Liu et al. 2013). Enhanced nutrient as well as moisture retention capacity, exchangeable cations, nutrient availability, and advanced nutrient turnover each contribute to higher growth of plants and biomass yield (Liu et al. 2013; Fellet et al. 2014; Ahmad et al. 2016). Table 4.4 lists some of the biochars used to reduce heavy metals in soil.

4.5.3 Microbial-Assisted Phytoremediation

Soil microbes associated with plants can often influence the uptake and availability of metals by plant species in the rhizobial zone where root secretions are made of various

TABLE 4.4
List of Some Biochar Used to Reduce Heavy Metals in Soil

Types of Biochar	Reduced Heavy Metal	References
Rice (straw, husk, and bran) biochar	Cd (71%), Zn (37%), Pb (60%)	Hayyat et al. (2016)
Bamboo and rice straw biochar	Zn (39.33%), Pb (27.64%), Cu (23.54%), Cd (23.25%)	Lu et al. (2017)
Hardwood biochar and corn straw	Zn, Cu	Chen et al. (2011)
Wheat straw biochar	Cd accumulation on the root	Rahim et al. (2024)
Sewage sludge biochar	Cr, Mn, Cu, and Zn (72.20%, 70.38%, 50.43%, 29.78%)	Zhou et al. (2017)
Rice straw biochar	Zn (61%), Pb (64%), Cd (76%)	Qasim et al. (2021)

substances that provide a place of nitrogen and carbon sources for microorganisms (Bais et al. 2006; Compant et al. 2010). Microbial chemotaxis to specific root-exuded chemical compounds is required for plant-driven microbial species selection as well as bacterial rhizosphere colonization for metal remediation (Doornbos et al. 2012). In the majority of plant–microbe relationships, roots release organic substances that effectively signal microorganisms, emphasizing the role of phytochemicals in useful microbe hiring via rhizobacterial colonialization (Lugtenberg and Kamilova 2009). Plant–microbial interaction is essential in shaping the rhizosphere because it benefits both hosts and microbes. Metal-tolerant useful microbes are commonly used as biostimulants to affect various metabolism and membrane permeability of plant roots and to enhance the development and growth of plants in metal-polluted soils by enabling mineral phytoavailability, modifying phytohormones, limiting ethylene synthesis, and blocking phytopatatoxicity (Glick 2010; Ma et al. 2011 a,b; Miransari 2011; Rajkumar et al. 2012; Ahemad and Kibret 2014). Bacteria such as *Arthrobacter* sp. and *Leifsonia* sp. (Actinobacteria), *Polaromonas* sp. and *Janthinobacterium* sp. (Betaproteobacteria), and *Janthinobacterium* sp. (Betaproteobacteria) have earlier been noted to speed up the solubilization and movement of Fe, Mn, and K for maintaining soil fertility (Abdulla 2009; Uroz et al. 2009).

Plant-associated microbes can produce phytohormones such as IAA, cytokinins, gibberellic acid, abscisic acid, and others under stress, modifying the hormone levels in plants and their stress response (Ma et al. 2011a; Ullah et al. 2015). Plant-growth-promoting bacterial strain is another important microbial group that aids plants in detoxifying toxic metals from contaminated sites (Seth 2012). As reciprocal soil microorganisms and free-living rhizobacteria, the above microbes can be immersed (Liao et al. 2003). Table 4.5 shows some important microbes used to reduce heavy metals in soil.

4.6 GENETIC ENGINEERING (USE OF TRANSGENIC PLANTS)

The hyperaccumulators are known for their tiny size and slow growth, which harms metal phytoextraction capacity and significantly limits the use of useful agronomic

TABLE 4.5
List of Important Microbes Used to Reduce Heavy Metal in Soil

Heavy Metals	Microbial Species	References
As	*Streptomyces* sp. *Bacillus* sp. *Acinetobacter lwoffii* *Bacillus licheniformis* *Debaryomyces hansenii* *Bacillus vietnamensis*	Das et al. (2018) and Khalid et al. (2020)
Cd	*G. versiforme* *Penicillium funiculosum* *Penicillium janthinellum* *P. indica* *P. fluorescens*	Zhang et al. (2019)
Zn	*R. irregularis* *Streptomyces pactum* *Bacillus safensis* *Kocuria rosea* *B. safensis* *P. fluorescens*	Raklami et al. (2021)
Ni	*Arthrobacter* sp. *Microbacterium oxydans* *Pseudomonas libanensis,* *Claroideoglomus claroideum*	Ma et al. (2019)
Cu	*Paenibacillusmacerans* NBRFT5 *Bacillus endophyticus* NBRFT4 *B. pumilus* NBRFT9	Alori et al. (2017)
Co	*Fungi (Rhizopus stolonifera)* *Algae (Oscillatoria* sp.*)*	Njoku et al. (2020)
Mn	*Scutelospore reticulate* *Glomus phaseous*	Alori et al. (2017)
Pb	*Trichoderma asperellum* *Trametes hirsuta*	Malik et al. (2020)
Fe	sp. *Geobacter* sp. *Shewanella*	Ahmed et al. (2020)
Al	*Brochothrix thermosphacta* *Vibrio alginolyticus*	Kurniawan et al. (2018)

procedures such as mechanical harvesting (Tong et al. 2004). Conventional breeding practices to boost trees and shrubs for metal removal through the plant–metals relationships have been suggested to solve the above limitations (Li et al. 2004). Regrettably, the success of this approach can sometimes be jeopardized regarding sexual inconsistencies caused by structural differences among plant parents. Biotechnology, by allowing direct genetic manipulation technologies, has the potential to eliminate this barrier (Kramer and Chardonnens 2001). According to research, manipulating relevant plant traits, such as metal tolerance, is a viable option for heavy metal phytoremediation in any contaminated location. Although utilizing bioenergy and biofuel plants to optimize the financial return of phytoremediation was proposed, there will still be widescale concerns about using gathered biomass as well as the fate of toxic materials in cultivated plants (Abhilash et al. 2011). On the other hand, recent advances in omics strategy enable the use of metabolomic, proteomic, metagenomic, genomic, and transcriptomic strategies to recognize characteristics that enhance the level of ground remediation methods and influence plant and microorganism sensitivity, deposition, and degradation ability against various inorganic and organic contaminants. Trans and cis genic tools are used to enhance contaminant absorption, transportation, as well as degeneration, plant development and vitality, root development, and abiotic stress tolerance in candidate plants. In the coming years, we could be able to tailor plants for site-specific remediation. Plant phytoremediation, or the removal or detoxification of dangerous chemicals from the environment, has been greatly enhanced because of genetic engineering. Transgenic plants are used in this strategy to overexpress specific genes associated with plant uptake, translocation, sequestering, and resilience to toxic compounds (Aken 2008). As a result, specific genes from microbes, trees, and animal life might be added into transgenic plants via gene transfer (DNA techniques) or *Agrobacterium tumefaciens*–mediated transition (Seth 2012). To increase Hg tolerance, *Arabidopsis thaliana* has an overexpression of the gene that codes for conveying mercuric ion reductase, whereas *Nicotiana tabaccum* has a yeast with an overexpression of the gene that codes for conveying mercuric ion reductase.

Metallothionein-expressing genes for Cd sensitivity were primarily designed for Hg and Cd removal from land (Rugh et al. 1998). To sequester more metals in cell compartments with low cell metabolism, transgenic plants with enhanced metal mobility and resilience methods such as vacuoles can be produced. Arsenic, for instance, complexes with phytochelatins (PCs) and glutathione (eGSH) in vacuoles (Dhankher et al. 2002). According to Seth (2012), the bacterial genes merA encode mercuric ion reductase, and merB encrypts organo-mercurial lyase, accelerating the tolerance of plants to Hg in transformants. As a consequence, transgenic plants with enhanced metal tolerance and carbon storage capacity are used satisfactorily for metal phytoremediation in polluted soils. Moreover, to create transgenic hyperaccumulator plants suitable for phytoremediation, a clearer picture of metal tolerance and cleansing processes in plants is required.

Table 4.6 summarizes the distribution of some heavy metals in hyperaccumulators in the tissue/cell.

4.7 CONCLUSIONS

Heavy metal contamination in soils has now become a massive public health and environmental concern in the present decade, because of the risk of food contamination and other human health consequences. Phytoremediation strategies may prove to be a valuable technique in overcoming this issue, such as some other physiochemical attempts to fix polluted soils appear to be financially unviable and produce less efficient outcomes. This review showed that many hyperaccumulator plants such as *Alyssum murale*, *Thlaspi caerulescens*, *Helianthus annuus*, *Brassica juncea*,

TABLE 4.6
Heavy Metal Distribution in Hyperaccumulators in the Tissue/Cell

Plant Species	Tissue/ Organ	Target Metal	Metal Accumulation by Plant (mg/kg)	References
Aeolanthus biformifolius	Cell wall	Cu	13,700	Chaney et al. (2010)
	Foliar samples	Mg	6,150	Antony et al. (2019)
		Al	85	
		S	2,080	
		K	22,800	
		Ca	16,900	
		Mn	399	
Eleocharis acicularis	Shoots	Cu	20,200	Sakakibara et al. (2011)
Haumaniastrum katangense	Roots	Cu	8,356	Sheoran et al. (2009)
		Mg	6,630	Antony et al. (2019)
		Al	2,170	
		P	921	
		S	898	
		K	6,280	
		Co	981	
		Ni	401	
		Cu	686	
		Zn	135	
		Ca	4,930	
		Mn	208	
		Fe	4,840	
Haumaniastrum katangense	Stems	Mg	5,260	Antony et al. (2019)
		Al	378	
		P	1,950	
		S	2,120	
		K	12,500	
		Ca	13,200	
		Co	983	
		Ni	740	
		Cu	95	
		Zn	147	
Ipomoea alpina	Roots	Cu	12,300	Yan et al. (2020)
Arabis paniculata	Shoots	Cd	8,400	Ullah et al. (2015)
Azolla pinnata	Shoot/Roots	Cd	740	Rai (2008)
Deschampsi acespitosa	Shoot/Roots	Cd	236.2	Kucharski et al. (2005)
Eleocharis acicularis	Shoots	Cd	239	Sakakibara et al. (2011)
Solanum photeinocarpum	Shoots	Cd	158	Zhang et al. (2011)

(Continued)

TABLE 4.6 (*Continued*)
Heavy Metal Distribution in Hyperaccumulators in the Tissue/Cell

Plant Species	Tissue/ Organ	Target Metal	Metal Accumulation by Plant (mg/kg)	References
Solanum lycopersicum	Leaves	Cd	133.45	Guo et al. (2020)
Rorippa globosa	Cell wall	Cd	100	Wei et al. (2008)
Rorippa globosa	Shoots	Cd	150	Wei (2006)
Thlaspi caerulescens	Cell wall	Cd	263	Lombi et al. (2001)
Thlaspi caerulescens	Roots	Cd	3,000	Sheoran et al. (2009)
Sedum alfredii	Roots	Cd	2,183	Jin et al. (2009)
Sedum plumbizincicola	Shoots	Cd	152.93	Huang et al. (2020)
Thlaspi caerulescens	Shoot/Roots	Cd	5,000	Koptsik (2014)
Betula occidentalis	Shoot/Roots	Pb	1,000	Koptsik (2014)
Brassica juncea	Shoot/Roots	Pb	10,300	Koptsik (2014)
Thlaspi rotundifolium	Roots	Pb	8,200	Lasat (2002)
Deschampsi acespitosa	Shoot/Roots	Pb	966.5	Kucharski et al. (2005)
Brassica nigra	Shoot/Roots	Pb	9,400	Koptsik (2014)
Euphorbia cheiradenia	Shoots	Pb	1,138	Chehregani and Ma-Layeri (2007)
Helianthus annuus	Shoot/Roots	Pb	5,600	Koptsik (2014)
Helianthus annuus	Shoots, Roots	Cd	65.7	Alaboudi et al. (2018)
Medicago sativa	Shoot/Roots	Pb	43,300	Koptsik (2014)
Achillea millefolium	Shoot/Roots	Hg	18.275	Jianxu et al. (2012)
Armoracia lapathifolia	Shoot/Roots	Hg	0.97	Aleksandra et al. (2008)
Cicer arietinum	Shoot/Roots	Hg	0.2	Jianxu et al. (2012)
Festuca rubra	Shoots	Hg	3.17	Luis et al. (2003, 2007)
Helianthus tuberosus	Shoot/Roots	Hg	1.89	Aleksandra et al. (2008)
Helianthus annuus	Shoots, Roots	Cd	65.7	Alaboudi et al. (2018)
Hordeum spp	Shoots	Hg	2.35	Luis et al. (2003, 2007)
Juncus maritimus	Shoot/Roots	Hg	0.315	Naser et al. (2011)
Lens culinaris	Shoots	Hg	1.4	Luis et al. (2003, 2007)
Lupinus polyphyllus	Shoots	Hg	0.20	Luis et al. (2003, 2007)
Marrubium vulgare	Shoots	Hg	13.8	Luis et al. (2003, 2007)
Pteris vittata	Shoot/Roots	Hg	91.98	Jianxu et al. (2012)
Pteris vittata	Shoots	Cd	216.5	Wan et al. (2017)
Pteris vittata	Roots	Hg	1601	Jian et al. (2009)
Rumex induratus	Shoots	Hg	6.45	Luis et al. (2003, 2007)

(*Continued*)

TABLE 4.6 (*Continued*)
Heavy Metal Distribution in Hyperaccumulators in the Tissue/Cell

Plant Species	Tissue/ Organ	Target Metal	Metal Accumulation by Plant (mg/kg)	References
Salix viminalis	Shoots	Hg	0.66	Wang et al. (2005)
Salix schwerinii	Shoots	Hg	0.55	Wang et al. (2005)
Macleaya cordata	Shoot/Roots	Hg	2.78	Jianxu et al. (2012)
Macleaya cordata	Roots	Cd	163.39	Nie et al. (2016)
Nephrolepis exaltata	Roots	Hg	2,357	Jian et al. (2009)
Triticum aestivum	Shoots	Hg	0.14	Luis et al. (2003, 2007)
Silene vulgaris	Shoot/Roots	Hg	4.25	Araceli et al. (2012)
Silene sendtneri	Shoots	Cd	2,156	Karalija et al. (2021)
Macadamia neurophylla	Roots	Mn	51,800	Sheoran et al. (2009)
Alyxia rubricaulis	Cell wall	Mn	11,500	Chaney et al. (2010)
Schima superba	Aboveground tissues	Mn	62,412	Yang et al. (2008)
Deschampsi acespitosa	Shoot/Roots	Zn	3,614	Kucharski et al. (2005)
Thlaspi calaminare	Roots	Zn	10,000	Sheoran et al. (2009)
Thlaspi caerulenscens	Cell wall	Zn	39,600	Chaney et al. (2010)
Eleocharis acicularis	Shoots	Zn	11,200	Sakakibara et al. (2011)
Pteris vittata	Gametophyte biomass	Cr	20,675	Kalve et al. (2011)
Alyssum caricum	Plant biomass	Ni	12,500	Li et al. (2003)
Alyssum heldreichii	Leaves	Ni	11,800	Bani et al. (2010)
Alyssum markgrafii	Leaves	Ni	19,100	Bani et al. (2010)
Alyssum murale	Leaves	Ni	4,730–20,100	Bani et al. (2010)
Alyssum murale	Plant biomass	Ni	15,000	Li et al. (2003)
Alyssum pterocarpum	Plant biomass	Ni	13,500	Li et al. (2003)
Alyssum serpyllifolium	Aboveground parts	Ni	10,000	Prasad (2005)
Alyssum bertolonii	Plant biomass	Ni	10,900	Li et al. (2003)
Isatis pinnatiloba	Aboveground parts	Ni	1,441	Altinozlu et al. (2012)
Phyllanthus serpentinus	Cell wall	Ni	38,100	Chaney et al. (2010)
Psychotria douarrei	Shoots	Ni	47,500	Cunningham and Ow (1996)
Berkheya coddii	Leaves	Ni	18,000	Mesjasz-Przyby et al. (2004)

(*Continued*)

TABLE 4.6 (*Continued*)
Heavy Metal Distribution in Hyperaccumulators in the Tissue/Cell

Plant Species	Tissue/ Organ	Target Metal	Metal Accumulation by Plant (mg/kg)	References
Thlaspi caerulescens	Shoot/Roots	Ni	2,740	Koptsik (2014)
Brassica juncea	Dry biomass	Au	10	Harris et al. (2009)
Haumaniastrum robertii	Cell wall	Co	10,200	Chaney et al. (2010)
Pteris biaurita	Roots	As	2,000	Srivastava et al. (2006)
Pteris cretica	Roots	As	1,800	Srivastava et al. (2006)
Pteris cretica	Fronds and roots	As	2,200–3,030	Zhao et al. (2002)
Pteris quadriaurita	Roots	As	2,900	Srivastava et al. (2006)
Pteris ryukyuensis	Roots	As	3,647	Srivastava et al. (2006)
Pteris vittata	Gametophyte biomass	As	8,331	Kalve et al. (2011)
Pteris vittata	Leaves and stems	As	1,000	Baldwin and Butcher (2007)
Corrigiola telephiifolia	Aboveground parts	As	2,110	Garcia-Salgado et al. (2012)
Eleocharis acicularis	Shoots	As	1,470	Sakakibara et al. (2011)

Xanthium strumarium L., and *Ipomoea carnea* can be used to eliminate large amounts of toxic metals, and other natural and inorganic pollutants, from the soil. Conventional phytoremediation techniques, on the other hand, are found to be less cost-effective if used on a large scale since naturally occurring hyperaccumulator trees and shrubs are slow-growing and produce low-utilizable terrestrial biomass. In this particular instance, researchers found that genetically engineered techniques that produces transgenic plants with increased biomass output, increased metal deposition, metal toxicity tolerance, and ability to adapt to a wide range of climatic conditions are more beneficial. Other phytoremediation methods, such as chemical-aided phytoextraction and microbial-supported phytoremediation, are also used by researchers to detoxify contaminated sites on a large scale.

4.8 RECOMMENDATIONS

Based on our study, the following recommendations may be helpful.

a. Because phytoremediation research is truly multidisciplinary, researchers from diverse backgrounds must be strongly encouraged to apply their skills and knowledge.
b. Existing plant varieties should be explored for a hyperaccumulation of various metals to develop novel and efficient metal hyperaccumulators.

c. Substantial and credible risk evaluation research should be carried out before using transgenic plants for phytoremediation in the field.
d. More phytoremediation studies should be carried out in the field with an accurate and objective cost–benefit analysis, taking into consideration the technology's highly green nature.
e. More research is necessary to understand the rhizosphere's relationships with metals, soil, bacteria, and plant roots.
f. Metal ion fate in plant cells should be made clearer using developments in spectroscopic and chromatographic methods that will contribute to the understanding of metal hyperaccumulation and sensitivity in plants.
g. Periodic meetings should be organized by an international body (such as IUPAC) to discuss and find solutions to the issues that the burgeoning technology of phytoremediation faces.
h. To develop innovative heavy metal phytoremediation approaches, more research is required to find the crucial host plant (hyperaccumulator) as well as microbial signaling pathways and their processes in facilitating this cross-interaction in the root system.

CONFLICTS OF INTEREST

The authors declare that they have no known competing financial interests or personal relationships that could have appeared to influence the work reported in this chapter.

REFERENCES

Abdulla, H., 2009. Bioweathering and biotransformation of granitic rock minerals by actinomycetes. *Microb. Ecol.* 58, 753–761.

Abhilash, P.C., Dubey, R.K., Tripathi, V., Srivastava, P., Verma, J.P., Singh, H.B., 2013. Adaptive soil management. *Curr. Sci.* 104, 1275–1276.

Abhilash, P.C., Powell, J., Singh, H.B., Singh, B., 2012. Plant-microbe interactions: Novel applications for exploitation in multipurpose remediation technologies. *Trends Biotechnol.* 30, 416–420.

Abhilash, P.C., Yunus, M., 2011. Can we use biomass produced from phytoremediation? *Biomass Bioenergy* 35, 1371–1372.

Adesodun, J.K., Atayese, M.O., Agbaje, T., Osadiaye, B.A., Mafe, O., Soretire, A.A., 2010. Phytoremediation potentials of sunflowers (*Tithonia diversifolia* and *Helianthus annuus*) for metals in soils contaminated with zinc and lead nitrates. *Water Air Soil. Pollut.* 207, 195–201.

Ahemad, M., Kibret, M., 2014. Mechanisms and applications of plant growth promoting rhizo bacteria: Current perspective. *J. King Saud Univ. Sci.* 26, 1–20.

Ahmad, N., Imran, M., Marral, M.W.R., Mubashir, M., Butt, B., 2016. Influence of biochar on soil quality and yield related attributes of wheat (*Triticum aestivum*L.). *J. Environ. Agric. Sci.* 7, 68–72.

Ainza, C., Trevors, J., Saier, M., 2010. Environmental mercury rising. *Water Air Soil Pollut.* 205, 47–48.

Aken, B.V., 2008. Transgenic plants for phytoremediation: Helping nature to cleanup environmental pollution. *Trend Biotech.* 26, 225–237.

Aleksandra, S.N., Galimska-Stypa, R., Kucharski, R., Zielonka, U., Małkowski, E., Gray, L., 2008. Remediation aspect of microbial changes of plant rhizosphere in mercury contaminated soil. *Environ. Monit.* Assess. 137, 101–109.

Ali, H., Khan, E., Sajad, M.A., 2013. Phytoremediation of heavy metals e Concepts and applications. *Chemosphere* 91, 869–881.

Alengebawy, A., Abdelkhalek, S.T., Qureshi, S.R. and Wang, M.Q., 2021. Heavy metals and pesticides toxicity in agricultural soil and plants: Ecological risks and human health implications. *Toxics 9*(3), 42.

Alori, E.T., Fawole, O.B., 2017. Microbial inoculants-assisted phytoremediation for sustainable soil management. *Phytorem. Manag. Environ. Contam.* 5, 3–17.

Altinozlu, H., Karagoz, A., Polat, T., Unver, I., 2012. Nickel hyper accumulation by natural plants in Turkish serpentine soils. *Turk. J. Bot.* 36, 269–280.

Antosiewicz, D.M., Escudě-Duran, C., Wierzbowska, E., Skłodowska, A., 2008. Indigenous plant species with the potential for the phytoremediation of arsenic and metals contaminated soil. *Water Air Soil Pollut.* 193, 197–210.

Araceli, P.S., Millan, R., Sierra, M.J., Alarcon, R., Garcia, P., Gil-Diaz, M., Vazquez, S., Lobo, M.C., 2012. Mercury uptake by *Silene vulgaris* grown on contaminated spiked soils. *J. Environ. Manag.* 95, 233–237.

Asensio, V., Flórido, F.G., Ruiz, F., Perlatti, F., Otero, X., Ferreira, T., 2018. Screening of native tropical trees for phytoremediation in copper-polluted soils. *Int. J. Phyto.* 20, 1456–1463. https://doi.org/10.1080/15226514.2018.1501341

Asgari Lajayer, B., Khadem Moghadam, N., Maghsoodi, M.R., Ghorbanpour, M., Kariman, K., 2019. Phytoextraction of heavy metals from contaminated soil, water and atmosphere using ornamental plants: Mechanisms and efficiency improvement strategies. *Environ. Sci. Pollut. Res.* 26, 8468–8484.

Aslam, M., Saeed, M.S., Sattar, S., Sajad, S., Sajjad, M., Adnan, M., Iqbal, M., Sharif, M.T., 2017. Specific role of proline against heavy metals toxicity in plants. *Int. J. Pure Appl. Biosci. 5*(6), 27–34.

Awofolu, O., 2005. A survey of trace metals in vegetation, soil and lower animal along some selected major roads in metropolitan city of Lagos. *Environ. Monit. Assess.* 105, 431–447.

Badri, D.V., Weir, T.L., vander Lelie, D. Vivanco, J.M., 2009. Rhizosphere chemical dialogues: Plant-microbe interactions. *Curr. Opin. Biotechnol.* 20, 642–650.

Bais, H.P., Weir, T.L., Perry, L.G., Gilroy, S., Vivanco, J.M., 2006. The role of root exudates in rhizosphere interactions with plants and other organisms. *Annu. Rev. Plant Biol.* 57, 233–266.

Baker, A.J.M., Reeves, R.D., Hajar, A.S.M., 1994. Heavy metal accumulation and tolerance in british populations of the metallophyte *Thlaspi caerulescens* J and CPresl (Brassicaceae). *New Phytol.* 127, 61–68.

Baldwin, P.R., Butcher, D.J., 2007. Phytoremediation of arsenic by two hyper accumulators in a hydroponic environment. *Microchem. J.* 85, 297–300.

Bandara, T., Xu, J., Potter, I.D., Franks, A., Chathurika, J.B.A.J., Tang, C., 2020. Mechanisms for the removal of Cd (II) and Cu (II) from aqueous solution and mine water by biochars derived from agricultural wastes. *Chemosphere* 254, 126745.

Bani, A., Pavlova, D., Echevarria, G., Mullaj, A., Reeves, R.D., Morel, J.L., Sulce, S., 2010. Nickel Hyper accumulation by the species of Alyssum and Thlaspi (Brassicaceae) from the ultramafic soils of the Balkans. *Bot. Serb.* 34, 3–14.

Basu, M., Guha, A.K., Ray, L., 2017. Adsorption behavior of cadmium on husk of lentil. *Process Safety Environ. Prot.* 106, 11–22.

Brendova, K., Zemanova, V., Pavlíkova, D., Tlustos, P., 2016. Utilization of biochar and activated carbon to reduce Cd, Pb and Zn phytoavailability and phytotoxicity for plants', *J. Environ. Manag.* 181, 637–645.

Bianucci, E., Godoy, A., Furlan, A., Peralta, J.M., Hernández, L.E., Carpena-Ruiz, R.O., Castro, S., 2018. Arsenic toxicity in soybean alleviated by a symbiotic species of Bradyrhizobium. *Symbiosis* 74, 167–176.

Bluskov, S., Arocena, J.M., Omotoso, O.O., Young, J.P., 2005. Uptake, distribution, and speciation of chromium in Brassica juncea. *Int. J. Phytorem.* 7(2), 153–165.

Bundschuh, J., Nath, B., Bhattacharya, P., Liu, C.W., Armienta, M.A., López, M.V.M., Lopez, D.L., Jean, J.S., Cornejo, L., Macedo, L.F.L., Filho, A.T., 2012. Arsenic in the human food chain: The Latin American perspective. *Sci. Total Environ.* 429, 92–106.

Chaney, R.L., Broadhurst, C.L., Centofanti, T., 2010. Phytoremediation of soil trace elements. In: Hooda, P.S. (Ed.), *Trace Elements in Soils*. Wiley, Chichester, pp. 311–352.

Chehregani, A., Malayeri, B.E., 2007. Removal of heavy metals by native accumulator plants. *Int. J. Agric. Biol.* 9, 462–465.

Cherian, S., Oliveira, M.M., 2005. Transgenic plants in phytoremediation: Recent advances and new possibilities. *Environ. Sci. Technol.* 39, 9377–9390.

Chen, J., Liu, Y.Q., Yan, X.W., Wei, G.H., Zhang, J.H., Fang, L.C., 2018. Rhizobium inoculation enhances copper tolerance by affecting copper uptake and regulating the ascorbate-glutathione cycle and phytochelatin biosynthesis-related gene expression in Medicago sativa seedlings. *Ecotoxicol. Environ. Safety* 162, 312–323.

Chen, X., Chen, G., Chen, L., Chen, Y., Lehmann, J., McBride, M.B., Hay, A.G., 2011. Adsorption of copper and zinc by biochars produced from pyrolysis of hardwood and corn straw in aqueous solution. *Bioresour. Technol.* 102(19), 8877–8884.

Chiang, P.N., Wang, M.K.K., Chiu, C.Y., Chou, S.Y., 2006. Effects of cadmium amendments on low-molecular-weight organic acid exudates in rhizosphere soils of tobacco and sunflower. *Environ. Toxicol.* 21, 479–488.

Choppala, G., Saifullah, Bolan, N., Bibi, S., Iqbal, M., Rengel, Z., Kunhikrishnan, A., Ok, Y.S., 2014. Cellular mechanisms in higher plants governing tolerance to cadmium toxicity. *Crit. Rev. Plant Sci.* 33, 374–391.

Compant, S., Clément, C., Sessitsch, A., 2010. Plant growth-promoting bacteria in the rhizo- and endosphere of plants: Their role, colonization, mechanisms involved and prospects for utilization. *Soil Biol. Biochem.* 42, 669–678.

Cunningham, S.D., Ow, D.W., 1996. Promises and prospects of phytoremediation. *Plant Physiol.* 110, 715–719.

Czarnecki, S., Düring, R.A., 2015. Influence of long-term mineral fertilization on metal contents and properties of soil samples taken from different locations in Hesse, Germany. *Soil* 1, 23–33.

Das, J., Sarkar, P., 2018. Remediation of arsenic in mung bean (Vigna radiata) with growth enhancement by unique arsenic-resistant bacterium Acinetobacter lwoffii. *Sci. Total Environ.* 624, 1106–1118.

Das, K., Das, S., Dhundasi, S., 2008. Nickel, its adverse health effects and oxidative stress. *Indian J. Med. Res.* 128, 412–425.

Dhankher, O.P., Li, Y.J., Rosen, B.P., Shi, J., Salt, D., Senecoff, J.F., Sashti, N.A., Meagher, R.B., 2002. Engineering tolerance and hyper accumulation of arsenic in plants by combining arsenate reductase and γ-glutamyl cysteine synthase expression. *Nat. Biotech.* 20, 1140–1150.

Dinh, N., Van Der Ent, A., Mulligan, D.R., Nguyen, A.V., 2018. Zinc and lead accumulation characteristics and in vivo distribution of Zn2+ in the hyperaccumulator Noccaea caerulescens elucidated with fluorescent probes and laser confocal microscopy. *Envt. Exp. Bot.* 147, 1–12.

Doornbos, R.F., vanLoon, L.C., Bakker, P.A.H.M., 2012. Impact of root exudates and plant defense signaling on bacterial communities in the rhizosphere. A review. *Agron. Sustain. Dev.* 32, 227–243.

Doty, S.L., Shang, Q.T., Wilson, A.M., Moore, A.L., Newman, L.A., Strand, S.E., 2007. Enhanced metabolism of halogenated hydrocarbons in transgenic plants containing mammalian P450 2E1. *Proc. Natl. Acad. Sci. USA* 97, 6287–6291.

Duda-Chodak, A., Baszczyk, U., 2008. The impact of nickel on human health. *J. Elementol.* 13, 685–696.

Duruibe, J.O., Ogwuegbuand, M.O.C., Egwurugwu, J.N., 2007. Heavy metal pollution and human biotoxic effects. *Int. J. Phys. Sci.* 2, 112–118.

Dushenkov, S., Vasudev, D., Kapolnik, Y., Gleba, D., Fleisher, D., Ting K.C., Ensley, B., 1997. *Environ. Sci. Technol. 31*(12); 3468–3476.

Edelstein, M., Ben-Hur, M. 2018. Heavy metals and metalloids: Sources, risks and strategies to reduce their accumulation in horticultural crops. *Sci. Hortic.* 234, 431–444.

Erakhrumen, A.A., 2007. Phytoremediation: An environmentally sound technology for pollution prevention, control and remediation in developing countries. *Educ. Res. Rev.* 2, 151–156.

Fageria, N.K., Baligar, V.C., Clark, R.B., 2002. Micronutrients in crop production. *Adv. Agron.* 77, 185–268.

Fageria, N.K., Filho, M.P.B., Moreira, A., Guimaraes, C.M., 2009. Foliar fertilization of crop plants. *J. Plant Nut.* 32, 1044–1064.

Fan, M., Liu, Z., Nan, L., Wang, E., Chen, W., Lin, Y., Wei, G., 2018. Isolation, characterization, and selection of heavy metal-resistant and plant growth-promoting endophytic bacteria from root nodules of Robinia pseudoacacia in a Pb/Zn mining area. *Microbiol. Res.* 217, 51–59.

Fellet, G., Marmiroli M., Marchiol L., 2014. Species grown on mine tailings amended with three types of biochar. *Sci. Total Environ.* 468, 598–608.

Filetti, F.M., Vassallo, D.V., Fioresi, M., Simoes, M.R., 2018. Reactive oxygen species impair the excitation-contraction coupling of papillary muscles after acute exposure to a high copper concentration. *Toxicol. Vitro* 51, 106–113.

Garcia-Salgado, S., Garcia-Casillas, D., Quijano-Nieto, M.A., Bonilla-Simon, M.M., 2012. Arsenic and heavy metal uptake and accumulation in native plant species from soils polluted by mining activities. *Water, Air, Soil Pollut.* 223, 559–572.

Gholizadeh, M., Hu, X., 2021. Removal of heavy metals from soil with biochar composite: A critical review of the mechanism. *J. Environ. Chem. Eng.* 9(5), 105830.

Ghosh, M., Singh, S.P., 2005. A review on phytoremediation of heavy metals and utilization of it's by products. *Appl. Ecol. Environ. Res.* 3, 1–18.

Girkina, N.T., Turnerb, B.L., Ostlec, N., Craigona, J., Sjögerstena, S., 2018. Root exudate analogues accelerate CO2 and CH4 production in tropical peat. *Soil Biol. Biochem.* 117, 48–55.

Glick, B.R., 2003. Phytoremediation: Synergistic use of plants and bacteria to clean up the environment. *Biotechnol.* 21: 383–93.

Glick, B.R., 2010. Using soil bacteria to facilitate phytoremediation. *Biotechnol. Adv.* 28, 367–374.

Gossuin, Y., Vuong, Q.L., 2018. NMR relaxometry for adsorption studies: Proof of concept with copper adsorption on activated alumina. *Sep. Purif. Technol.* 202, 138–143.

Gou, Z.H., Miao, X.F., 2010. Growth changes and tissues anatomical characteristics of giant reed (*Arundo donax* L,) in contaminated soil with arsenic, cadmium and lead. *J. Cent. South Univ. Technol.* 17, 770–777.

Gulati, K., Banerjee, B., BalaLall, S., Ray, A., 2010. Effects of diesel exhaust, heavy metals and pesticides on various organ systems: Possible mechanisms and strategies for prevention and treatment. *Indian J. Exp. Biol.* 48, 710–721.

Gupta, P., Rani, R., Chandra, A., Kumar, V., 2018. Potential applications of Pseudomonas sp.(strain CPSB21) to ameliorate Cr6+ stress and phytoremediation of tannery effluent contaminated agricultural soils. *Sci. Rep.* 8(1), 4860.

Harris, T., Naidoo, K., Nokes, J., Walker, T., Orton, F., 2009. Indicative assessment of the feasibility of Ni and Au phytomining in Australia. *J. Clean. Prod.* 17, 194–200.

Hasan, S.M.M., Akber, M.A., Bahar, M.M., Islam, M.A., Akbor, M.A., Siddique, M.A.B, Islam, M.A., 2021. Chromium contamination from tanning industries and Phytoremediation potential of native plants: A study of savar tannery industrial estate in Dhaka, Bangladesh. *Bull. Environ. Contam. Toxicol. 106*(6), 1024–1032.

He, Z.L., Yang, X.E., StoVella, P.J., 2005. Trace elements in agroecosystems and impacts on the environment. *J. Trace Elem. Med. Biol.* 19, 125–140.

Hemavathy, R.R.V., Kumar, P.S., Suganya, S., Swetha, S., Varjani, S.J., 2019. Modelling on the removal of toxic metal ions from aquatic system by different surface modified Cassia fistula seeds. *Bioresour. Technol.* 281, 1–9.

Hess, R., Schmid, B., 2002. Zinc supplement overdose can have toxic effects. *J. Paediatr. Haematol. Oncol.* 24, 582–584.

Hocaoğlu-Özyiğit, A., Genç, B.N., 2020. Cadmium in plants, humans and the environment. *Front. Life Sci. Related Technol. 1*(1), 12–21.

Imran, M., Rehim, A., 2016. Zinc fertilization approaches for agronomic bio-fortification and estimated human bioavailability of zinc in maize grain. *Arch. Agron. Soil Sci.* https://doi.org/10.1080/03650340.2016.1185660

Imran, M., Kanwal, S., Hussain, S., Maqsood, M.A., Aziz, T., 2015. Efficacy of zinc application methods for concentration and estimated bioavailability of zinc in grains of rice grown on a calcareous soil. *Pak. J. Agric. Sci.* 52, 169–175.

Iqbal, M.P., 2012. Lead pollution—a risk factor for cardiovascular disease in Asian developing countries. *Pak. J. Pharm. Sci.* 25, 289–294.

Islam, M.S., Ahmed, M.K., Al-mamun, M.H., 2015a. Metal speciation in soil and health risk due to vegetables consumption in Bangladesh. *Environ. Monit. Assess.* 187, 288–302.

Islam, M.S., Ahmed, M.K., Al-Mamun, M.H., Islam, S.M.A. 2019. Sources and ecological risk of heavy metals in soils of different land uses in Bangladesh. *Pedosphere 29*(5), 665–675.

Islam, M.S., Ahmed, M.K., Al-Mamun, M.H., Masunaga, S. 2015b. Potential ecological risk of hazardous elements in different land-use urban soils of Bangladesh. *Sci. Total Environ.* 512–513, 94–102.

Islam, M.S., Ahmed, M.K., Al-mamun, M.H., Raknuzzaman, M., 2015c. The concentration, source and potential human health risk of heavy metals in the commonly consumed foods in Bangladesh. *Ecotox. Environ. Saf.* 122, 462–469.

Jadia, C.D., Fulekar, M.H., 2008. Phytoremediation: The application of vermi compost to remove zinc, cadmium, copper, nickel and lead by sunflower plant. *Environ. Eng. Manage. J.* 7, 547–558.

Jaishankar, M., Tseten, T., Anbalagan, N., Mathew, B.B., Beeregowda, K.N., 2014. Toxicity, mechanism and health effects of some heavy metals. *Interdiscip Toxicol. 7*(2), 60–72.

Järup, L., 2003. Hazards of heavy metal contamination. *Br. Med. Bull.* 68, 167–182.

Jeffery, S., Verheijen, F.G.A., Van DerVelde, M., Bastos, A.C., 2011. A quantitative review of the effects of biochar application to soils on crop productivity using meta-analysis. *Agric. Ecosyst. Environ.* 144, 175–187.

Jha, Y., Subrmanian, R.B., Mishra, K.K., 2017. Role of plant growth promoting rhizobacteria in accumulation of heavy metal in metal contaminated soil. *Emerg. Life Sci. Res.* 3, 48–56.

Ji, K., Kim, J., Lee, M., Park, S., Kwon, H., Cheong, H., Jang, J., Kim, D., Yu, S., Kim, Y., Lee, K., Yang, S., Jhung, Ik, Yang, W., Paek, D., Hong, Y., Choi, K., 2013. Assessment of exposure to heavy metals and health risks among residents near abandoned metal mines in Goseong. *Korea. Environ. Pollut.* 178, 322–328.

Jia, J., Liu, C., Wang, L., Liang, X., Chai, X., 2018. Double functional polymer brush grafted cotton fiber for the fast visual detection and efficient adsorption of cadmium ions. *Chem. Eng. J.* 347, 631–639.

Jian, C., Safwan, S., Fengxiang, X.H., David, L.M., Charles, A.W., Zhimin, Y., Yi, S., 2009. Bioaccumulation and physiological effects of mercury in *Pteris vittata* and *Nephrolepis exaltata*. *Ecotoxicology* 18, 110–121.

Jianxu, W., Xinbin, F., Christopher, W.N.A., Ying, X., Lihai, S., 2012. Remediation of mercury contaminated sites: A review. *J. Hazard. Mater.* 221–222, 1–18.

Jin, X.F., Liu, D., Islam, E., Mahmood, Q., Yang, X.E., He, Z.L., Stoffella, P.J., 2009. Effects of zinc on root morphology and antioxidant adaptations of cadmium-treated *Sedum alfredii* H. *J. Plant Nutr.* 32, 1642–1656.

Kalve, S., Sarangi, B.K., Pandey, R.A., Chakrabarti, T., 2011. Arsenic and chromium hyper accumulation by an ecotype of *Pteris vittata*: Prospective for phytoextraction from contaminated water and soil. *Curr.* Sci. 100, 888–894.

Kcil, A., Erust, C., Ozdemiroglu, S., Fonti, V., Beolchini, F., 2015. A review of approaches and techniques used in aquatic contaminated sediments: Metal removal and stabilization by chemical and biotechnological processes. *J. Cleaner Prod.* 86, 24–36.

Khalid, S., Shahid, M., Niazi, N.K., Murtaza, B., Bibi, I., Dumat, C., 2017. A comparison of technologies for remediation of heavy metal contaminated soils. *J. Geochem. Explor.* 182, 247–268.

Khan, M.A., Ahmad, I., ur Rahman, I., 2007. Effect of environmental pollution onheavy metals content of *Withanias omnifera*. *J. Chin. Chem. Soc.* 54, 339–343.

Khan, S., Cao, Q., Zheng, Y.M., Huang, Y.Z., Zhu, Y.G., 2008. Health risks of heavy metals in contaminated soils and food crops irrigated with wastewater in Beijing, China. *Environ. Pollut.* 152, 686–692.

Koptsik, G.N., 2014. Problems and prospects concerning the phytoremediation of heavy metal polluted soils: A review. *Eurasian Soil Sci.* 47, 923–939.

Kos, B., Grčman, H., Leštan, D., 2003. Phytoextraction of lead, zinc and cadmium from soil by selected plants. *Plant Soil Environ.* *49*(12), 548–553.

Kramer, U., 2010. Metal hyper accumulation in plants. *Ann. Rev. Plant Biol.* 61,517–534.

Kramer, U., Chardonnens, A.N., 2001. The use of transgenic plants in the bioremediation of soils contaminated with trace elements. *Appl. Microbiol. Biotechnol.* 55, 661–72.

Kucharski, R., Sas Nowosielska, A., Makowski, E., Japenga, J., Kuperberg, J.M., Pogrzeba, M., Krzyzak, J., 2005. The use of indigenous plant species and calcium phosphate for the stabilization of highly metal polluted sites in southern Poland. *Plant Soil* 273, 291–305.

Kurniawan, S.B., Purwanti, I.F., Titah, H.S., 2018. The effect of pH and aluminium to bacteria isolated from aluminium recycling industry. *J. Ecol. Eng.* *19*(3), 154–161.

Lange, B., Van Der Ent, A., Baker, A.J.M., Echevarria, G., Mahy, G., Malaisse, F., Meerts, P., Pourret, O., Verbruggen, N., Faucon, M.P., 2017. Copper and cobalt accumulation in plants: A critical assessment of the current state of knowledge. *N. Phyto.* 213, 537–551.

Lasat, M.M., 2002. Phytoextraction of toxic metals: A review of biological mechanisms. *J. Environ. Qual.* 31, 109–120.

Lei, M., Zhang, Y., Khan, S., Qin, P., & Liao, B., 2010. Pollution, fractionation and mobility of Pb, Cd, Cu, and Zn in garden and paddy soils from a Pb/Zn mining area. *Environ. Monitor. Assessm.* 168, 215–222.

Lei, S., Shi, Y., Qiu, Y., Che, L., Xue, C., 2019, Performance and mechanisms of emerging animal-derived biochars for immobilization of heavy metals. *Sci. Total Environ.* 646, 1281–1289.

Li, J.H., Lu, Y., Yin, W., Gan, H.H., Zhang, C., Deng, X.L., Lian, J., 2009. Distribution of heavy metals in agricultural soils near a petrochemical complex in Guangzhou, China. *Environ. Monit. Assess.* 153, 365–375.

Li, L., Zhang, Y., Ippolito, J.A., Xing, W., Qiu, K., Yang, H., 2020. Lead smelting effects heavy metal concentrations in soils, wheat, and potentially humans. *Envt. Pol.* 257, 113641. https://doi.org/10.1016/j.envpol.2019.113641

Li, T.Q., Yang, X.E., Long, X.X., 2004. Potential of using Sedum alfredii Hance for phytoremediating multi-metal contaminated soils. *J Soil Water* Conserv. 18, 79–83.

Li, Y., Bai, P., Yan, Y., Yan, W., Shi, W., Xu, R., 2019, Removal of Zn^{2+}, Pb^{2+}, Cd^{2+}, and Cu^{2+} from aqueous solution by synthetic clinoptilolite. *Micropor. Mesopor. Mater.* 273, 203–211.

Li, Y.M., Chaney, R., Brewer, E., Roseberg, R., Angle, J.S., Baker, A., Reeves, R., Nelkin, J., 2003. Development of a technology for commercial phytoextraction of nickel: Economic and technical considerations. *Plant Soil* 249, 107–115.

Liao, J.P., Lin, X.G., Cao, Z.H., Shi, Y.Q., Wong, M.H., 2003. Interactions between arbuscular mycorrhizae and heavy metals under a sand culture experiment. *Chemosphere* 50, 847–853.

Liao, Z., Fan, J., Lai, Z., Bai, Z., Wang, H., Cheng, M., Zhang, F., Li, Z., 2023. Response network and regulatory measures of plant-soil-rhizosphere environment to drought stress. *Adv. Agron.* 180, 93.

Liu, L., Li, W., Song, W., Guo, M., 2018. Remediation techniques for heavy metal-contaminated soils: Principles and applicability. *Sci. Total Environ.* 633, 216–219.

Liu, X., Zhang, A., Ji, C., Joseph, S., Bian, R., Li, L., Paze Ferreiro, J., 2013. Biochar's effect on crop productivity and the dependence on experimental conditions a meta-analysis of literature data. *Plant Soil* 373, 583–594.

Lombi, E., Zhao, F.J., Dunham, S.J., McGrath, S.P., 2001. Phytoremediation of heavy metal-contaminated soils: Natural Hyper accumulation versus chemical enhanced phytoextraction. *J. Environ. Qual.* 30, 1919–1926.

Lu, H., Zhang, W., Yang, Y., Huang, X., Wang, S., Qiu, R., 2012. Relative distribution of Pb^{2+} sorption mechanisms by sludge-derived biochar. *Water Res.* 46, 854–862.

Lu, K., Yang, X., Gielen, G., Bolan, N., Ok, Y.S., Niazi, N.K., Xu, S., Yuan, G., Chen, X., Zhang, X., Liu, D., 2017. Effect of bamboo and rice straw biochars on the mobility and redistribution of heavy metals (Cd, Cu, Pb and Zn) in contaminated soil. *J. Environ. Manag.* 186, 285–292.

Lugtenberg, B., Kamilova, F., 2009. Plant-growth-promoting rhizobacteria. *Annu. Rev. Microbiol.* 63, 541–556.

Luis, R., Lopez-Bellido, F., Carnicer, A., Alcalde-Morano, V., 2003. Phytoremediation of mercury-polluted soils using crop plants. *Fresenius Environ. Bull.* 12, 967–971.

Luis, R., Rincon, J., Asencio, I., Rodriguez-Castellanos, L., 2007. Capability of selected crop plants for shoot mercury accumulation from polluted soils: Phytoremediation perspectives. *Int. J. Phytoremediat.* 9, 1–13.

Ma, Y., Rajkumar, M., Luo, Y.M., Freitas, H., 2011a. Inoculation of endophytic bacteria on host and non-host plants e effects on plant growth and Ni uptake. *J. Hazard. Mater.* 195, 230–237.

Ma, Y., Rajkumar, M., Rocha, I., Oliveira, R.S., Freitas, H., 2015b. Serpentine bacteria influence metal translocation and bioconcentration of *Brassica juncea* and *Ricinus communis* grown in multi-metal polluted soils. *Front. Plant Sci.* 5, 757.

Ma, Y., Rajkumar, M., Oliveira, R.S., Zhang, C., Freitas, H., 2019. Potential of plant beneficial bacteria and arbuscular mycorrhizal fungi in phytoremediation of metal-contaminated saline soils. *J. Hazard. Mater.* 379, 120813.

Macek, T., Mackova, M., Kas, J., 2000. Exploitation of plants for the removal of organics in environmental remediation. *Biotechnol.* 18, 23–34.

Maharajan, T., Chellasamy, G., Tp, A.K., Ceasar, S.A., Yun, K., 2023. The role of metal transporters in phytoremediation: A closer look at Arabidopsis. *Chemosphere* 310, 136881.

Malik, A., Butt, T.A., Naqvi, S.T.A., Yousaf, S., Qureshi, M.K., Zafar, M.I., Farooq, G., Nawaz, I. and Iqbal, M., 2020. Lead tolerant endophyte Trametes hirsuta improved the growth and lead accumulation in the vegetative parts of Triticum aestivum L. *Heliyon*, 6, 7.

Meers, E., Slycken, S.V., Adriaensen, K., Ruttens, A., Vangronsveld, J., Laing, G.D., Witters, N., Thewys, T., Tack, F.M.G., 2010. The use of bio-energy crops (*Zea mays*) for 'phytoremediation' of heavy metals on moderately contaminated soils: A field experiment. *Chemosphere* 78, 35–41.

Memon, A.R., Schr€oder, P., 2009. Implications of metal accumulation mechanisms to phytoremediation. *Environ. Sci. Poll. Res.* 16, 162–175.

Mensah, E., Kyei-Baffour, N., Ofori, E. and Obeng, G., 2009. Influence of human activities and land use on heavy metal concentrations in irrigated vegetables in Ghana and their health implications. In: *Appropriate Technologies for Environmental Protection in the Developing World: Selected Papers from ERTEP 2007*, July 17–19 2007, Ghana, Africa, pp. 9–14.

Mesjasz-Przyby, O.J., Nakonieczny, M., Migula, P., Augustyniak, M., Tarnawska, M.M., Reimold, W.U., Koeberl, C., Przyby, O.W., Owacka, E.G., 2004. Uptake of cadmium, lead, nickel and zinc from soil and water solutions by the nickel hyperaccumulator *Berkheya coddii. Acta Biol.Cracov. Ser. Bot.* 46, 75–85.

Milic, D., Lukovic, J., Ninkov, J., Zeremski-Skoric, T., Zoric, L., Vasin, J., Milic, S., 2012. Heavy metal content in halophytic plants from inland and maritime saline areas. *Cent. Eur. J. Biol.* 7, 307–317.

Miransari, M., 2011. Soil microbes and plant fertilization. *Appl. Microbiol. Biotechnol.* 92, 875–885.

Mishra, S., Dwivedi, S.P., Singh, R.B., 2010. A review on epigenetic effect of heavy metal carcinogenesis on human health. *Open Nutraceut. J.* 3, 188–193.

Mukhopadhyay, S., Maiti, S.K., 2010. Phytoremediation of metal enriched mine waste: A review. *Global J. Environ. Res.* 4, 135–150.

Naser, Z., Liu, J.S., Wang, Q.C., Liang, Z.Z., 2011. Mercury contamination due to zinc smelting and chlor-alkali production in NE China. *Appl. Geochem.* 26, 188–193.

Ojuederie, O.B., Babalola, O.O., 2017. Microbial and plant-assisted bioremediation of heavy metal polluted environments: A review. *Int. J. Environ. Res. Public Health* 14(12), 1504.

Oliveira, V.H.D., Tibbett, M., 2018. toxicity and transport of Cd and Zn in *Populus trichocarpa, Environ. Exp. Botany* 155, 281–292.

Padmavathiamma, P.K., Li, L.Y., 2007. Phytoremediation technology: Hyper accumulation metals in plants. *Water Air Soil Pollut.* 184, 105–126.

Pajevic, S., Borisev, M., Nikolic, N., Arsenov, D.D., Orlovic, S., Zupunski, M., 2016. Phytoextraction of heavy metals by fast-growing trees: A review. In: Tsao, D. (Ed.), *Phytoremediation.* Springer, Heidelberg, pp. 29–64.

Papa, S., Bartoli, G., Pellegrino, A., Fioretto, A., 2010. Microbial activities and trace element contents in an urban soil. *Environ. Monit. Assess.* 165, 193–203.

Parmar, S., Singh, V., 2015. Phytoremediation approaches for heavy metal pollution: A review. *J. Plant Sci. Res.* 2(2), 139.

Pasricha, S., Mathur, V., Garg, A., Lenka, S., Verma, K., Agarwal, S., 2021. Molecular mechanisms underlying heavy metal uptake, translocation and tolerance in hyperaccumulators-an analysis: Heavy metal tolerance in hyperaccumulators. *Environ. Challenges*, 4, 100197.

Patra, D.K., Pradhan, C., Patra, H.K., 2020. Toxic metal decontamination by phytoremediation approach: Concept, challenges, opportunities and future perspectives. *Environ. Technol. Innov.* 18, 100672.

Paze Ferreiro, J., Lu, H., Fu, S., Mendez, A., Gasco, G., 2014. Use of phytoremediation and biochar to remediate heavy metal polluted soils: A review. *Solid Earth.* 5, 65–75.

Peng, W., Li, X., Song, J., Jiang, W., Liu, Y., Fan, W., 2018. Bioremediation of cadmium- and zinc-contaminated soil using *Rhodobacter sphaeroides. Chemosphere* 197, 33–41

Pereira, B.F.F., DeeAbreu, C.A., Herpin, U., DeeAbreu, M.F., Berton, R.S., 2010. Phytoremediation of lead by jack beans on a rhodic hapludox amended with EDTA. *Sci. Agric.* 67, 308–318.

Popova, L.P., Maslenkova, L.T., Ivanova, A., Stoinova, Z., 2012. Role of salicylic acid in alleviating heavy metal stress. In: Parvaiz, A., Prasad, M.N.V. (Ed.), *Environmental Adaptations and Stress Tolerance of Plants in the Era of Climate Change*. Springer, New York, pp. 447–466.

Popova, L.P., Maslenkova, L.T., Yordanova, R.Y., Ivanova, A.P., Krantev, A.P., Szalai, G., 2009. Exogenous treatment with salicylic acid attenuates cadmium toxicity in pea seedlings. *Plant Physiol Biochem.* 47, 224–231.

Qasim, B., Razzak, A.A. and Rasheed, R.T., 2021. Effect of biochar amendment on mobility and plant uptake of Zn, Pb and Cd in contaminated soil. *IOP Conf. Ser. Earth Environ. Sci. 779*(1), 012082.

Raklami, A., El Gharmali, A., Ait Rahou, Y., Oufdou, K., Meddich, A., 2021. Compost and mycorrhizae application as a technique to alleviate Cd and Zn stress in Medicago sativa. *Int. J. Phytoremed. 23*(2), 190–201.

Prasad, M.N.V., 2005. Nickel ophilous plants and their significance in phyto technologies. *Braz. J. Plant Physiol.* 17, 113–128.

Proshad, R., Islam, M.S., Kormoker, T., 2018. Assessment of heavy metals with ecological risk of soils in the industrial vicinity of Tangail district, Bangladesh. *Int. J. Adv. Geosci.* 6, 108–116.

Rahman, M.A., Rahman, M.M., Reichman, S.M., Lim, R.P., Naidu, R., 2014. Heavy metals in Australian grown and imported rice and vegetables on sale in Australia: Health hazard. *Ecotoxicol. Environ. Saf.* 100, 53–60.

Rai, P.K., 2008. Phytoremediation of Hg and Cd from industrial effluents using an aquatic free floating macrophyte *Azolla pinnata. Int. J. Phytoremediat.* 10, 430–439.

Rai, R., Agrawal, M., Agrawal, S.B., 2016. Impact of heavy metals on physiological processes of plants: With special reference to photosynthetic system. *Plant Resp. Xenobiot.* 127–140. https://doi.org/10.1007/978-981-10-2860-1_6

Rajakaruna, N., Tompkins, K.M., Pavicevic, P.G., 2006. Phytoremediation: An affordable green technology for the clean-up of metal contaminated sites in SryLanka. *Cey. J. Sci. (Biol. Sci.)* 35(1), 25–39.

Rajkumar, M., Sandhya, S., Prasad, M.N.V., Freitas, H., 2012. Perspectives of plant-associated microbes in heavy metal phytoremediation. *Biotechnol. Adv.* 30, 1562–1574.

Ramanjaneyulu, A.V., Neelima, T.L., Madhavi, A., Ramprakash, T., 2017. Phytoremediation: An overview. In: Humberto, R.M., Ashok, G.R., Thakur, K., Sarkar, N.C. (Eds.), *Applied Botany*. American Academic Press, New York, pp. 42–84.

Rani, K.S., Srinivas, B., GouruNaidu, K., Ramesh, K.V. 2018. Removal of copper by adsorption on treated laterite. *Mater. Today: Proc. 5*(1), 463–469.

Ranieri, E., Moustakas, K., Barbafieri, M., Ranieri, A.C., Herrera Melián, J.A., Petrella, A., Tommasi, F., 2020. Phytoextraction technologies for mercury-and chromium-contaminated soil: A review. *J. Chem. Tech. Biotech.* 95, 317–327. https://doi.org/10.1002/jctb.6008

Raskin, I., Ensley, B.D., 2000. Phytoremediation of Toxic Metals: Using Plants to Clean Up the Environment. John Wiley & Sons, Inc., New York.

Rayu, S., Karpouzas, D.G., Singh, B.K., 2012. Emerging technologies in bioremediation: Constraints and opportunities. *Biodegradation 23*(6), 917–926.

Rehim, A., ZafareuleHye, M., Imran, M., Ali, M.A., Hussain, M., 2014. Phosphorus and zinc application improves rice productivity. *Pak. J. Sci.* 66, 134–139.

Remy, E., Duque, P., 2016. Assessing tolerance to heavy-metal stress in Arabidopsis thaliana seedlings. In *Environmental Responses in Plants: Methods and Protocols*, pp. 197–208. https://doi.org/10.1007/978-1-4939-3356-3_16

Rizvi, A., Khan, M.S., 2018. Heavy metal induced oxidative damage and root morphology alterations of maize (Zea mays L.) plants and stress mitigation by metal tolerant nitrogen fixing *Azotobacter chroococcum. Ecotoxicol. Environ. Safety* 157, 9–20.

Romero-Cano, L.A., Garcia-Rosero, H., Gonzalez-Gutierrez, L.V., Baldenegro-Perez, L.A., Carrasco-Marín, F., 2017, Functionalized adsorbents prepared from fruit peels: Equilibrium, kinetic and thermodynamic studies for copper adsorption in aqueous solution. *J. Cleaner Prod.* 162, 195–204.

Roy, M., McDonald, L.M., 2013. Metal uptake in plants and health risk assessments in metal-contaminated smelter soils. *Land Degrad. Dev.* https://doi.org/10.1002/ldr.2237

Rugh, C.L., Seueoff, J.F., Meagher, R.B., Merkle, S.A., 1998. Development of transgenic yellow poplar for mercury phytoremediation. *Nat. Biotech.* 16, 925–928.

Saglam, A., Yetişsin, F., Demiralay, M., Terzi, R., 2016. Copper stress and responses in plants. In *Plant Metal Interaction*, pp. 21–40. https://doi.org/10.1016/B978-0-12-803158-2.00002-3

Saifullah, M.E., Qadir, M., DeeCaritat, P., Tack, F.M.G., Laing, G.D., Zia, M.H., 2009. *Chemosphere* 74, 1279–1291.

Sakakibara, O.Y., Ha, N.T.H., Sano, S., Sera, K., 2011. Phytoremediation of heavy metal contaminated water and sediment by *Eleocharis acicularis*. *Clean: Soil, Air, Water* 39, 735–741.

Salem, H.M., Eweida, E.A., Farag, A., 2000. Heavy metals in drinking water and their environmental impact on human health. In: *ICEHM2000*, Cairo University, Egypt, pp. 542–556.

Sarker, S.S., Akter, S., Siddique, M.A.B., Rahman, K.M.J., Nahar, S., Sharmin, S.A., 2024. Chromium and arsenic bioaccumulation and biomass potential of pink morning glory (Ipomoea carnea Jacq.). *Environ. Sci. Pollut. Res. 31*(2), 2187–2197.

Sarwar, N., Ishaq, W., Farid, G., Shaheen, M.R., Imran, M., Geng, M., Hussain, S., 2015. Zinc cadmium interactions: Impact on wheat physiology and mineral acquisition. *Ecotox. Environ. Saf.* 122, 528–536.

Sarwar, N., Saifullah, M.S.S., Zia, M.H., Naeem, A., Bibi, S., Farid, G., 2010. Role of plant nutrients in minimizing cadmium accumulation by plant. *J. Sci. Food Agric.* 90, 925–937.

Segura, A., Rodríguez-Conde, S., Ramos, C., Ramos, J.L., 2009. Bacterial responses and interactions with plants during rhizoremediation. *Microb. Biotechnol.* 2, 452–464.

Seth, C.S., 2012. A review on mechanisms of plant tolerance and role of transgenic plants in environmental clean up. *Bot. Rev.* 78, 32–62.

Seth, C.S., Chaturvedi, P.K., Misra, V., 2007. Toxic effect of arsenic and cadmium alone and in combination on Giant duckweed (*Spirodela polyrrhiza* L.) in response to its accumulation. *Environ. Toxicol.* 22, 539–549.

Shabani, N., Sayadi, M.H., 2012. Evaluation of heavy metals accumulation by two emergent macrophytes from the polluted soil: An experimental study. *Environmentalist* 32, 91–98.

Shahid, M., Dumat, C., Khalid, S., Schreck, E., Xiong, T., Niazi, N.K., 2017. Foliar heavy metal uptake, toxicity and detoxification in plants: A comparison of foliar and root metal uptake. *J. Hazard. Mater.* 325, 36–58.

Sheoran, V., Sheoran, A.S., Poonia, P., 2009. Phytomining: A review. *Min. Eng.* 22, 1007–1019.

Shi, T., Ma, J., Wu, X., Ju, T., Lin, X., Zhang, Y., Li, X., Gong, Y., Hou, H., Zhao, L., Wu, F., 2018. Inventories of heavy metal inputs and outputs to and from agricultural soils: A review, *Ecotoxicol. Environ. Safety* 164, 118–124.

Siddiqui, M.H., Kumar, A., Kesari, K.K., Arif, J.M., 2009. Biomining—a useful approach toward metal extraction. *Am.-Euras. J. Agron.* 2, 84–88.

Silva, I.R., Novais, R.F., Jham, G.N., Barros, N.F., Gebrim, F.O., Nunes, F.N., Neves, J.C.L., Leite, F.P., 2004. Responses of eucalypt species to aluminum: The possible involvement of low molecular weight organic acids in the Al tolerance mechanism. *Tree Physiol.* 24, 1267–1277.

Skuza, L., Szu´cko-Kociuba, I., Filip, E., Bozek, I., 2022. Natural molecular mechanisms of plant hyperaccumulation and hypertolerance towards heavy metals. *Int. J. Mol. Sci.* 23, 9335. https://doi.org/10.3390/ijms23169335

Souza, E.C., Vessoni-Penna, T.C., de Souza Oliveira, R.P., 2014. Bio surfactant-enhanced hydrocarbon bioremediation: An overview. *Int. Biodeterior. Biodegrad.* 89, 88–94.

Souza, L.A., Piotto, F.A., Nogueirol, R.C., Azevedo, R.A., 2013. Use of non-hyperaccumulator plant species for the phytoextraction of heavy metals using chelating agents. *Sci. Agric.* 70, 290–295.

Srivastava, M., Ma, L.Q., Santos, J.A.G., 2006. Three new arsenic hyper accumulating ferns. *Sci. Total Environ.* 364, 24–31.

Strom, L., Owen, A.G., Godbold, D.L., Jones, D.L., 2002. Organic acid mediated p mobilization in the rhizosphere and uptake by maize roots. *Soil Biol.* Biochem. 34, 703–710.

Sytar, O., Brestic, M., Taran, N., Zivcak, M., 2016. Plants used for biomonitoring and phytoremediation of trace elements in soil and water. In: Ahmad, P. (Ed.), *Plant Metal Interaction.* Elsevier, Amsterdam, pp. 361–384.

Syed, A., Zeyad, M.T., Shahid, M., Elgorban, A.M., Alkhulaifi, M.M., Ansari, I.A., 2021. Heavy metals induced modulations in growth, physiology, cellular viability, and biofilm formation of an identified bacterial isolate. *ACS Omega 6*(38), 25076–25088.

Tappero, R., Peltier, E., Gräfe, M., Heidel, K., Ginder-Vogel, M., Livi, K.J.T., Rivers, M.L., Marcus, M.A., Chaney, R.L., Sparks, D.L., 2007. Hyperaccumulator Alyssum murale relies on a different metal storage mechanism for cobalt than for nickel. *New Phytologist 175*(4), 641–654.

Tang, Y.T., Qiu, R.L., Zeng, X.W., Ying, R.R., Yu, F.M., Zhou, X.Y., 2009. Lead zinc, cadmium hyper accumulation and growth stimulation in *Arabis paniculata* Franch. *Environ. Exp. Bot.* 66, 126–134.

Tangahu, B.V., Abdullah, S.R.S., Basri, H., Idris, M., Anuar, N., Mukhlisin, M., 2011. A review on heavy metals (As, Pb, and Hg) uptake by plants through phytoremediation. *Int. J. Chem. Eng.* https://doi.org/10.1155/2011/939161

Thornton, I., 1981. Geochemical aspects of the distribution and forms of heavy metals in soils. In: Lepp, N.W. (Ed.), *Effect of Heavy Metal Pollution on Plants: Metals in the Environment*, vol. II. Applied Science Publishers, London and New Jersey, pp. 1–34.

Tong, Y.P., Kneer, R., Zhu, Y.G., 2004. Vacuolar compartmentalization: A second-generation approach to engineering plants for phytoremediation. *Trends Plant Sci.* 9, 7–9.

Tripathi, R.D., Srivastava, S., Mishra, S., Singh, N., Tuli, R., Gupta, D.K., Maathuis, F.J.M., 2007. Arsenic hazards: Strategies for tolerance and remediation by plants. *Trends Biotechnol.* 25, 158–165.

Ullah, A., Heng, S., Munis, M.F.H., Fahad, S., Yang, X., 2015. Phytoremediation of heavy metals assisted by plant growth promoting (PGP) bacteria: A review. *Environ. Exp. Botany* 117, 28–40.

Uroz, S., Calvaruso, C., Turpault, M.P., Sarniguet, A., de Boer, W., Leveau, J.H.J., 2009. Efficient mineral weathering is a distinctive functional trait of the bacterial genus Collimonas. *Soil Biol. Biochem.* 41, 2178–2186.

Usman, K., Al-Ghouti, M.A., Abu-Dieyeh, M.H., 2018. Phytoremediation: Halophytes as promising heavy metal hyper accumulators. In: Saleh, H.E.M., Aglan, R. (Eds.), *Heavy Metals.* Intech Open, United Kingdom, pp. 201–217.

van der Ent, A., Malaisse, F., Erskine, P.D, Mesjasz-Przybyłowicz, J., Przybyłowicz, W.J., Barnabas, A.D., Sośnicka, M., Harris, H.H., 2019. Abnormal concentrations of Cu–Co in Haumaniastrum katangense, Haumaniastrum robertii and Aeolanthus biformifolius: Contamination or hyperaccumulation? *Metallomics 11*(3), 586–596, https://doi.org/10.1039/c8mt00300a

Van Der Ent, A., Tang, Y.-T., Sterckeman, T., Echevarria, G., Morel, J.-.L., Qiu, R.-.L., 2018. Nickel hyperaccumulation mechanisms: A review on the current state of knowledge. *Plant Soil* 423, 1–11. https://doi.org/10.1007/s11104-017-3539-8

Vishnoi, S.R., Srivastava, P.N., 2008. Phytoremediation-green for environmental clean. In: *The 12th World Lake Conference,* Jaipur, India, pp. 1016–1021.

Wang, G., Zeng, F., Song, P., Sun, B., Wang, Q., Wang, J., 2022. Effects of reduced chlorophyll content on photosystem functions and photosynthetic electron transport rate in rice leaves. *J. Plant Physiol.* 272, 153669.

Wang, J., Shi, L., Zhai, L., Zhang, H., Wang, S., Zou, J., Shen, Z., Lian, C., Chen, Y., 2021. Analysis of the long-term effectiveness of biochar immobilization remediation on heavy metal contaminated soil and the potential environmental factors weakening the remediation effect: A review. *Ecotoxicol. Environ. Safety* 207, 111261.

Wang, Y., Stauffer, C., Keller, C., 2005. Changes in Hg fractionation in soil induced by willow. *Plant Soil* 275, 67–75.

Waqas, M., Khan, A.L., Kang, S.M., Kim, Y.H., Lee, I.J., 2014. Phytohormone-producing fungal endophytes and hardwood-derived biochar interact to ameliorate heavy metal stress in soybeans. *Biol. Fertilit. Soils* 50, 1155–1167.

Wei, S., daSilva, J.A.T., Zhou, Q., 2008. Agro-improving method of phyto extracting heavy metal contaminated soil. *J. Hazard. Mater.* 150, 662–668.

Wiszniewska, A., Muszyńska, E., Kołton, A., Kamińska, I., Hanus-Fajerska, E., 2019. In vitro acclimation to prolonged metallic stress is associated with modulation of antioxidant responses in a woody shrub Daphne jasminea. *Plant Cell Tissue Organ Cult.* 139, 339–357.

Wu, F.B., Chen, F., Wei, K.G., Zhang, P., 2004. Effects of Cadmium on free amino acids, glutathione, and ascorbic acid concentration in two barley genotypes (*Hordeum vulgare* L.) differing in cadmium tolerance. *Chemosphere* 57, 447–454.

Wuana, R.A., Okieimen, F.E., 2011. Heavy metals in contaminated soils: A review of sources, chemistry, risks and best available strategies for remediation. *ISRN Ecology* 2011, 1–20.

Xu, J., Yin, H.X., Li, X., 2009. Protective effects of proline against cadmium toxicity in micro propagated hyperaccumulator, *Solanum nigrum L. Plant Cell. Rep.* 28, 325–333.

Yadav, R., Arora, P., Kumar, S., Chaudhury, A., 2010. Perspectives for genetic engineering of poplars for enhanced phytoremediation abilities. *E*cotoxicology 19, 1574–1588.

Yan, A., Wang, Y., Tan, S.N., Mohd Yusof, M.L., Ghosh, S., Chen, Z., 2020. Phytoremediation: A promising approach for revegetation of heavy metal-polluted land. *Front. Plant Sci.* 11, 359.

Yan, F., Schubert, S., Mengel, K., 1996. Soil pH increase due to biological decarboxylation of organic anions. *Soil Biol. Biochem.* 28, 617–624.

Yang, S.X., Deng, H., Li, M.S., 2008. Manganese uptake and accumulation in a woody hyperaccumulator, *Schima superba. Plant Soil Environ.* 54, 441–446.

Yeboah, J.O., Shi, G., Shi, W., 2021. Effect of heavy metal contamination on soil enzymes activities. *J. Geosci. Environ. Prot.* 9(6), 135–154.

Yu, J., Huang, Z., Chen, T., Qin, D., Zeng, X., Huang, Y., 2012. Evaluation of ecological risk and source of heavy metals in vegetable-growing soils in Fujian province, China. *Environ. Earth Sci.* 65, 29–37.

Yuan, G.L., Sun, T.H., Han, P., Li, J., Lang, X.X., 2014. Source identification and ecological risk assessment of heavy metals in topsoil using environmental geochemical mapping: Typical urban renewal area in Beijing, China. *J. Geochem. Explor.* 136, 40–67.

Yuan, S., Hong, M., Li, H., Ye, Z., Gong, H., Zhang, J., Huang, Q., Tan, Z., 2020. Contributions and mechanisms of components in modified biochar to adsorb cadmium in aqueous solution. *Sci. Total Environ.* 733, 139320.

Zhang, X., Xia, H., Li, Z., Zhuang, P., Gao, B., 2011. Identification of a new potential Cd-hyperaccumulator *Solanum photeinocarpum* by soil seed bank-metal concentration gradient method. *J. Hazard. Mater.* 189, 414–419.

Zhang, H., Xu, F., Xue, J., Chen, S., Wang, J., Yang, Y., 2020. Enhanced removal of heavy metal ions from aqueous solution using manganese dioxide-loaded biochar: Behavior and mechanism. *Sci. Rep.* *10*(1), 6067.

Zhang, X., Xia, H., Li, Z., Zhuang, P., Gao, B., 2011. Identification of a new potential Cd-hyperaccumulator *Solanum photeinocarpum* by soil seed bank-metal concentration gradient method. *J. Hazard. Mater.* 189, 414–419.

Zhao, F.J., Dunham, S.J., McGrath, S.P., 2002. Arsenic Hyper accumulation by different fern species. *New Phytol.* 156, 27–31.

Zhou, D., Liu, D., Gao, F., Li, M., Luo, X., 2017. Effects of biochar-derived sewage sludge on heavy metal adsorption and immobilization in soils. *Int. J. Environ. Res. Public Health 14*(7), 681.

Zheng, L., Peng, D., Meng, P., 2018. Promotion effects of nitrogenous and oxygenic functional groups on cadmium (II) removal by carboxylated corn stalk, *J. Clean. Prod.* 201, 609–623.

5 Geochemical Evaluation of Soil Contamination from Rat-Hole Coal Mining Areas in Jaintia Hills, Meghalaya, India

Ravi Kumar Umrao and Resmi S

5.1 INTRODUCTION

Jaintia Hills, situated in the northeastern state of Meghalaya, India, is renowned for its picturesque landscapes, abundant greenery, and cultural heritage deeply rooted in nature. However, beneath this natural beauty lies an extensive network of coal seams that have attracted mining activities for decades. The unrestrained expansion of coal mining has raised concerns about the depletion of natural resources and the severe environmental consequences it causes. Rat-hole mining, a method prevalent in this region due to its low cost and labour-intensive nature, has garnered attention for its detrimental impact on soil quality and overall ecosystem health. Areas such as Lumthari, Khliehriat, Sutnga, and Jowai in the Jaintia Hills serve as hubs for coal mining activities. Khliehriat, situated in the East Jaintia Hills, is known as the "Coal Capital" and hosts numerous coal mines. Sutnga, also located in the East Jaintia Hills, is emblematic of the spread of mining activities into previously untouched areas, raising concerns about the sustainability of such expansion. The concentration of mining activities in these areas has led to significant environmental concerns.

Jaintia Hills boasts a substantial deposit of sub-bituminous Tertiary coal with unique physicochemical characteristics, such as high sulphur, volatile matter, vitrinite, and low ash contents (Chabukdhara and Singh, 2016). These characteristics, along with the enrichment of environmentally sensitive elements such as Fe, Mg, Bi, Al, V, Cu, Cd, Ni, Pb, and Mn, are associated with more severe environmental impacts during mining and utilisation in coal-based industries (Chabukdhara and Singh, 2016).

Although several studies have explored the impact of coal mining on water quality, acid mine drainage, aquatic biodiversity, and ecorestoration in the Jaintia Hills District (Swer and Singh, 2003; Pyrbot et al., 2019; Singh, 2019), a comprehensive geochemical evaluation of soil contamination in coal mining areas is lacking. Pyrbot et al. (2019) reported adverse effects on water resources, leading to acidic streams with a pH of 3–5, rendering them unsuitable for human use and devoid of aquatic life.

DOI: 10.1201/9781003442554-6

Streams and rivers face contamination from acid mine drainage, heavy metal leaching, organic enrichment, and silting by coal and sand particles (Swer and Singh, 2003; Pyrbot et al., 2019; Singh, 2019). Based on a community survey, Singh (2019) highlighted that soil degradation affects agricultural land and leads to a decline in crop productivity, prompting farmers to abandon traditional agriculture for alternative livelihoods. However, geochemical evaluations of soil contamination, specifically assessments of the impact of heavy metal pollution, are lacking in the coal mining areas of the Jaintia Hills District, Meghalaya. This book chapter aims to fill this gap by providing a thorough geochemical evaluation of soil contamination in and around rat-hole coal mining activities, addressing factors contributing to soil pollution, assessing contamination magnitude, identifying primary pollutants, and discussing potential sources of heavy metal contamination in soils.

5.2 STUDY AREA AND GEOLOGICAL SETUP

The study area encompasses the Jaintia Hills district in Meghalaya, India, as illustrated in Figure 5.1, and is covered by the Survey of India toposheets 83C/7 and 83C/11. The topography of the Jaintia Hills is characterised by rolling hills, dense forests, and a network of rivers and streams, which play vital roles in maintaining the

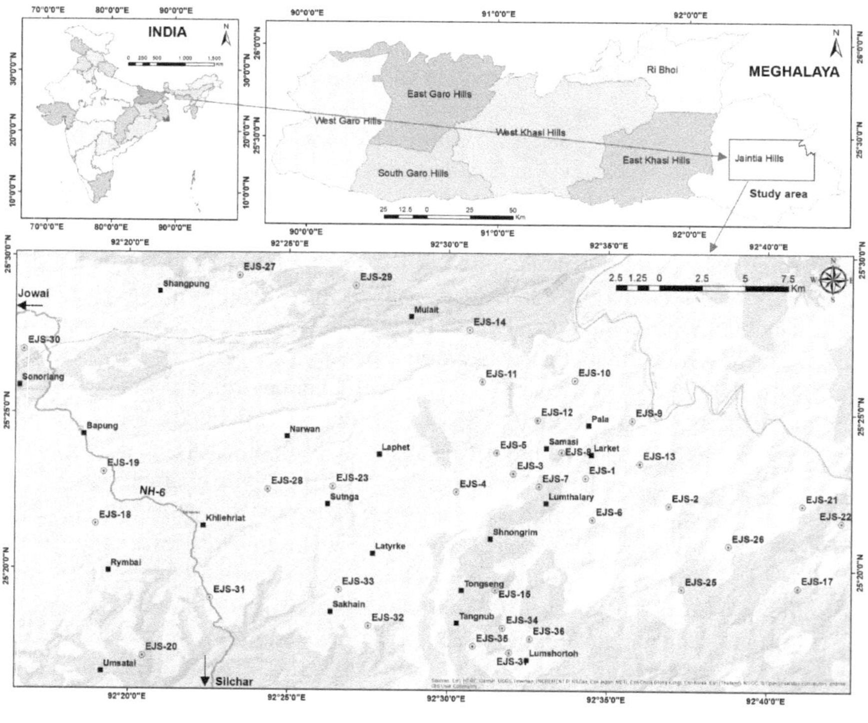

FIGURE 5.1 Soil sample location map in parts of the Jaintia Hills District, Meghalaya, India.

region's ecological balance. The interplay of geological, climatic, and anthropogenic factors makes this a compelling case study for assessing the geochemical impact of rat-hole coal mining on soil contamination.

The Shillong Plateau exhibits diverse geological features, including the Assam–Meghalaya gneissic complex, the Shillong Group of metasedimentary rocks, and various intrusions, such as granitic plutons, ultramafic-alkaline-carbonatite complexes, and the Sylhet trap. Additionally, it is adorned with deposits ranging from Cretaceous-Palaeogene sediments to recent alluvial materials (Sadiq et al., 2014a, 2014b; 2018; Sadiq and Umrao, 2017, 2020; Umrao and Sardar, 2022).

Geologically, the rocks exposed in the study area belong to the Assam–Meghalaya gneissic complex and sedimentary rocks from the Jaintia and Barail Groups. The gneissic complex consists of migmatite rocks from the Precambrian, mainly exposed in the northern and northwestern parts of the area. Tertiary sediments (the sediments of the Shella and Kopili formations of the Jaintia Group and Barail Group) cover migmatites and granites in the area. Tertiary sedimentation began with the deposition of the lower Sylhet sandstone, also known as the Therria sandstone, which hosts the majority of coal seams in the area (Raja Rao, 1981). The Shella Formation, characterised by alternating limestone and sandstone units, indicates sea level oscillations, resulting in shallow marine and deltaic or estuarine environments (Raja Rao, 1981; Umrao and Sardar, 2022). The Kopili Formation, from the upper Eocene period, represents a transition to a deeper marine environment with the accumulation of shales (Umrao and Sardar, 2022). The Barail Group shelf facies include sandstone, shale, carbonaceous shale, and minor coal seams (GSI Misc. Publ., 2009).

The coal in the Shella Formation is believed to have formed over stable shelf conditions, particularly in the Lakadong sandstone member, which is known for its superior quality compared to other Tertiary coals in Northeast India (Raja Rao, 1981). The coal has a low moisture content (0.7%–1.9%), relatively low volatile matter content (37.5%–42.1%), high carbon content (82.3%–84.3%), and low oxygen content (4.0%–7.6%). However, a high sulphur content, primarily organic sulfur, categorises coal as low-quality (Barooah and Baruah, 1996). This detailed geological overview sets the stage for understanding the soil contamination dynamics resulting from rat-hole coal mining activities in the region.

5.3 MATERIALS AND METHODS

Thirty-seven soil samples were systematically collected from the coal mining belt in the Jaintia Hills district, Meghalaya (Figure 5.1), to conduct a comprehensive geochemical analysis for contamination assessment. All samples were collected from below the humus layer in the B horizon of the natural soil. The geochemical composition of the soil provides valuable insights into the elemental behaviour associated with weathering, environmental changes, and potential anthropogenic contamination. After collection, the samples were air-dried under sunlight and processed to achieve a particle size of −120 mesh. The elemental analysis was performed using a PANalytical X-ray fluorescence (XRF) instrument at the Chemical Laboratory, GSI,

NER, Shillong, following the standard protocol available at the GSI OCBIS portal. The selected elements for analysis included Al, Fe, Ti, Mn, Co, Cr, Cu, Ni, Pb, Sr, V, and Zn, aiming to study potential soil contamination in and around the coal mining belt of the Jaintia Hills district, Meghalaya.

5.3.1 Statistical Analysis

This study employed various statistical analyses, including the Pearson correlation coefficient, principal component analysis (PCA), and cluster analysis, to elucidate relationships among the trace elements. Correlation analysis involves scrutinising the correlations between two variables to measure their degree of association. Specifically, Pearson correlation analysis was conducted to explore the correlations among the heavy metals.

PCA, a widely utilised statistical tool, involves multivariate analysis to eliminate data redundancy. This approach provides a more profound understanding of the interrelationships within a matrix by substituting a large dataset with a reduced number of components known as principal components (PCs). These components explain most of the variance in the dataset. The load factor indicates the relationship between the variables and each factor (Giri et al., 2013; Resmi et al., 2019; Soltani-Gerdefaramarzi et al., 2021). PCA, employed in this study, aimed to discern the sources of heavy metals in soil, utilising varimax factor loading to enhance the interpretability of the results.

Cluster analysis, another statistical method, categorises variables based on their similarity or degree of affinity. This method groups closely associated metals that may share a common source. The dendrogram for cluster analysis was prepared using the Agglomerative Hierarchical Clustering (AHC) method, contributing to a comprehensive understanding of metal associations.

5.3.2 Soil Pollution Assessment Indices

The environmental impact of metals and the pollution level in the soil samples can be measured through indices such as the enrichment factor (EF), geoaccumulation index (I_{geo}), contamination factor (CF), pollution load index (PLI), Nemerow pollution index ($PI_{Nemerow}$), and modified degree of contamination (mC_d).

5.3.2.1 Enrichment Factor

The EF is an effective and widely used tool for assessing the degree of enrichment and facilitating the comparison of contamination levels across various environmental media (Benhaddya and Hadjel, 2013; Resmi et al., 2019). It is instrumental in distinguishing elements originating from geogenic and anthropogenic processes (Soltani-Gerdefaramarzi et al., 2021). The EF methodology relies on the normalisation technique, where the metal concentration is normalised as a ratio to another constituent or a reference element in the soil. Al, Fe, Mn, Si, and Ti are commonly used as reference elements (Lu et al., 2009). This study used Fe as the reference material for normalising the metal concentration.

The EF is calculated for the elements using Eq. (5.1):

$$\mathrm{EF} = \frac{\left(C_{\mathrm{m}}/C_{\mathrm{Fe}}\right)_{\mathrm{sample}}}{\left(C_{\mathrm{m}}/C_{\mathrm{Fe}}\right)_{\mathrm{background}}} \tag{5.1}$$

where C_{m} and C_{Fe} are the metal and Fe concentrations in the soil sample and background, respectively.

In this study, the background values utilised for comparison were derived from the average concentrations of elements in the shale, as documented by Turekian and Wedepohl (1961). As Zhang and Liu (2002) emphasised, an EF value within the range of 0.5–1.5 signifies that the trace metals are entirely sourced from crustal origins. Conversely, an EF value exceeding 1.5 indicates a substantial proportion of trace metals originating from non-crustal materials, such as biota or pollution drainages. The following five contamination categories are generally recognised based on the EF (Giri et al., 2013):

EF < 2: Deficiency to low contamination
EF = 2–5: Moderate contamination
EF = 5–20: Considerable contamination
EF = 20–40: High contamination
EF > 40: Severe contamination

5.3.2.2 Geoaccumulation Index (I_{geo})

The level of metal contamination in soil and sediments can also be determined by the I_{geo} proposed by Müller (1969). The I_{geo} was calculated using Eq. (5.2), proposed by Muller (1969).

$$I_{geo} = \log_2\left[\frac{C_n}{1.5B_n}\right] \tag{5.2}$$

where C_n is the metal (n) concentration recorded in the sample of the study area, B_n is the background value of the corresponding metal (n) (Turekian and Wedepohl, 1961), and a factor of 1.5 is used because of possible variations in the background data due to lithological effects. Muller (1969) classified I_{geo} values into seven categories (Table 5.1).

TABLE 5.1
Classification Scheme of the Geoaccumulation Index

Class	Value	Classification
0	≤0	Uncontaminated
1	0–1	Uncontaminated to moderately contaminated
2	1–2	Moderately contaminated
3	2–3	Moderately to highly contaminated
4	3–4	Highly contaminated
5	4–5	Highly to very highly contaminated
6	≥5	Very Highly contaminated

5.3.2.3 Contamination Factor

The CF was calculated based on Pekey et al. (2004) using the ratio between the metal content in the sediment at a sample location and the background values (Eq. 5.3).

$$\mathrm{CF} = \frac{C_n}{B_n} \tag{5.3}$$

where C_n is the concentration of metal n in the sample, and B_n is the background value of the corresponding metal (n). The CF is categorised into four classes: $\mathrm{CF} < 1$, low contamination; $1 \leq \mathrm{CF} < 3$, moderate contamination; $3 \leq \mathrm{CF} < 6$, considerable contamination; and $\mathrm{CF} > 6$, very high contamination (Pekey et al., 2004).

5.3.2.4 Pollution Load Index

The PLI proposed by Tomlinson et al. (1980) was used in this study to measure the PLI in soil samples. The PLI was calculated using formulae based on individual elements' CF values (Eq. 5.4).

$$\mathrm{PLI} = \left(\mathrm{CF}_1 \times \mathrm{CF}_2 \times \cdots \times \mathrm{CF}_n\right)^{1/n} \tag{5.4}$$

where CF is the contamination factor, and "n" is the number of metals. This empirical index provides a simple, comparative means for assessing trace/heavy metal pollution levels. The PLI is classified as a $\mathrm{PLI} \leq 1$, which represents no metal pollution, and $\mathrm{PLI} > 1$, which represents metal pollution in soil (Tomlinson et al., 1980).

5.3.2.5 Nemerow Pollution Index

The $\mathrm{PI}_{\mathrm{Nemerow}}$ assesses the overall degree of soil pollution and includes the contents of all analysed heavy metals (Gong et al., 2008). $\mathrm{PI}_{\mathrm{Nemerow}}$ is calculated based on the formula given in Eq. (5.5).

$$\mathrm{PI}_{\mathrm{Nemerow}} = \sqrt{\frac{\left(\frac{1}{n}\sum_{i-1}^{n}\mathrm{PI}\right)^2 + \mathrm{PI}_{\max}^2}{n}} \tag{5.5}$$

where PI is the calculated value for the single pollution index, $\mathrm{PI}_{\max}$ is the maximum value for the single pollution index of all heavy metals, and n is the number of heavy metals. Cheng et al. (2007) classified $\mathrm{PI}_{\mathrm{Nemerow}}$ into five grades: (1) $\mathrm{PI}_{\mathrm{Nemerow}} < 0.7$ (safe domain soil), (2) $0.7 \leq \mathrm{PI}_{\mathrm{Nemerow}} < 1$ (precaution domain soil), (3) $1 \leq \mathrm{PI}_{\mathrm{Nemerow}} < 2$ (slight pollution), (4) $2 \leq \mathrm{PI}_{\mathrm{Nemerow}} < 3$ (moderate pollution), and (5) $\mathrm{PI}_{\mathrm{Nemerow}} > 3$ (high pollution).

5.3.2.5.1 Modified Degree of Contamination

The $\mathrm{mC_d}$ index is the modified degree of contamination index of Hakanson (Hakanson, 1980), and this index was first used by Abrahim and Parker (2008). $\mathrm{mC_d}$ allows the assessment of overall heavy metal soil contamination by summing all the single pollution indices (C^i_f) to estimate overall metal contamination at a location. mC_d is calculated using Eq. (5.6).

$$mC_d = \left(\sum_{i=1}^{n} C^i_f \right) / n \tag{5.6}$$

where C^i_f is the content of an individual heavy metal, and n is the number of analysed heavy metals. mC_d was classified into seven categories according to the degree of contamination: $mC_d < 1.5$, very low; $1.5 \le mC_d < 2$, low; $2 \le mC_d < 4$, moderate; $4 \le mC_d < 8$, high; $8 \le mC_d < 16$, very high; $16 \le mC_d < 32$, strongly high; and $mC_d \ge 32$, ultrahigh.

5.4 RESULTS AND DISCUSSIONS

5.4.1 Heavy Metal Concentrations

The concentrations of heavy metals, including Al, Fe, Ti, Mn, Co, Cr, Cu, Ni, Pb, Sr, V, and Zn, exhibited significant variability within the range of values observed in the study area. The concentrations ranged between 63,932 and 129,717 mg/kg for Al, 31,558 and 226,971 mg/kg for Fe, 4,315 and 15,762 mg/kg for Ti, 77 and 11,617 mg/kg for Mn, 4 and 44 mg/kg for Co, 103 and 428 mg/kg for Cr, 4 and 36 mg/kg for Cu, 10 and 247 mg/kg for Ni, 14 and 57 mg/kg for Pb, 37 and 118 mg/kg for Sr, 62 and 764 mg/kg for V, and 20 and 168 mg/kg for Zn (Table 5.2). The geometric mean concentrations were 95,123.26 mg/kg for Al, 88,981.23 mg/kg for Fe, 7,425.20 mg/kg for Ti, 631.07 mg/kg for Mn, 14.16 mg/kg for Co, 171.07 mg/kg for Cr, 20.75 mg/kg for Cu, 50.94 mg/kg for Ni, 27.70 mg/kg for Pb, 57.67 mg/kg for Sr, 221.88 mg/kg for V, and 74.60 mg/kg for Zn. The geometric mean concentration distribution revealed a sequential order of heavy metals in the study area: Fe>Al>Ti>Mn>V>Cr>Zn>Sr>Ni>Pb>Cu>Co. Comparisons with background values from shale (Turekian and Wedepohl, 1961) indicated that geometric mean concentrations of Al, Fe, Ti, Cr, Pb, and V were 1.19, 1.89, 1.61, 1.90, 1.38, and 1.71 times greater than the corresponding average shale values, respectively. In contrast, the Mn, Co, Cu, Ni, Zn, and Sr concentrations were below their respective background values.

Examining the coefficient of variability (CV) highlighted the most discriminating factor for describing variability among the heavy metals. The toxic metal content was ranked as Cr>Zn>Ni>Pb>Cu>Co, with moderate variations indicated by CV values ranging from 34.00% to 80.73%, except for Mn, which exhibited high variation at 153.76%. Lower CV values suggested natural sources, whereas higher CV values indicated potential anthropogenic influences (Zhang et al., 2007). The concentration of Mn, which is greater than other metals and is distributed randomly, suggests that Mn is susceptible to external factors, possibly originating from anthropogenic sources.

The box plot visually illustrates the distribution of heavy metal concentrations in the soil (Figure 5.2). Elevated concentrations of elements beyond what could be attributed to geogenic sources suggested potential anthropogenic causes. Geogenic

TABLE 5.2

Descriptive Statistics of Heavy Metals in Soil Samples ($n = 37$) and Background Values of Average Shale

Element	Range	Minimum	Maximum	GeoMean	Median	CV/%	Skewness	Kurtosis	Background Value (BG)	% of Samples above BG
Al	65,784.66	63,932.31	129,716.97	95,123.26	99,603.16	18.33	0.09	−0.83	80,000	81.08
Fe	195,412.81	31,558.31	226,971.12	88,981.23	90,632.99	49.31	1.30	1.26	47,200	94.59
Ti	11,447.28	4,315.20	15,762.48	7,425.20	6,892.34	32.95	1.32	1.66	4,600	97.30
Mn	11,539.32	77.45	11,616.76	631.07	464.67	153.76	2.58	7.42	850	40.54
Co	40.00	4.00	44.00	14.16	13.00	63.87	0.65	−0.66	19	35.14
Cr	325.00	103.00	428.00	171.07	160.00	40.63	1.65	2.58	90	100.00
Cu	32.00	4.00	36.00	20.75	24.00	34.00	−0.74	0.30	45	0.00
Ni	237.00	10.00	247.00	50.94	47.00	80.73	1.65	2.57	68	35.14
Pb	43.00	14.00	57.00	27.70	27.00	38.44	0.78	−0.13	20	72.97
Sr	81.00	37.00	118.00	57.67	56.00	35.88	1.20	0.80	170	0.00
V	702.00	62.00	764.00	221.88	208.00	65.33	1.64	2.10	130	89.19
Zn	148.00	20.00	168.00	74.60	81.00	42.51	0.26	−0.46	95	40.54

Note: Concentration values in mg/kg.
Source: After Turekian and Wedepohl (1961).

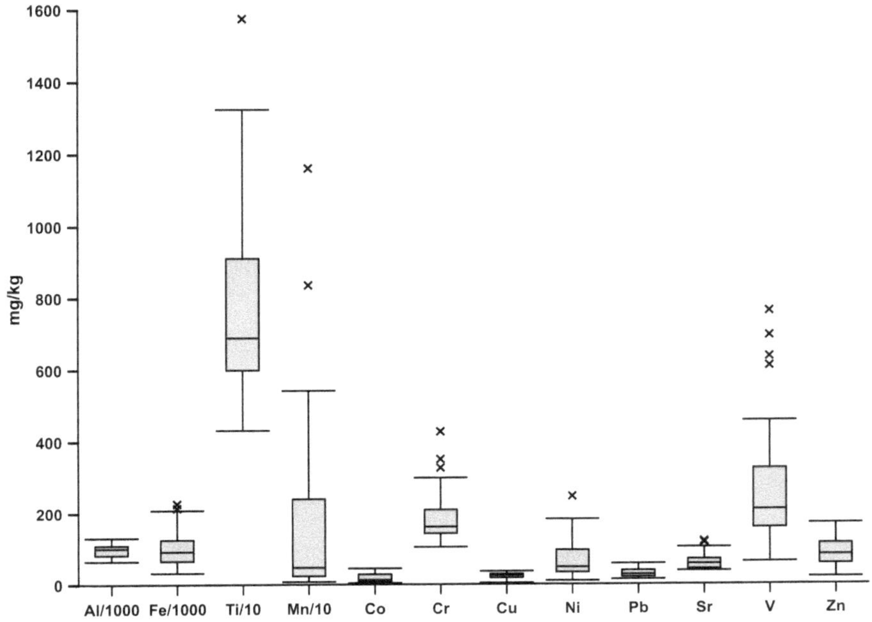

FIGURE 5.2 Box plot of the metal concentrations in the soil samples (to display all the elements on the *a*-axis, the concentrations of Fe and Al are divided by 1,000, and those of Ti and Mn are divided by 10).

influences were substantiated by the prevalence of diverse sedimentary rock exposures in the study area, particularly from the Jaintia and Barail Groups, comprising coal, black shale, shale, sandstone, and limestone. Notably, rocks such as black shale are enriched in rare earth elements, Zr, V, Y, and Th metals and contain diagenetic pyrite, contributing significantly to soil contamination (Umrao and Sardar, 2022). Consequently, the higher concentrations of V, Ti, Al, Cr, and Pb could be attributed to the presence of black shale and Kopili shale from the Jaintia Group in the study area.

5.4.2 Multivariate Statistical Analysis

Multivariate statistical analyses were employed to assess the factors influencing heavy metal concentrations and ascertain their potential sources. Pearson correlation analysis, PCA, and cluster analysis were conducted for comprehensive evaluation. The results of the correlation analysis are presented in Table 5.3, revealing significant positive correlations (r=0.94–0.5) among most elements, signifying a common origin, interdependence, and similar behaviour during transportation. Notably, Al and Ti exhibited less significant negative and positive correlations, with Ti demonstrating a moderate negative correlation. These variations in correlation can be attributed to various processes, including physical, chemical, biological, and anthropogenic factors (Resmi et al., 2019). Strong positive correlations were observed between Fe and

TABLE 5.3
Pearson Correlation Matrix for the Heavy Metal Concentrations in the Soil Samples

	Al	Fe	Ti	Mn	Co	Cr	Cu	Ni	Pb	Sr	V	Zn
Al	1											
Fe	0.15	1										
Ti	0.10	−0.48	1									
Mn	−0.15	**0.86**	−0.46	1								
Co	−0.11	**0.81**	−0.62	**0.80**	1							
Cr	0.29	0.37	−0.22	0.24	0.38	1						
Cu	0.38	0.11	−0.38	−0.09	0.19	0.19	1					
Ni	0.00	**0.90**	−0.57	**0.90**	**0.87**	0.50	0.01	1				
Pb	0.15	0.46	−0.44	0.46	**0.57**	0.48	0.13	**0.58**	1			
Sr	−0.20	0.40	−0.45	0.41	0.47	−0.03	0.10	0.42	**0.54**	1		
V	0.04	**0.91**	−0.55	**0.87**	**0.84**	**0.53**	0.00	**0.94**	**0.52**	0.35	1	
Zn	0.14	**0.66**	−0.68	**0.52**	**0.78**	0.44	0.45	**0.66**	**0.57**	**0.54**	**0.67**	1

Note: Significant positive correlations ($r = 0.94$ to 0.5) is in bold.

Mn, Co, Ni, V, and Zn elements. Co demonstrated robust positive correlations with Ni, V, and Zn and a moderate positive correlation with Pb. The moderate correlation between Pb and Zn may suggest a minimal impact of vehicular movement on soil contamination. At the same time, Cr, Cu, and Sr displayed poor-to-moderate correlations with other metals. Additionally, Pb exhibited a moderately positive correlation with Sr, V, and Zn.

5.4.3 Principal Component Analysis

PCA offers a comprehensive insight into the interrelationships within the variance matrix, streamlining a vast dataset into a smaller set of PCs. These PCs, which account for most of the dataset's variation, are determined through varimax factor loading (Soltani-Gerdefaramarzi et al., 2021). The PCA results, post-varimax rotation, are depicted in Table 5.4. Three prominent factor models were identified, each possessing eigenvalues surpassing 1, collectively explaining 78.54% of the total variance (Table 5.4).

PC1, representing 53.17% of the variance, prominently features Fe, Mn, Co, Ni, and V as closely associated metals (Figure 5.3a). These elements exhibit strong positive loading factors (>0.8), while Pb and Zn also demonstrate substantial loadings (>0.52) in PC1. This association is attributed to geogenic processes, suggesting that the source of these metals stems from the geochemical weathering of parent rock material (Lu et al., 2009; Sahoo et al., 2017). A noteworthy positive loading for Al

TABLE 5.4
Component Matrix for PC Factor Loading of Heavy Metals

Element	Factor 1	Factor 2	Factor 3
Al	−0.043	**0.855**	0.010
Fe	**0.884**	0.111	0.241
Ti	−0.397	0.049	**−0.727**
Mn	**0.909**	−0.190	0.148
Co	**0.810**	−0.037	0.459
Cr	0.504	**0.629**	0.029
Cu	−0.205	0.518	**0.699**
Ni	**0.949**	0.041	0.236
Pb	0.521	0.167	0.458
Sr	0.298	−0.380	**0.675**
V	**0.955**	0.095	0.183
Zn	0.539	0.210	**0.711**
Initial eigenvalue	6.381	1.722	1.322
% of total variance	53.171	14.352	11.019
% of cumulative variance	53.171	67.523	78.542

Note: Factor loadings >0.6 refer to values in bold.

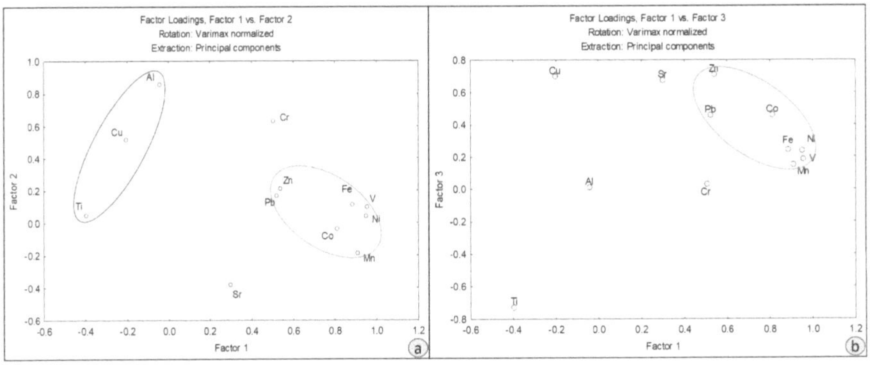

FIGURE 5.3 (a and b): Factor analysis of elements indicating different origins.

and Cr was observed in the second PC, PC2, accounting for 14.35% of the total variance (Table 5.3). The metals with a loading factor of 2 in PC2 may be linked to anthropogenic origins (Resmi et al., 2019). The third PC, PC3, which contributed 11.02% of the variance, included Cu, Sr, and Zn as strong positive loading factors (>0.67), with Ti exhibiting a robust negative loading factor (−0.7) (Figure5.3b). The presence of Cu, Sr, and Zn in PC3, characterised by relatively lower concentrations and uniform distributions in the study area, suggests a likely association with biogenic sources (Resmi et al., 2019).

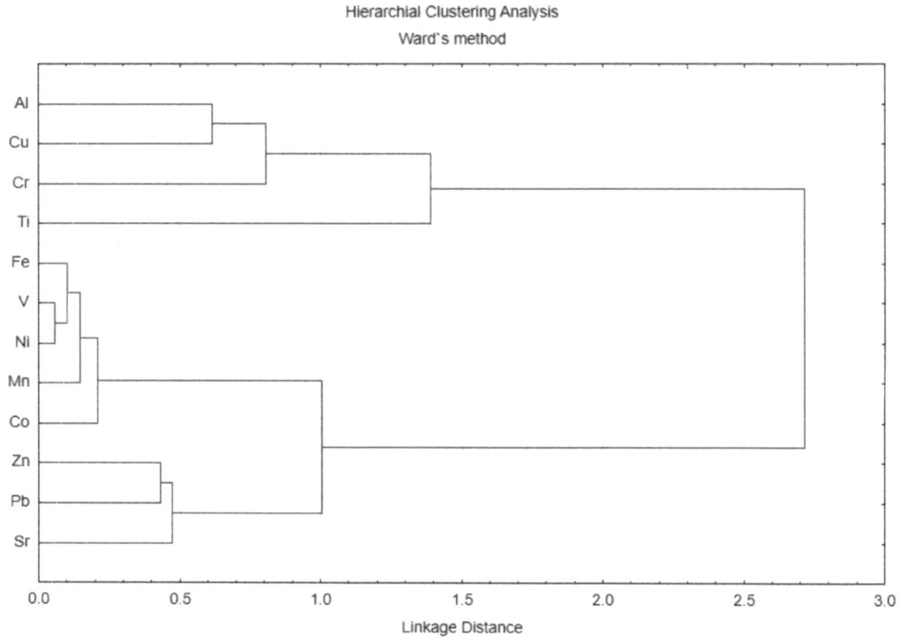

FIGURE 5.4 Dendrograms of cluster analysis for heavy metals in soil samples.

The cluster analysis results, as illustrated in the dendrogram in Figure 5.4, serve to categorise closely associated metals originating from the same sources. In the cluster analysis diagram, the distance between clusters indicates the degree of relationship among elements, with smaller distances signifying strong relationships and greater spreads suggesting weaker relationships (Soltani-Gerdefaramarzi et al., 2021). Evaluating the dendrogram from AHC, Fe, Ni, Mn and Co are grouped as strong clusters, implying similar sources of origin for these metals. The second minor cluster encompasses Zn, Pb and Sr, indicating a shared origin for these metals. The close relationship between the two clusters suggests the potential mixing of two nearly similar sources influencing the origin of these metals. The third set comprises Al, Cu, and Cr with a robust connection to Ti identical to the one observed in PCA (Figure 5.3a). The relationships among the elements within this cluster are weak, as evidenced by greater distances from the other groups in the dendrogram.

5.4.4 Assessment of Heavy Metal Contamination

The assessment of heavy metal contamination in soil samples involved comparing their concentrations with background values from uncontaminated counterparts, specifically the average shale values from Turekian and Wedepohl (1961) adopted in this study.

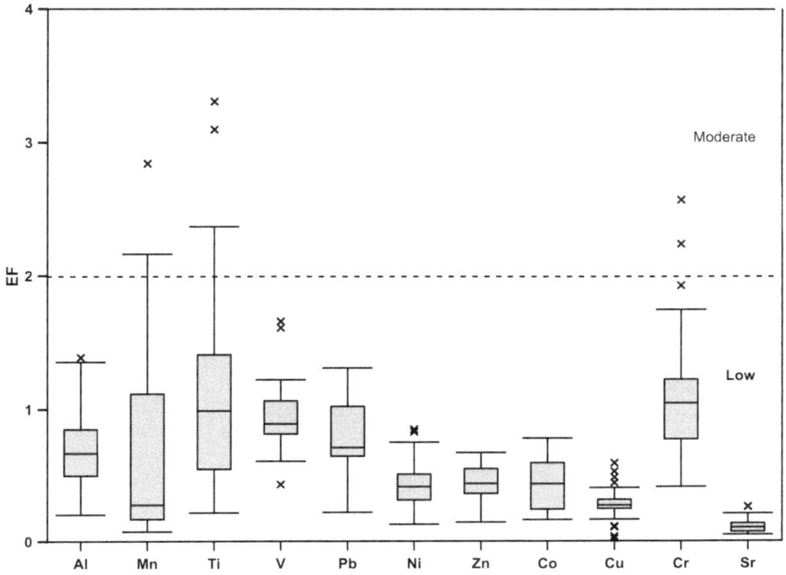

FIGURE 5.5 Box plot of enrichment factors for heavy metals in soil samples.

TABLE 5.5
Statistical Parameters of the Enrichment Factor (EF)

Statistical Parameter	Al	Mn	Ti	V	Pb	Ni	Zn	Co	Cu	Cr	Sr
Minimum	0.21	0.07	0.22	0.44	0.22	0.13	0.15	0.16	0.02	0.42	0.05
Maximum	1.39	2.84	3.31	1.66	1.31	0.85	0.68	0.79	0.60	2.57	0.26
Median	0.67	0.28	0.99	0.89	0.71	0.42	0.44	0.44	0.27	1.05	0.10

Figure 5.5 presents the EF for heavy metals in the soil, while Table 5.5 provides the statistical EF data. The findings indicate low-to-moderate metal contamination in the study area. Specifically, Al, V, Pb, Ni, Zn, Co, Cu, and Sr fall under the category of EF < 2, suggesting a deficiency in low contamination. Moreover, Mn, Ti, and Cr indicate low-to-moderate enrichment of metals. The EF values for Al (0.21–1.39, median: 0.67), Mn (0.07–2.84, median: 0.28), Ti (0.22–3.31, median: 0.99), V (0.44–1.66, median: 0.89), Pb (0.22–1.31, median: 0.71), Ni (0.13–0.85, median: 0.42), Zn (0.15–0.68, median: 0.44), Co (0.16–0.79, median: 0.44), Cu (0.02–0.60, median: 0.27), Cr (0.42–2.57, median: 1.05), and Sr (0.05–0.26, median: 0.10) collectively suggest low-to-moderate pollution in the soil. Areas near Samasi and Mulait exhibit moderate contamination for Mn, Tongseng, Mulait, and Sakhain for Ti and Samasi and Sutnga for Cr metals. The order of the median EFs of the heavy metals was as follows: Cr>Ti>V>Pb>Al>Zn>Co>Ni>Mn>Cu>Sr.

Table 5.6 provides the statistical data for the I_{geo} calculated for the trace elements in the studied soil samples (Figure 5.6). The I_{geo} results revealed varying degrees

TABLE 5.6
Statistical Parameters of the Geoaccumulation Index (I_{geo})

Statistical Parameter	Fe	Al	Mn	Ti	V	Pb	Ni	Zn	Co	Cu	Cr	Sr
Minimum	−1.166	−0.908	−4.041	−0.677	−1.653	−1.100	−3.350	−2.833	−2.833	−4.077	−0.390	−3.604
Maximum	1.681	0.112	3.188	1.192	1.970	0.926	1.276	0.237	0.627	−0.907	1.665	−1.931
Median	0.356	−0.269	−1.456	−0.002	0.093	−0.152	−1.118	−0.815	−1.132	−1.492	0.245	−3.006

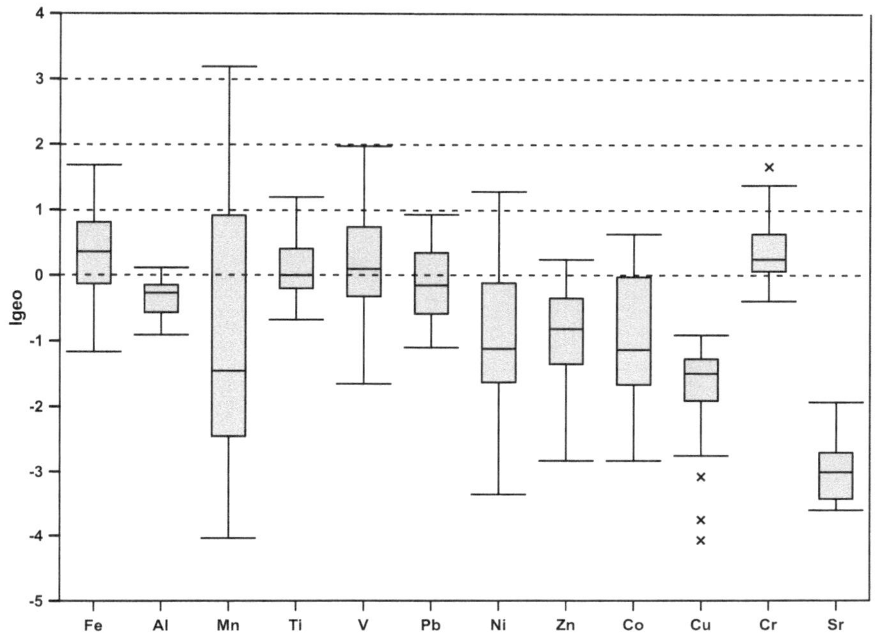

FIGURE 5.6 Box plot of geoaccumulation index for heavy metals in soil samples.

of metal contamination in the soil. Across the 12 elements analysed, the I_{geo} values ranged from −4.1 to 3.2, with Cu and Mn showing the lowest and highest I_{geo} values, respectively. The elements Cu and Sr, with a maximum $I_{geo} \leq 0$, indicate uncontaminated soil. Al, Pb, Zn, and Co, with I_{geo} values between ≤ 0 and 1, suggest uncontaminated to moderately contaminated soil. Fe, Al, Mn, Ti, V, Ni, and Cr, with I_{geo} values between ≤ 0 and 3, indicate uncontaminated to moderately to highly contaminated soil for these metals. Mn exhibits the highest I_{geo} variation from −4 to 3.2. Specific areas, such as Samasi, Lumthalary, Larket, Pala, Mulait, and Lumshortoh, show contamination ($I_{geo} > 1$) for Fe, Mn, and V. Contamination for Ti and Ni ($I_{geo} > 1$) is observed near Mulait and Samasi, respectively. In contrast, Cr contamination ($I_{geo} > 1$) is noted near Samasi, Pala, Larket, Sutnga, and Lumshortoh.

Table 5.7 summarises the statistical parameters of the CF, PLI, $PI_{Nemerow}$, and mC_d. Mn exhibited the highest CF value (Figure 5.7), indicating very high contamination (CF > 6) near the Samasi, Lumthalary, and Mulait areas. Considerable contamination (3 < CF < 6) is observed for Fe, Mn, V, Ni, and Cr in and around the Samasi, Lumthalary, Pala, Larket, Mulait, Sutnga, and Lumshortoh areas. The remaining parts of the study area show moderate-to-low contamination for these metals. Al, Zn, Pb, Co, Cu, and Sr exhibited moderate-to-low contamination levels in all soil samples.

The PLI was computed using Eq. (5.3) for Pb, Ni, Zn, Co, Cu, and Cr in this study, resulting in PLI values ranging from 0.35 to 1.80 (Table 5.7; Figure 5.8). Seventeen of the 37 samples indicated the presence of metal pollution in the soil around the Samasi, Lumthalary, Pala, Larket, Mulait, Tangnub, and Lumshortoh areas.

TABLE 5.7

Statistical Parameters of the Contamination Factor (CF), Pollution Load Index (PLI), Nemerow Pollution Index ($PI_{Nemerow}$), and Modified Degree of Contamination (mC_d)

Statistical Parameter	CF (Fe)	CF (Al)	CF (Mn)	CF (Ti)	CF (V)	CF (Pb)	CF (Ni)	CF (Zn)	CF (Co)	CF (Cu)	CF (Cr)	CF (Sr)	PLI	$PI_{Nemerow}$	mC_d
Minimum	0.67	0.80	0.09	0.94	0.48	0.70	0.15	0.21	0.21	0.09	1.14	0.12	0.35	0.93	0.60
Maximum	4.81	1.62	13.67	3.43	5.88	2.85	3.63	1.77	2.32	0.80	4.76	0.39	1.80	3.70	2.18
Median	1.92	1.25	0.55	1.50	1.60	1.35	0.69	0.85	0.68	0.53	1.78	0.19	0.81	1.52	1.01

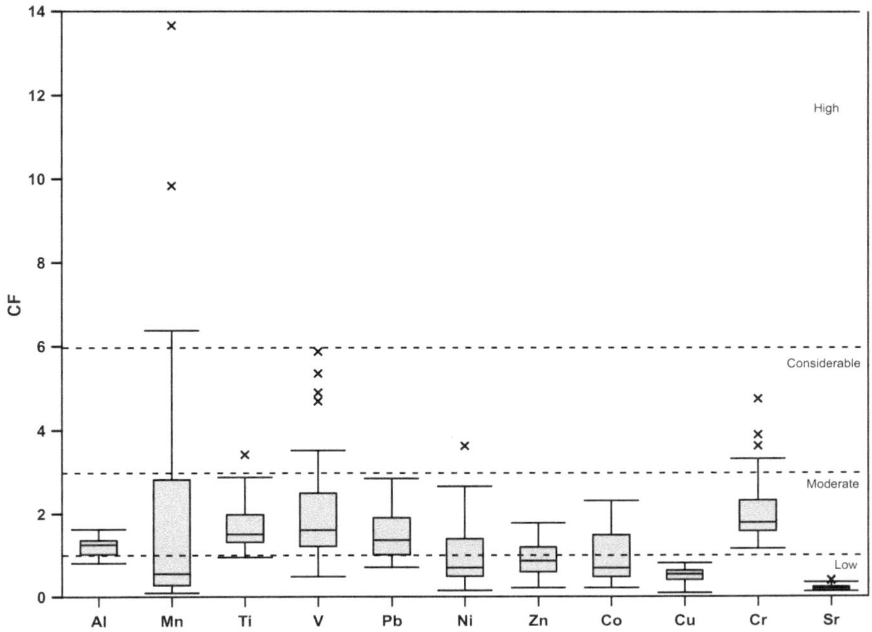

FIGURE 5.7 Box plot of the contamination factors of heavy metals in the soil samples.

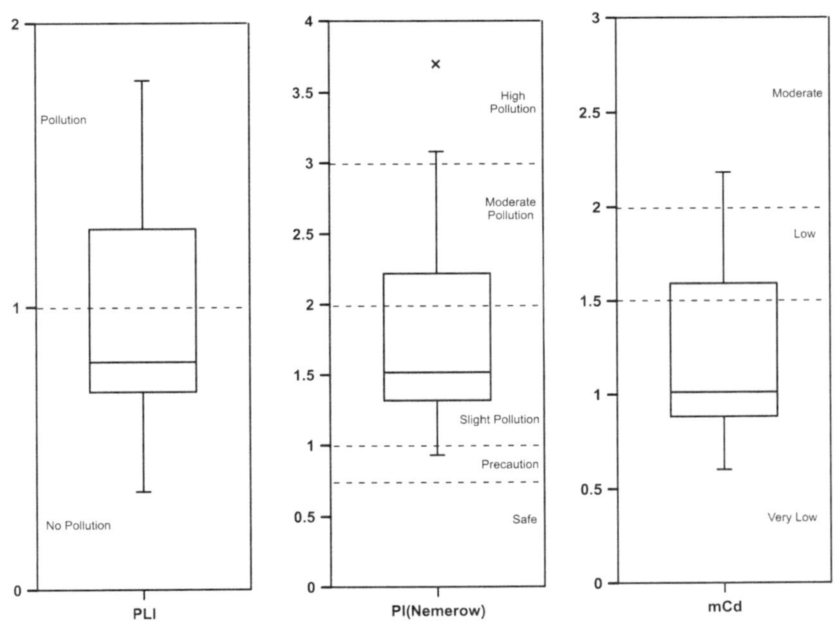

FIGURE 5.8 Box plots of the pollution load index, Nemerow pollution index, and modified degree of contamination for heavy metals in the soil samples.

The $PI_{Nemerow}$ was calculated using Eq. (5.4) for Pb, Ni, Zn, Co, Cu, and Cr to assess the overall degree of pollution in the soil samples. A $PI_{Nemerow}$ ranging from 0.93 to 3.70 (Table 5.7; Figure 5.8) indicated high pollution near the Pala and Larket areas, moderate pollution near the Samasi, Lumthalary, Mulait, Sutnga, and Lumshortoh areas, and slight pollution in the soil samples from the remaining areas.

The mC_d, calculated using Eq. (5.5) for Pb, Ni, Zn, Co, Cu, and Cr, varied from 0.60 to 2.18, indicating very low-to-moderate pollution in the soil samples (Table 5.7; Figure 5.8). Moderate contamination ($2 \leq mC_d < 4$) was observed near the Mulait and Samasi areas, while low contamination ($1.5 \leq mC_d < 2$) was noted around the Samasi, Lumthalary, Larket, and Lumshortoh areas. The remaining areas exhibited very low contamination levels.

The assessment of heavy metal contamination in the soil samples employed the EF, I_{geo}, CF, PLI, $PI_{Nemerow}$, and mC_d indices. The results indicated a generally low-to-moderate contamination level in the area, with a few outliers indicating high soil contamination. Heavy metal contamination was not detected in the soil samples from the gneissic complex/migmatite rocks in the Shangpung and Umsatai areas.

The contamination in the soil is predominantly concentrated in and around the Samasi, Lumthalary, Pala, Larket, Mulait, Tangnub, Sutnga, and Lumshortoh areas, with Mn, Ti, V, and Cr being the most concerning metals. These areas primarily expose sedimentary rocks such as coal, black shale, shale, sandstone, and limestone from the Jaintia and Barail Groups (Umrao and Sardar, 2022). Rat-hole mining of coal, which is prevalent in the study area, leads to unattended dumps of overburdened rocks, including sandstone, shale, and carbonaceous shale, near the mines, contributing significantly to soil contamination (Sahu and Goyal, 2010; Pyrbot et al., 2019).

The geochemical evaluation of soil samples, considering various indices for metal contamination and understanding the composition of surrounding rocks, suggests that higher concentrations of heavy metals primarily correspond to geogenic causes. However, coal dumps and haphazard overburden from rat-hole mines also emerge as notable contributors to soil contamination in the area. This interpretation is supported by the substantially lower concentrations of heavy metals such as Cu, Cr, Pb, Co, and Zn, primarily associated with anthropogenic activities.

5.5 CONCLUSIONS

Comprehensive geochemical and statistical analyses were conducted to assess heavy metal contamination in the soil of rat-hole coal mining-affected areas in the Jaintia Hills district of Meghalaya. Elevated Al, Fe, Ti, Mn, Cr, Pb, and V concentrations were observed, aligning with the background average shale values. The CV highlighted moderate variation, ranking heavy metals in the sequence of Cr>Zn>Ni>Pb>Cu>Co. Multivariate statistical analysis revealed strong correlations among Fe, Mn, Co, Ni, V, and Zn. In contrast, other metals exhibited moderate-to-poor correlations. PCA revealed three components explaining 78.54% of the total variance. The AHC dendrogram indicated a shared source of origin for two clusters, suggesting a strong relationship and the mixing of two similar sources. The combined interpretation of PCA and AHC suggested a predominant role of geogenic activity in influencing elemental concentrations in soil, with minimal impact from anthropogenic or biogenic activities. The pollution indices (EF, I_{geo}, CF) identified Fe, Mn, Ti, V, and Cr as potential pollutants, while Pb, Ni, Zn, Co, Cu, and Sr indicated low-to-moderate levels of soil contamination. The collective impact of various metals on soil pollution was assessed using PLI and $PI_{Nemerow}$. The PLI values ranged from 0.35 to 1.80, indicating the presence of nonpolluted to polluted soil. The $PI_{Nemerow}$ values varied from 0.93 to 3.70, ranging from precautionary to high pollution in the soil samples. The mC_d ranged from 0.60 to 2.18, indicating very low-to-moderate pollution in the soil samples. Overall, the geochemical assessment using these indices suggested a low-to-moderate contamination level in the study area, with some outliers indicating high soil contamination, particularly in and around the Samasi, Lumthalary, Pala, Larket, Mulait, Tangnub, Sutnga, and Lumshortoh areas, where Mn, Ti, V, and Cr were identified as the most concerning metals. The soil's primary source of metal concentration was attributed to the geochemical weathering of parent rock materials and associated processes, mainly sedimentary rocks such as coal, black shale, shale, sandstone, and limestone. Sedimentary pyrite in these rocks significantly contributed to soil and water contamination. In addition to geogenic causes, coal dumps and haphazard overburden from rat-hole mines were identified as additional contributors to soil contamination in the area.

ACKNOWLEDGEMENTS

The authors are thankful to the Director General of the Geological Survey of India for providing the necessary support for the study. The first author also thanks all individuals for their help in sampling and processing work and Chemical Division, GSI, NER, Shillong, for the analysis of samples. The authors are thankful to the anonymous reviewer(s) for constructive comments, which helped to improve this chapter in its present form.

REFERENCES

Abrahim, G.M.S. and Parker, R.J. (2008) Assessment of heavy metal enrichment factors and the degree of contamination in marine sediments from Tamaki Estuary, Auckland, New Zealand, *Environmental Monitoring and Assessment*, 136, 227–238.

Benhaddya, M.L. and Hadjel, M. (2014) Spatial distribution and contamination assessment of heavy metals in surface soils of Hassi Messaoud, Algeria, *Environmental Earth Sciences*, 71, 1473–1486. https://doi.org/10.1007/s12665-013-2552-3

Barooah, P.K. and Baruah, M.K. (1996) Sulphur in Assam coal, *Fuel Processing Technology*, 46, pp. 8, 3–97.

Chabukdhara, M. and Singh, O.P. (2016) Coal mining in northeast India: an overview of environmental issues and treatment approaches, *International Journal of Coal Science & Technology*, 3, 87–96.

Cheng, J.L., Shi, Z. and Zhu, Y.W. (2007) Assessment and mapping of environmental quality in agricultural soils of Zhejiang Province, China, *Journal of Environmental Sciences*, 19, 50–54.

Geological Survey of India, Misc. Pub. (2009) *Geology and Mineral Resources of Meghalaya, No. 30 part IV, v.2(ii)*, 53.

Giri, S., Singh, A.K. and Tewary, B.K. (2013) Source and distribution of metals in bed sediments of Subarnarekha River, India, *Environmental Earth Sciences*, 70(7), 3381–3392. https://doi.org/10.1007/s12665-013-2404-1

Gong, Q., Deng, J., Xiang, Y., Wang, Q. and Yang, L. (2008) Calculating pollution indices by heavy metals in ecological geochemistry assessment and a case study in parks of Beijing, *Journal of China University of Geosciences*, 19, 230–241.

Hakanson, L. (1980) An ecological risk index for aquatic Pollution control: A sedimentological approach, *Water Research*, 14, 975–1001.

Lu, X., Wang, L., Lei, K., Huang, J. and Zhai, Y. (2009) Contamination assessment of copper, lead, zinc, manganese and nickel in street dust of Baoji, NW China, *Journal of Hazardous Materials*, 161, 1058–62.

Müller, G. (1969) Index of geoaccumulation in sediments of the Rhine River, *Geo Journal*, *2*(3), 108–118.

Pekey, H., Karakas, D., Ayberk, S., Tolun, L. and Bakoglu, M. (2004) Ecological risk assessment using trace elements from surface sediments of I´zmit Bay (Northeastern Marmara Sea) *Turkey, Marine Pollution Bulletin*, 48, 946–953.

Pyrbot, W., Shabong, L. and Singh, O.P. (2019) Neutralization of acid mine drainage contaminated water and ecorestoration of stream in a coal mining area of East Jaintia Hills, Meghalaya, *Mine Water and the Environment*, 38, 551–555. https://doi.org/10.1007/s10230-019-00601-9

Raja Rao, C.S. (1981) *Coalfields of India—Bulletin Series A: Coalfields of North Eastern India: Vol. 1: Coalfields of North Eastern India*. Geological Survey of India, Calcutta, p. 124.

Resmi S., Tripathi, SK, Barik, S. and Sengupta, D. (2019) Trace metals concentration in the river sediments of Western Sundarban creeks and its environmental impact, *Indian Journal of Geosciences*, *73*(2), 143–156.

Sadiq, M., Ranjith A. and Umrao, R.K. (2014b) REE mineralization in the carbonatites of the Sung Valley Ultramafic-Alkaline-Carbonatite Complex, Meghalaya, India, *Open Geosciences*, *6*(4), 457–475.

Sadiq, M. and Umrao, R.K. (2017) Significant occurrence of REE in titaniferous lateritic bauxite (TLB) capping in Sung Valley Complex, Meghalaya, India, *Indian Journal of Geosciences*, *71*(2), 387–394.

Sadiq, M. and Umrao, R.K. (2020) Nb-Ta-rare earth element mineralization in titaniferous laterite cappings over Sung Valley ultramafic rocks in Meghalaya, India, *Ore Geology Reviews*, *120*(5), 103439. https://doi.org/10.1016/j.oregeorev.2020.103439

Sadiq, M., Umrao, R.K. and Dutta, J.C. (2014a) Occurrence of rare earth elements in parts of Nongpoh granite, Ri-Bhoi district, Meghalaya, *Current Science*, *106*(2), 162–165.

Sadiq, M., Umrao, R.K., Sharma, B.B., Chakraborti, S., Bhattacharyya, S. and Kundu, A. (2018) Mineralogy, geochemistry and geochronology of mafic magmatic enclaves and their significance in evolution of Nongpoh granitoids, Meghalaya, NE India. In: Sensarma, S. and Storey, B. C. (Eds.), *Large Igneous Provinces from Gondwana and Adjacent Regions* (Vol. 463, pp. 171–198). Geological Society of London Special Publication. https://doi.org/10.1144/SP463.2

Sahoo, P.K., Tripathy, S., Panigrahi, M.K. and Equeenuddin, S.M. (2017) Anthropogenic contamination and risk assessment of heavy metals in stream sediments influenced by acid mine drainage from a northeast coalfield, India. *Bulletin of Engineering Geology and the Environment,* 76, 537–552. https://doi.org/10.1007/s10064-016-0975-2

Sahu, B P. and Goel, N.P. (2010) *Social and Environmental Impact Assessment of Opencast Mining in Meghalaya: A Case Study of Jaintia Hills*. Regency Publication, New Delhi.

Singh, O.P. (2019) Technical report of project entitled study on mining affected areas and its impact on livelihood Meghalaya, Basin Management Agency Shillong, Meghalaya, p. 226.

Soltani-Gerdefaramarzi, S., Ghasemi, M. and Ghanbarian, B. (2021) Geogenic and anthropogenic sources identification and ecological risk assessment of heavy metals in the urban soil of Yazd, central Iran, *PLoS One, 16*(11), e0260418.

Swer, S. and Singh, O.P. (2003) Coal mining impacting water quality and aquatic biodiversity in Jaintia Hills District of Meghalaya, *ENVIS Bulletin on Himalayan Ecology*, 11, 26–33.

Tomlinson, D.L., Wilson, J.G., Harris, C.R. and Jeffrey, D.W. (1980) Problems in the assessment of heavy-metal levels in estuaries and the formation of a pollution index, *Helgoländer Meeresuntersuchungen*, 33, 566–575. https://doi.org/10.1007/BF02414780

Turekian, K.K. and Wedepohl, K.H. (1961) Distribution of the elements in some major units of the earth's crust, *Bulletin of the Geological Society of America,* 72, 175–192.

Umrao, R.K. and Sardar, S.K. (2022) Significant occurrences of REE in Tertiary sequence of East Jaintia Hills, Meghalaya, India. *Journal of the Geological Society of India*, 98, 621–626. https://doi.org/10.1007/s12594-022-2036-8

Zhang, J. and Liu, C.L. (2002) Riverine composition and estuarine geochemistry of particulate metals in China – weathering features, anthropogenic impact and chemical fluxes. *Estuarine, Coastal and Shelf Science, 54*(6), 1051–1070.

Zhang, X.Y., Sui, Y.Y., Zhang, X.D., Meng, K. and Herbert, S.J. (2007) Spatial variability of nutrient properties in black soil of northeast China, *Pedosphere*, 17, 19–29.

6 Biochar for the Remediation of Hydrocarbons-Contaminated Soil
A Review

Subhash Chandra and Isha Medha

6.1 INTRODUCTION

The issue of soil contamination with hydrocarbons has become a worldwide concern. The pervasive and escalating use of petroleum products and by-products in society renders their discharge into the environment unavoidable. The main origin of hydrocarbons in the soil is derived from the by-products of petrochemical industries, crude oil extraction, as well as oil spills and emissions resulting from the operation of heavy diesel engines in the mining industry. In recent years, the petrochemical industries have increased their production of petroleum products due to rapid industrialization, urbanization, and rising energy demands. In 2022, the worldwide output of petroleum and oils reached over 4.4 billion tonnes, and it is projected to experience significant growth in the following years. According to Sihag [1], between 1.7 and 8.8 million metric tonnes of oil and petroleum waste are estimated to be released into the environment annually, originating from both natural and anthropogenic sources. Hydrocarbon pollutants can accumulate in different plant species and can be carried to surface and groundwater resources. Hydrocarbon pollutants in the soil pose a significant risk to the environment, ecology, and human health. Furthermore, long and branched chain alkanes and alkenes make up the bulk of hydrocarbon pollutants in soil. Aromatic hydrocarbons are very persistent in soil environments, and how they get up there is dependent on factors including microbial abundance and soil physiochemistry. The fact that these pollutants tend to have a lengthy half-life in the soil makes them problematic. Several in-situ and ex-situ methods can be used to eliminate hydrocarbon pollutants from the soil, including air stripping, thermal cracking, soil flushing, bioremediation, and phytoremediation procedures [2,3]. Among these approaches, bioremediation and phytoremediation have become highly popular because to their simplicity and cost-effectiveness [4]. The effectiveness of bioremediation and phytoremediation approaches relies on the physiochemical state of the polluted soil. Typically, soil that is deteriorated and contains hydrocarbon

DOI: 10.1201/9781003442554-7

pollutants is unable to support the growth of microorganisms and flora. Under these conditions, incorporating soil amending material to enhance the physiochemical properties of the soil may be a method to enhance the efficiency of in-situ remediation approaches. Biochar (BC) has been increasingly popular in recent years as a soil-amending material due to its exceptional efficacy and sustainable characteristics. BC is a black porous and functional carbon material derived through the pyrolysis of biomass waste in the temperature range of 300°C–800°C [5,6]. In the past few years, several studies have been done regarding the application of BC for the remediation of hydrocarbon-contaminated soil. For instance, Zhen [7] reported the application of BC and rhamnolipid for the bioremediation of petroleum-hydrocarbon-contaminated soil. Their research demonstrated that the use of BC and rhamnolipid modified biochar reduced the harmful effects of petroleum hydrocarbons on Spartina anglica by enhancing plant growth, including increased plant height, root vitality, and total chlorophyll content. As a result, the effectiveness of the phytoremediation method was improved. In a separate study conducted by Zhang [8], BC was employed as a soil ameliorating substance to enhance the efficacy of the phytoremediation technique. Their investigation has shown that the inclusion of BC in the soil enhanced the growth of the plantation, hence enhancing the effectiveness of soil remediation through phytoremediation approach [8]. Another study by Zhang [9] concluded that the application of BC alone and in combination with petroleum-degrading bacteria substantially reduced the level of petroleum contaminants in the soil at the end of the 60 days of remediation study in the soil contaminated with total petroleum products. In fact, the presence of BC facilitated the proliferation of microbes and enhanced their ability to break down petroleum products. Consequently, the combined use of BC and microbes yielded the most favorable outcomes by the conclusion of the remediation period. This chapter provides a thorough analysis of the different sources of hydrocarbon pollutants in soil, their behavior and movement in the soil environment, different approaches to soil remediation using ex-situ and in-situ techniques, and the significance of using BC in the soil to enhance the bioremediation process.

6.2 HYDROCARBON-CONTAMINATED SOIL

6.2.1 Distinguished Sources of Hydrocarbon Contaminants

Hydrocarbons can be introduced into the environment via a multitude of pathways. Uncontrolled and intentional discharges of petroleum hydrocarbon pollutants by human activities—including but not limited to transportation and storage, tank leakages, accidental oil spills, effluent discharge from the petrochemical industry, fugitive emissions, ruptures in aged underground pipelines, warfare and political instability, acts of sabotage, and natural disasters, have detrimental effects on the terrestrial and marine ecosystem [3]. Petroleum hydrocarbons have the potential to alter both the physical and chemical characteristics of soil, including its texture, compaction, structural state, saturation hydraulic conductivity, penetration resistance, and mineral and heavy metal concentration and composition. One of the main ways that petroleum hydrocarbon contaminants end up in soil is by seepage from natural oil resources that have been discovered during deep inland and offshore explorations. The crude oil industry is one of the main contributors to the environmental contamination caused by petroleum hydrocarbons, according to the EPA Toxic Release Inventory report (2005). Worldwide, oil tanker accidents and human error are responsible for the release of an estimated 1.7–8.8 million metric tonnes of petroleum hydrocarbons into the ocean each year.

6.2.2 Fate and Transport of Hydrocarbon Contaminants in Soil

Hydrocarbons are intricate combinations of contaminants that include n-alkanes, branched chain alkenes, and polyaromatic hydrocarbons (PAHs). When hydrocarbons are discharged into the soil atmosphere, they experience several alterations in their physiochemical characteristics, which are influenced by their molecular weight and solubility. The alteration in their physicochemical properties often initiates by a weathering mechanism that involves dispersion, dissolution, sorption, and photodegradation. Upon entering the soil, hydrocarbons undergo the initial physical phase of volatilization, during which the lighter and more volatile component of the hydrocarbon escapes into the atmosphere by evaporation. While the semivolatile and heavier part sinks into the soil, however, the process of volatilization continues in the subsurface regions [10]. Contrary to the physical process of volatilization, which merely alters the volatile fraction of the hydrocarbon to evaporate, the chemical weathering process is crucial in defining fate and transport of these contaminants in the soil. Firstly, the solubility of hydrocarbon pollutants in water or aqueous solution plays a crucial role in initiating the chemical weathering process. The dissolving affinity of these pollutants in soil aqueous solution is crucial in determining their destiny inside the soil matrix. Typically, hydrocarbons with long chains and aromatic structures have a lower tendency to dissolve in water. In addition, hydrocarbon oils with high molecular weight and viscosity have lower solubility in aqueous solutions of soil compared to hydrocarbons with lower viscosity. It was reported that unlike long-chain hydrocarbons, short-chain hydrocarbons having chain-length less than C8 usually dissolve in water and can undergo alterations in their physicochemical properties by chemical processes and microbial actions [11]. The subsequent weathering processes that cause physicochemical changes in hydrocarbons are sorption and desorption, which occur inside the soil matrix. During this process, different portions of the hydrocarbons have a tendency to be adsorbed onto the soil particles. The sorption processes involve the distribution of hydrocarbons into the organic matter (OM) of the soil, their movement into the macro, micro, and nanopores of the soil through diffusion, and the formation of bond between hydrocarbons and OM through van der Waals and electrostatic interactions. Reportedly, it has been discovered that the sorption of hydrocarbons in the soil is influenced by its physiochemical characteristics. For example, the sorption of hydrocarbon in soil is enhanced by higher levels of OM, clay content, and hydrophilicity [12]. In other words, soil with favorable physicochemical properties may efficiently adsorb hydrocarbons and modify their chemical properties, while degraded soils are unable to effectively adsorb hydrocarbon pollutants in the soil matrix. Furthermore, the presence of aged PAH in the soil is a significant determinant of the fate of hydrocarbons. These PAH compounds are resistant to degradation by both chemical processes and microbiological activity [13].

6.2.3 Effect of Hydrocarbon Contaminants in the Soil

6.2.3.1 Effect on Soil Physicochemical Properties

Hydrocarbons in the soil have a substantial impact on the physicochemical and biological characteristics of the soil. Hydrocarbon pollutants in the soil lead to harmful geoenvironmental issues and negatively impact the soil's water environment by releasing poisonous substances. Hydrocarbon pollutants in the soil hinder air diffusion and water flow in the pores because of their hydrophobic nature, leading to alterations in soil physical properties as plastic limit and permeability. Furthermore, changes in soil qualities can impact chemical variables such as pH, total organic carbon, and soil nutrients, which in turn influence the quantity of microbes in the soil [14,15]. An example is that the existence of hydrocarbon pollutants in the soil increases soil pH, OM, and total organic carbon content while decreasing nitrate, phosphate, and gravimetric water potential [16]. Vishnu and coworkers found that crude oil hydrocarbon pollutants in the soil lowered the soil pH by reacting with soil minerals, while also raising the soil's electrical conductivity and total organic carbon levels. The Fourier-Transform Infrared Spectroscopy analysis of the soil contaminated with crude oil showed the existence of hydroxyl (–OH), Si–H, and C=C groups, which may disrupt and modify the breakdown of minerals and the absorption of nutrients in the soil [17]. Another study conducted by Cheraghi and colleagues found that the pH of the hydrocarbon-contaminated soil decreased, mineral leaching decreased, and total petroleum products and PAHs were highly concentrated [18]. Wang and colleagues obtained marsh soil samples from oil fields and conducted an analysis of their physiochemical parameters. According to the soil examination, the soil samples were found to be highly contaminated with petroleum compounds. The petroleum pollutants in the soil significantly impacted both the soil temperature and soil water content. Both measurements showed significant seasonal variation; soil water content was at its lowest in summer and highest in autumn. Hydrocarbon pollutants raise the soil pH by accumulating salt and decrease the soil's ability to exchange anions and cations [19]. Thus, it can be estimated that the presence of hydrocarbons in the soil substantially reduces its productivity by varying soil pH, nutrients, and OM content. Moreover, the presence of hydrocarbons in the soil also has a detrimental effect on soil microenvironment and enzymes, which has been elaborated in the next section.

6.2.3.2 Effect on Soil Microbial Ecosystem

The soil's microbial activity and growth are significantly affected by the presence of hydrocarbon contaminants in the soil as these contaminants also affect the soil's physicochemical properties (discussed in the previous section). Alrumman and colleagues found that various microbial biomasses were dramatically reduced in soil spiked with kerosene and diesel as pollutants in incubated samples (incubated at 4°C for 2 weeks). The study revealed significant variations in soil enzymatic activity in soils polluted with kerosene and diesel hydrocarbons [20]. Soil dehydrogenase activity was increased in soil contaminated with kerosene but decreased in soil contaminated with diesel hydrocarbons compared to uncontaminated soil. The soil phosphatase activity was reduced in the soil contaminated with kerosene and increased in the soil contaminated with diesel hydrocarbons compared to the control soil [20]. A 10-week lab investigation revealed that oil pollution of the soil led to a notable initial rise in all biological indices examined. When PAHs were present, biomass-C, respiration, protease activity, and heterotrophic counts increased dramatically, but urease activity decreased. The presence of oil and PAHs initially blocked N-mineralization [21]. A separate investigation revealed that soil contaminated with high levels of hydrocarbon pollutants had a notable reduction in both the Shannon and evenness indices of the soil's microbial community. Upon analysis, it was discovered that the soil heavily polluted with hydrocarbons (>20,000 mg/kg) had a considerable decrease in both microbial alpha and beta diversities compared to soil with moderate contamination levels [22]. According to Labud et al., a study conducted over a period of 6 months found that sandy soil with 5%–10% hydrocarbon contaminants (such as diesel oil, petrol, and petrol) had a significant negative effect on microbial biomass. This impact was particularly evident in soil polluted with gasoline. The presence of petroleum and diesel in the soil increased microbial respiration [23]. Guo and colleagues gathered soil samples from the vicinity of oil production wells and locations where oil spills occurred. Their findings demonstrated a significant decrease in the abundance of cultivable bacteria and fungi and a decrease in urease activity as the concentration of total petroleum hydrocarbons increased. Nevertheless, they also observed a direct relationship between the overall amount of hydrocarbons, the presence of soil OM, and the promotion of fungi–bacteria–urease activity at a low level of hydrocarbon pollution [24]. A separate study examined the impact of prolonged exposure to total petroleum hydrocarbons on the population of algae in soil, the amount of microorganisms present, and the activity of enzymes in the soil. Their investigation found that the levels of microbial biomass, soil enzyme activity, and microalgae decreased in areas with medium-to-high pollution levels, explicitly ranging from 5,200 to 21,430 mg/kg of soil. Nevertheless, minimal pollution prompted algae growth without impacting the microbial biomass and enzymes [25]. Therefore, it can be inferred that the existence of hydrocarbon pollutants in the soil enhances specific microbial growth as a result of an increase in the organic carbon portion. However, long-term studies have indicated a significant decline in soil enzymatic activities due to elevated levels of hydrocarbon contaminants.

6.3 REMEDIATION OF HYDROCARBON-CONTAMINATED SOIL

6.3.1 Physical and Chemical Remediation Techniques

This encompasses the use of various physical and chemical techniques to remove and recover hydrocarbon contaminants from the soil. The physical and chemical processes mentioned are primarily conducted in situ, utilizing methods such as physical barriers, heat treatment, air stripping, and soil flushing.

6.3.1.1 Physical Barrier and Surface Capping

The physical barrier is a method employed to limit the vertical and horizontal mobility of hydrocarbon pollutants in the soil. Geomembrane walls, plastic sheets, and verified barriers are employed as vertical barriers to impede the downward migration of pollutants in the soil, thereby safeguarding groundwater reserves from contamination [26]. In addition, vertical trenches occasionally dug around the region that contained the hydrocarbon-contaminated soil. These trenches are then lined with materials such as cement and bentonite to limit the lateral spread of contaminants. The surface capping technique is placing an impermeable cover over the contaminated soil to prevent water from infiltrating the soil, reducing the leaching and chemical interaction of hydrocarbon pollutants in the soil. Surface capping material is commonly applied in multiple layers over the soil and strengthened by the growth of vegetation to prevent soil erosion caused by intense rainfall and surface runoff [27].

6.3.1.2 Air Stripping

Air stripping is a method that involves the use of high-pressure air to flow through a soil column that is polluted. When hydrocarbon-contaminated soil is exposed to pressured air, the volatile components of hydrocarbons in the soil become vaporized and are released from the soil along with the departing air mass [2]. This technology, primarily designed to eliminate dissolved hydrocarbon and biodegradable hydrocarbon fractions from the soil, is both cost-effective and time-efficient. The flushing air mass occasionally contains a significant amount of oxygen to facilitate the aerobic breakdown of volatile fractions of hydrocarbons in the soil [28].

6.3.1.3 Soil Washing

Soil washing is a method of physically treating soil by injecting a washing fluid, typically an extractant, into the soil. The extractant eliminates the volatile and semivolatile components of hydrocarbons found in both the larger and smaller portions of the soil. Afterward, the extractant is brought back to the surface using extraction pumps [29]. Therefore, it eliminates the volatile components from the soil. Nevertheless, this approach incurs a greater expense for treatment due to the utilization of high-pressure vacuum pumps and also carries the risk of extractant containing hydrocarbons infiltrating the groundwater reserves.

6.3.1.4 Stabilization

Stabilization can be achieved using ex-situ and in-situ methods, which are designed to limit the mobility of hydrocarbon pollutants in the soil. One way to achieve this is by incorporating various stabilizers, including pozzolanic materials, clay-based minerals, and cementitious material, into the soil [30,31]. The stabilizers in the soil function by impeding the passage of pollutants and transforming them into a less hazardous state through physical and chemical interactions.

6.3.1.5 Immobilization

Immobilization is a chemical process that involves adding various sorbents to the soil to increase the sorption of volatile aromatic hydrocarbons in the soil matrix. The process functions by enhancing the responsiveness of pollutants with the sorbents while reducing their ability to dissolve. The pollutants are rendered immobile through a range of chemical reactions such as complexation, precipitation, and electrostatic retention [32,33]. Immobilization can be performed in two ways: ex-situ and in-situ. It serves as a proactive barrier to prevent the leakage of toxins from the polluted sites.

6.3.1.6 Chemical Oxidation and Reduction

This approach is an in-situ method that utilizes chemical oxidizing agents, including hydroxides, potassium permanganate, potassium dichromate, and hydrogen peroxide. Upon entering soil that contains hydrocarbon contaminants, these oxidizing agents rapidly convert the volatile portion of those pollutants into a harmless form, such as carbon dioxide (CO_2) and water (H_2O) [34–36]. The efficacy of this approach relies on the distribution of these oxidizing agents over the designated area in the lower layers of the surface, as well as the duration of contact between the oxidizing agents and the contaminants. Increased dispersion and prolonged contact time result in greater degradation of contaminants.

6.3.1.7 Activated Carbon

Activated carbon is a hydrophobic material and is known for its large surface area, structure with tiny pores, and ability to absorb a lot of substances [37,38]. Unlike graphite, this type of carbon is characterized by its irregularly stacked or randomly arranged sheets or groups of atoms. Activated carbon has the ability to effectively absorb hydrocarbon pollutants found in the soil because of its high adsorption capacity. The activated carbon contains micro and meso holes that serve as pathways for the retention of hydrocarbon pollutants on its surface [39]. This retention occurs by interparticle diffusion, where the contaminants move from the solution to the surface of the activated carbon. The elimination of impurities in the soil is regulated by a dynamic balance between the process of adsorption, degradation, and the entry of pollutants, all of which take place on the activated carbon.

6.3.2 Biochar for the Remediation of Hydrocarbon-Contaminated Soil

6.3.2.1 What Is Biochar?

BC is a porous and amorphous carbon material. It is produced as the primary result of pyrolysis, which is the process of heating different types of biomass waste [40]. BC is often defined by its porous surface, which has a high specific surface area. It also contains surface functional groups, micro and macro pores, and a significant amount of π-electrons [41,42]. BC can be produced using varieties of feedstock, some of which are mentioned in Table 6.1. BC can be produced from different pyrolysis methods, the details of which are discussed in the subsequent section [43–51].

6.3.2.2 Production Methods of Biochar

BC production is achieved using different pyrolysis methods, including slow pyrolysis, fast pyrolysis, gasification, and hydrothermal carbonization. The choice of each method is contingent upon the nature of the goods produced (BC and bio-oil), the desired quantity of production, the demand for product derivatives, and the overall economic viability of the process.

Slow pyrolysis involves the extended heating of biomass in an environment with limited oxygen, between a temperature range of 300°C and 800°C [52]. Slow pyrolysis is characterized by a low heating rate of 2–7°C/min and a lengthy residence period for both the solid and vapor phases, typically exceeding 1 h. BC is often created by a

TABLE 6.1
Feedstock for Biochar Production

S. No.	Feedstock Used	Production Conditions	References
1	Corncob waste	Pyrolysis temperature: 500°C and 700°C Duration: 2 h	[43]
2	Rice straw waste	Pyrolysis temperature: 400°C, 500°C, 600°C, and 700°C Duration: 60, 90, and 120 min	[41]
3	Maize waste	Pyrolysis temperature: 1,200°C Duration: 40 min	[44]
4	Sugarcane baggase	Pyrolysis temperature: 350°C, 450°C, and 550°C	[45]
5	Sesame straw waste	Pyrolysis temperature: 700°C Duration: 4 h	[46]
6	Oak wood waste	Pyrolysis temperature: 650°C Duration: 15 h	[47]
7	Sugarcane straw	Pyrolysis temperature: 700°C Duration: 1 h	[48]
8	Pig manure	Pyrolysis temperature: 400°C Duration: 1 h	[49]
9	Sludge waste	Pyrolysis temperature: 550°C Duration: 2 h	[50]
10	Dairy manure	Pyrolysis temperature: 450°C Duration: 4 h	[51]

process called slow pyrolysis, which involves employing temporary pit kilns, mound kilns, brick kilns, metal kilns, and other similar devices. The slow pyrolysis process primarily produces BC, but it also generates different amounts and types of syngas as by-products [53]. The production of BC using slow pyrolysis can be done using tube furnace and microwave heating furnace as shown in Figure 6.1.

The initial experiment of fast pyrolysis in the late 1970s garnered significant interest, and since then it has undergone remarkable development for the generation of bio-oil as the main product, along with some BC [54]. Fast pyrolysis involves rapidly heating biomass to temperatures between 450°C and 600°C, causing it to evaporate and condense into a dark brown liquid. This process occurs at a high heating rate of 100°C/min and has a short vapor residence time of approximately 2 s to limit secondary reactions [55]. The calorific value of this liquid is 50% of that of mineral oils or petroleum fuels, as stated by Bridgwater and Peacocke in 2000. Fast pyrolysis is conducted in pressurized reactors, resulting in a bio-oil yield of up to 80% based on the dry feed. Additionally, char and gas are produced as by-products through cogeneration.

Gasification is a method of producing syngas as the main output, along with some char particles, ash, and tar. These by-products have the potential to be used as fuel in the future [56]. Gasification is a thermal process in which biomass is heated to a temperature as high as 1,000°C, resulting in the partial oxidation of fuel elements [57].

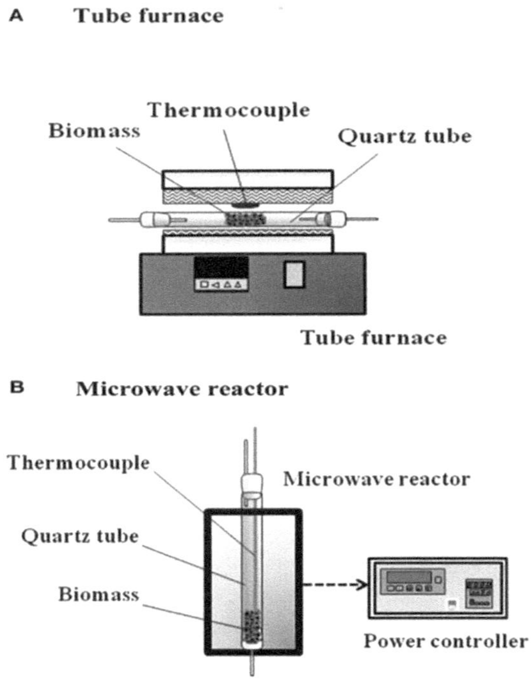

FIGURE 6.1 (a) Schematic diagram of tube furnace for biochar production; (b) schematic diagram of microwave reactor. (Adapted from Ref. [1] under Commons Creative License.)

Hydrothermal carbonization is a highly effective method of converting biomass into hydrochar, which is a distinct form of BC. This approach differs from other pyrolysis techniques. The hydrothermal treatment for biomass conversion to BC offers several advantages. Firstly, it eliminates the need for energy-intensive drying processes. Additionally, it boasts a high conversion efficiency and is environmentally friendly. Furthermore, this method requires relatively low operating temperatures compared to other thermal processes. The hydrothermal conversion process involves heating biomass to temperatures between 180°C and 350°C in the presence of sub-critical water, while maintaining a pressure greater than atmospheric pressure, inside a steam-filled reactor. In pyrolysis, biomass undergoes a sequence of simultaneous processes, including hydrolysis, dehydration, decarboxylation, aromatization, and recondensation, during which it is decomposed. This process results in the formation of hydrochar, a carbon-rich and dense-energy substance.

6.3.2.3 Biochar of Remediation of Contaminated Soil

The widespread existence of hydrocarbon pollutants in the soil is a significant risk to both the environment and human well-being. The bioremediation technique is an effective and cost-efficient method for stabilizing hydrocarbons in the soil. Bioremediation is a method of remediation that employs either the microbiological approach or a plant-based strategy (phytoremediation), and occasionally a mix of both, to eliminate or stabilize petroleum-based toxins in the soil [58]. BC, a porous and black carbon substance generated from biomass, is commonly used as a soil-ameliorating material. BC in the soil enhances both the physiochemical quality of the soil and the quantity of microorganisms [59,60]. BC, a versatile carbon substance, possesses a remarkable ability to adsorb both inorganic and organic hydrocarbon pollutants on its surface. There are various factors that affect the removal of hydrocarbon contaminants in the soil. Majority of them are soil pH, temperature, oxidation-reduction potential, water content, microbial abundance, and soil OM [61–63]. Soil with aerobic conditions, a pH level between 7 and 9 (neutral to slightly alkaline), a high temperature range of 30°C–40°C, a high oxidation potential, and a high number of microbes are typically thought to have the highest effectiveness in degrading hydrocarbons in a soil–water environment [61–63]. BC in the soil can create the conditions mentioned above because of its porous structure, which enables it to retain water in its pores. Additionally, the alkaline nature of BC increases the pH of the soil to a neutral or alkaline level, while the abundance of organic functional groups enhances the soil's oxidation potential. The BC's surface possesses a large specific surface area and an abundance of micropores, which in turn enhance the presence of microorganisms within the soil containing BC particles. BC generated at temperatures above 500°C exhibits a significant specific surface area and is inherently hydrophobic. Therefore, the utilization of elevated temperatures produced BC into the soil, which can efficiently break down hydrocarbon pollutants. BC in the soil can remove hydrocarbon pollutants by a range of chemical reactions, including precipitation, electrostatic contact, and oxidation by oxygenated functional groups found on the surface of the BC [64]. Ren [65] studied the effect of corncob-derived BC application in petroleum products contaminated soil. The team reported that the application of corncob-derived BC substantially removed the oil contaminants from the soil and improved microbial immobilization rate. Zhang [66] studied the effects of 12 different types of BCs on the degradation of polyaromatic hydrocarbons in the soil. Based on their findings, BC was found to promote microbial growth on a genus level, and it had a significant impact on the fungal community on multiple levels, including class and phylum. The researchers went on to prove that BC's inorganic element contaminants promoted microbial development, which in turn accelerated the biodegradation of polyaromatic hydrocarbons in soil [66]. Kong [67] studied the application of sawdust and wheat-straw-derived BC for the remediation of PAHs contaminated soil based on a lab-based incubation study. A substantial reduction in the content of PAHs in the soil has been observed following the incubation of the soil with BC. The suggested mechanism posits that the use of BC altered the soil's habitat for microbial development, hence promoting the microbial engagement in the biodegradation of PAHs [67]. Another study by Han [68] examined the impact of using BC in conjunction with phytoremediation to clean up soil contaminated with petroleum. Their findings indicated that the use of BC as a soil amendment, in combination with Rye grass, a plant species used for phytoremediation, effectively eliminated Total Petroleum Hydrocarbon from the soil [68]. Thus, it can be inferred from the data discussed that the BC application in the hydrocarbon-contaminated soil works by enhancing the microbial abundance that further promotes biodegradation of the pollutants. In addition, the functional characteristics of the BC particles in the soil include adsorption of hydrocarbon contaminants on its surface and degradation of them through various chemical interactions such as oxidation of aromatic rings, reduction of branched chain to simple linear chains, and reduction reaction on its surface.

6.3.2.4 Mechanism of Hydrocarbon Removal in the Soil Using Biochar

The underlying mechanism by which BC accelerates the degradation of hydrocarbons in the soil is the biostimulation process. The microbes in the soil stimulate by forming biofilms on the substrate surface. This biofilm formation on the substrate is affected by various factors such as surface area, pore density, surface roughness, free energy, surface charge, and hydrophobicity [69]. BC is typically characterized by having porous surface (mix of micro, meso, and macro pores), surface charge, surface functional groups, high surface area, and hydrophobic nature [69]. Having these characteristics, BC application in the soil substantially attracts the microbial growth through biofilm formation. The microbial abundance in the hydrocarbon-contaminated soil promotes the degradation of liable fractions of hydrocarbons in the soil. A recent study by [70] concluded that the co-application of BC in the hydrocarbon-contaminated soil substantially increased the hydrocarbon removal at the end of 30 days of incubation by 35%–89% compared to control (8%–41%). The study further confirmed the involvement of microbes in hydrocarbon removal in the soil by studying the CO_2 emission, which usually increased due to increase in the microbial respiration and biodegradation of contaminants in the soil [70]. In addition, the coexistance of BC with microbes in the soil promotes the biodegradation process by adsorbing the these contaminants on the biochar surface where the microbes present. Physical intervention of BC and microbes further accelerates the degradation process. Physical interaction between microbial cell and BC is done by van der Waal's force and hydrogen bonding, in which, the microbial cells get transferred from the bulk solution to the BC's surface followed by adhesion and colonization on BC's surface by secreting multiple polymeric substances to form extracellular biofilm [71–73]. The detailed mechanism of hydrocarbon contaminant's degradation by the synergistic effect of BC and microbes in the soil can be visualized in Figure 6.2.

6.4 CONCLUSIONS AND FUTURE PROSPECTS

Hydrocarbon pollutants are introduced into the soil through numerous human activities. The existence of hydrocarbons in the soil is a cause for concern due to the elevated levels of petroleum products, which can result in aromatic hydrocarbon toxicity, decreased microbial and enzymatic functions, and ultimately diminish soil productivity. Another issue regarding the existence of hydrocarbons in the soil is that these pollutants might infiltrate groundwater sources. There are multiple methods for removing hydrocarbons from soil, including physical, chemical, and biological approaches. Out of all the methods available, the bioremediation strategy, is the most efficient and cost-effective option to treat polluted soil in its original location.

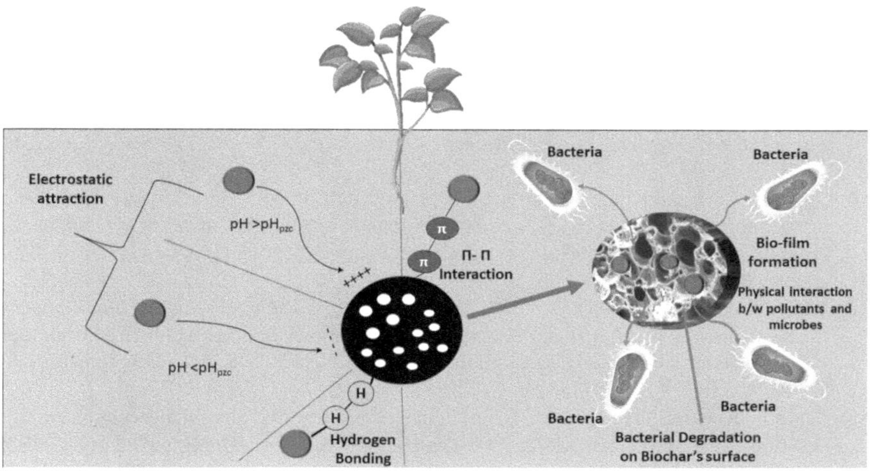

FIGURE 6.2 Mechanism of the degradation of petroleum hydrocarbons in the soil with the application of biochar.

Nevertheless, the soil's high content of aromatic hydrocarbons and its unfavorable physiochemical features restrict microbial activities in the soil. The application of BC in the soil facilitates the cleanup process through a variety of approaches. BC in the soil boosts the physiochemical properties, adsorbs pollutants in the soil pores, and stimulates microbial activities to facilitate the biodegradation of these contaminants. Combining BC with phytoremediation could be a highly successful method for bioremediating soil contaminated with hydrocarbons.

Further research could be conducted using BC-based composite materials or modified BC to investigate their impact on the remediation of hydrocarbon-contaminated soils. Another study might be conducted by integrating BC with in-situ microbial culture to remediate contaminated soil and evaluate the impact of modified BC on the microbial community and abundance. Furthermore, it is imperative to conduct extended field experiments using modified BCs to assess the effectiveness of BC in long-term remediation of polluted soil under real field conditions. These studies should also investigate the various aspects that could potentially influence the efficiency of this approach.

REFERENCES

1. Sihag S, Pathak H, Jaroli DP. Factors affecting the rate of biodegradation of polyaromatic hydrocarbons. *Int J Pure Appl Biosci*. 2014;2:185–202.
2. Srinivasan A, Chowdhury P, Viraraghavan T. *Air Stripping in Industrial Wastewater Treatment*. Encyclopedia of Life Support Systems. University of Regina, Canada, 2008.
3. Innocent Chukwunonso, Ossai, Ahmed A, Hassan A. Remediation of soil and water contaminated with petroleum hydrocarbon: A review. *Environ Technol Innov*. 2020;100526. doi:10.1016/j.eti.2019.100526

4. Agamuthu P, Tan YS, Fauziah SH. Bioremediation of hydrocarbon contaminated soil using selected organic wastes. *Proc Environ Sci.* 2013;18:694–702. doi:10.1016/j.proenv.2013.04.094
5. Shao P, Yin H, Li Y, Cai Y, Yan C, Yuan Y, Dang Z. Remediation of Cu and As contaminated water and soil utilizing biochar supported layered double hydroxide: Mechanisms and soil environment altering. *J Environ Sci (China).* 2023;126:275–286. doi:10.1016/j.jes.2022.05.025
6. Chandra S, Bhattacharya J. Influence of temperature and duration of pyrolysis on the property heterogeneity of rice straw biochar and optimization of pyrolysis conditions for its application in soils. *J Clean Prod.* 2019;215:1123–1139. doi:10.1016/j.jclepro.2019.01.079
7. Zhen M, Chen H, Liu Q, Song B, Wang Y, Tang J. ScienceDirect Combination of rhamnolipid and biochar in assisting phytoremediation of petroleum hydrocarbon contaminated soil using Spartina anglica. *J Environ Sci.* 2019;85:107–118. doi:10.1016/j.jes.2019.05.013
8. Saum L, Jiménez MB, Crowley D. Influence of biochar and compost on phytoremediation of oil-contaminated soil. *Int J Phytoremed.* 2018;20:54–60. https://doi.org/10.1080/15226514.2017.1337063
9. Zhang B, Zhang L, Zhang X. Bioremediation of petroleum hydrocarbon-contaminated soil by petroleum-degrading bacteria immobilized on biochar. *RSC Adv.* 2019;9:35304–35311. doi:10.1039/c9ra06726d
10. Essaid HI, Bekins BA, Herkelrath Wi N, Delin GN. Crude oil at the Bemidji Site: 25 years of monitoring, modeling, and understanding. *Groundwater.* 2011;49:706–726.
11. Hemptinne JC De, Delepine H, Jose C, Jose J. Aqueous solubility of hydrocarbon mixtures. *Oil Gas Sci Technol.* 1998;53:409–419.
12. Truskewycz A, Gundry TD, Khudur LS, Kolobaric A, Taha M, Aburto-medina A, Ball AS, Shahsavari E. Petroleum hydrocarbon contamination in terrestrial ecosystems. *Fate Microbial Respons.* 2019;1–20. doi:10.3390/molecules24183400
13. Providenti MA, Lee H, Trevors JT. Selected factors limiting the microbial degradation of recalcitrant compounds. *J Ind Microbiol.* 1993;12:379–395.
14. Akinwumi I, Diwa D, Obianigwe N. Effects of crude oil contamination on the index properties, strength and permeability of lateritic clay. *Int J Appl Sci Eng Res.* 2014;3:816–824.
15. Sutton NB, Maphosa F, Morillo JA, Al-soud WA, Langenhoff AAM, Grotenhuis T, Rijnaarts HHM, Smidt H. Impact of long-term diesel contamination on soil microbial. *Appl Environ Microbiol.* 2013;79:619–630. doi:10.1128/AEM.02747-12
16. Masakorala K, Yao J, Chandankere R, Liu H, Liu W, Cai M. A combined approach of physicochemical and biological methods for the characterization of petroleum hydrocarbon-contaminated soil. *Environ Sci Pollut Res.* 2014;21:454–463. doi:10.1007/s11356-013-1923-3
17. Vishnu CPDA, Purna VJ, Rao C. Investigation of physical and chemical characteristics on soil due to crude oil contamination and its remediation. *Appl Water Sci.* 2019;9:1–10. doi:10.1007/s13201-019-0970-4
18. Cheraghi M, Sobhanardakani S, Lorestani B, Merrikhpour H, Mosaed HP. Biochemical and physical characterization of petroleum hydrocarbon contaminated soils in Tehran. *J Chem Heal Risks.* 2015;5:199–208.
19. Wang Y, Feng J, Lin Q, Lyu X, Wang X, Wang G. Effects of crude oil contamination on soil physical and chemical properties in momoge wetland of China. *Chinese Geogr Sci.* 2013;23:708–715. doi:10.1007/s11769-013-0641-6
20. Alrumman SA, Standing DB, Paton GI. Effects of hydrocarbon contamination on soil microbial community and enzyme activity. *J King Saud Univ Sci.* 2015;27:31–41. doi:10.1016/j.jksus.2014.10.001

21. Unbehaun H, Dittler B, Kühne G, Wagenführ A. The impact of hydrocarbon remediation (diesel oil and polycyclic aromatic hydrocarbons) on enzyme activities and microbial properties of soil. *Acta Biotechnol.* 2000;20:313–333. doi:10.1002/abio.370200312
22. Gao H, Wu M, Liu H, Xu Y, Liu Z. Effect of petroleum hydrocarbon pollution levels on the soil microecosystem and ecological function. *Environ Pollut.* 2022;293. doi:10.1016/j.envpol.2021.118511
23. Labud V, Garcia C, Hernandez T. Effect of hydrocarbon pollution on the microbial properties of a sandy and a clay soil. *Chemosphere.* 2007;66:1863–1871. doi:10.1016/j.chemosphere.2006.08.021
24. Guo H, Yao J, Cai M, Qian Y, Guo Y, Richnow HH, Blake RE, Doni S, Ceccanti B. Effects of petroleum contamination on soil microbial numbers, metabolic activity and urease activity. *Chemosphere.* 2012;87:1273–1280. doi:10.1016/j.chemosphere.2012.01.034
25. Megharaj M, Singleton I, McClure NC, Naidu R. Influence of petroleum hydrocarbon contamination on microalgae and microbial activities in a long-term contaminated soil. *Arch Environ Contam Toxicol.* 2000;38:439–445. doi:10.1007/s002449910058
26. Ossai IC, Ahmed A, Hassan A, Hamid FS. Remediation of soil and water contaminated with petroleum hydrocarbon: A review. *Environ Technol In*nov. 2020;17:1–42. doi:10.1016/j.eti.2019.100526
27. Mulligan CN, Yong RN, Gibbs BF. Remediation technologies for soils and groundwater: an evaluation. *Eng Geol.* 2001;60:193–207. doi: 10.1016/S0013-7952(00)00101-0
28. Abdel-Moghny T, Mohamed RSA, El-Sayed E, Aly SM, Snousy MG. Removing of hydrocarbon contaminated soil via air flushing enhanced by surfactant. *Appl Petrochemical Res.* 2012;2:51–59. doi:10.1007/s13203-012-0008-4
29. Godheja J, Shekhar SK, Siddiqui SA, Modi D. Xenobiotic compounds present in soil and water: a review on remediation strategies. *J Environ Anal Toxicol.* 2016;6:2161–0525.
30. Banaszkiewicz K, Marcinkowski T. Use of cement-fly ash-based stabilization techniques for the treatment of waste containing aromatic contaminants. *E3S Web Conf.* 2017. https://doi.org/10.1051/e3sconf/20172200009
31. Bates E, Hills C. *Stabilization and Solidification of Contaminated Soil and Waste: A Manual of Practice*. Edward Bates and Colin Hills; 2015.
32. Bolan N, Kunhikrishnan A, Thangarajan R, Kumpiene J, Park J, Makino T, Beth M, Scheckel K. Remediation of heavy metal (loid) s contaminated soils – To mobilize or to immobilize ? *J Hazard Mater.* 2014;266:141–166. doi:10.1016/j.jhazmat.2013.12.018
33. Vu TH, Tran MV. A review on immobilisation of toxic wastes using geopolymer technique. In *Proc 4th Congrès Int Géotechnique-Ouvrages-Structures CIGOS*, Ho Chi Minh City, Vietnam, 26–27 October 2017; pp. 299–309.
34. Kluck C, Achari G. Chemical oxidation techniques for in situ remediation of hydrocarbon impacted soils. *Environ Eng.* 2004;1–8.
35. Besha AT, Bekele DN, Naidu R. Recent advances in surfactant-enhanced In-Situ Chemical Oxidation for the remediation of non-aqueous phase liquid contaminated soils and aquifers. *Environ Technol Innov.* 2018;9:303–322. doi:10.1016/j.eti.2017.08.004
36. Asgari AR, Abizadeh R, Mahvi AH, Nasseri S, Dehghani MH, Nazmara S, Yaghmaeian K. Remediation of total petroleum hydrocarbons using combined in-vessel composting and oxidation by activated persulfate. *Glob J Environ Sci Manag.* 2017;3:373–384. doi:10.22034/gjesm.2017.03.04.004
37. Aljuboury DADA, Palaniandy P, B AAH, Feroz S. Treatment of petroleum wastewater by conventional and new technologies: A review. *Glob NEST.* 2017. https://doi.org/10.30955/gnj.002239
38. Berg E. Adsorption of organic andinorganic compounds onactivated carbon and biochar. Degree Project, UMEA University; 2017.
39. Bansal RC, Goyal M. *Activated Carbon Adsorption.* CRC Press; 2005.

40. Weber K, Quicker P. Properties of biochar. *Fuel.* 2018;217:240–261. doi:10.1016/j.fuel.2017.12.054
41. Chandra S, Bhattacharya J. Influence of temperature and duration of pyrolysis on the property heterogeneity of rice straw biochar and optimization of pyrolysis conditions for its application in soils. *J Clean Prod.* 2019;215:1123–1139. doi:10.1016/j.jclepro.2019.01.079
42. Wang J, Wang S. Preparation, modification and environmental application of biochar: A review. *J Clean Prod.* 2019;227:1002–1022. doi:10.1016/j.jclepro.2019.04.282
43. Xiao F, Bedane AH, Xiaojun J, Mann MD, Pignatello JJ. Science of the Total Environment Thermal air oxidation changes surface and adsorptive properties of black carbon (char/biochar). *Sci Total Environ.* 2018;618:276–283. doi:10.1016/j.scitotenv.2017.11.008
44. Lydia P, Carsten W, Mueller C, Rumpel A, Rka Ã, Alexandra CW. Multi-technique approach to assess the fate of high-temperature biochar in soil and to quantify its effect on soil organic matter composition. *Org Geochem.* 2017;122:177–186.
45. Cross A, Sohi SP. Soil biology & biochemistry: The priming potential of biochar products in relation to labile carbon contents and soil organic matter status. *Soil Biol Biochem.* 2011;43:2127–2134. doi:10.1016/j.soilbio.2011.06.016
46. Park J, Sik Y, Kim S, Cho J, Heo J, Delaune RD, Seo D. Chemosphere Competitive adsorption of heavy metals onto sesame straw biochar in aqueous solutions. *Chemosphere.* 2016;142:77–83. doi:10.1016/j.chemosphere.2015.05.093
47. López-cano I, Roig A, Cayuela ML, Alburquerque JA, Sánchez-monedero MA. Biochar improves N cycling during composting of olive mill wastes and sheep manure. *Waste Manag.* 2016;49:553–559. doi:10.1016/j.wasman.2015.12.031
48. Melo LCA, Puga AP, Coscione AR, Beesley L, Abreu CA, Camargo OA. Sorption and desorption of cadmium and zinc in two tropical soils amended with sugarcane-straw-derived biochar. *J Soil Sediments.* 2016;16:226–234. doi:10.1007/s11368-015-1199-y
49. Kołodynska D, Wnetrzak R, Leahy JJ, Hayes MHB, Kwapinski W, Hubicki Z. Kinetic and adsorptive characterization of biochar in metal ions removal. *Chem Eng J.* 2012;197:295–305. doi:10.1016/j.cej.2012.05.025
50. Lu H, Zhang W, Yang Y, Huang X, Wang S, Qiu R. Relative distribution of Pb2+ sorption mechanisms by sludge-derived biochar. *Water Res.* 2012;46:854–862. doi:10.1016/j.watres.2011.11.058
51. Cao X, Ma L, Liang Y, Gao B, Harris W. Simultaneous immobilization of lead and atrazine in contaminated soils using dairy-manure biochar. *Environ Sci Technol.* 2011;4884–4889.
52. Laird D, Fleming P, Wang B, Horton R, Karlen D. Biochar impact on nutrient leaching from a Midwestern agricultural soil. *Geoderma.* 2010;158:436–442. doi:10.1016/j.geoderma.2010.05.012
53. Brownsort P, Mašek O. Biomass pyrolysis processes: Performance parameters and their influence on biochar system benefits. MSc dissertation, University of Edinburgh, Edinburgh, UK, 2009.
54. Bridgwater A, Peacocke G. Fast pyrolysis processes for bomass. *Renew Sustain Energy Rev.* 2000;4:1–73.
55. Bridgwater A V. Review of fast pyrolysis of biomass and product upgrading. *Biomass Bioenergy.* 2012;38:68–94. doi:10.1016/j.biombioe.2011.01.048
56. Couto N, Rouboa A, Silva V, Monteiro E, Bouziane K. Influence of the biomass gasification processes on the final composition of syngas. *Energy Proc.* 2013;36:596–606. doi:10.1016/j.egypro.2013.07.068
57. C. Higman, M. van der B. Burgt. *Gasification.* Elsevier; 2003.
58. Riser-Roberts E. *Remediation of Petroleum Contaminated Soils: Biological, Physical, and Chemical Processes.* CRC Press; 2020.

59. Gupta S, Kua HW. Factors determining the potential of biochar as a carbon capturing and sequestering construction material: critical review. *J Mater Civ Eng*. 2017;29:04017086.
60. Nanda S, Reddy SN, Mitra SK, Kozinski JA. The progressive routes for carbon capture and sequestration. *Energy Sci Eng*. 2016;6:99–122. doi:10.1002/ese3.117
61. Das N, Chandran P. Microbial degradation of petroleum hydrocarbon contaminants: An overview. *Biotechnol Res Int*. 2011. doi: 10.4061/2011/941810
62. Bartha R, Bossert I. The treatment and disposal of petroleum wastes in petroleum microbiology. *Pet Microbiol*. 1984;553–578.
63. Yuniati MD. Bioremediation of petroleum-contaminated soil: A review. *IOP Conf Ser Earth Environ Sci*. 2018. pp. 1755–1315.
64. Ahmad M, Rajapaksha AU, Lim JE, Zhang M, Bolan N, Mohan D, Vithanage M, Lee SS, Ok YS. Biochar as a sorbent for contaminant management in soil and water: A review. *Chemosphere*. 2014;99:19–33. doi:10.1016/j.chemosphere.2013.10.071
65. Ren H, Wei Z, Wang Y, Deng Y, Li M, Wang B. Effects of biochar properties on the bioremediation of the petroleum-contaminated soil from a shale-gas field. *Environ Sci Pollut Res*. 2020;36427–36438.
66. Zhang G, He L, Guo X, Han Z, Ji L, He Q, Han L. Mechanism of biochar as a biostimulation strategy to remove polycyclic aromatic hydrocarbons from heavily contaminated soil in a coking plant. *Geoderma*. 2020;375. doi:10.1016/j.geoderma.2020.114497
67. Kong L, Gao Y, Zhou Q, Zhao X, Sun Z. Biochar accelerates PAHs biodegradation in petroleum-polluted soil by biostimulation strategy. *J Hazard Mater*. 2018;343:276–284. doi:10.1016/j.jhazmat.2017.09.040
68. Han T, Zhao Z, Bartlam M, Wang Y. Combination of biochar amendment and phytoremediation for hydrocarbon removal in petroleum-contaminated soil. *Environ Sci Pollut Res*. 2016;21219–21228. doi:10.1007/s11356-016-7236-6
69. Rummel CD, Jahnke A, Gorokhova E, Kühnel D, Schmitt-Jansen M. Impacts of biofilm formation on the fate and potential effects of microplastic in the aquatic environment. *Environ Sci Technol Lett*. 2017;4:258–267. doi:10.1021/acs.estlett.7b00164
70. Guo J, Yang S, He Q, Chen Y, Zheng F, Zhou H, Hou C, Du B, Jiang S, Li H. Improving benzo(a)pyrene biodegradation in soil with wheat straw-derived biochar amendment: Performance, microbial quantity, CO2 emission, and soil properties. *J Anal Appl Pyrolysis*. 2021;156:105132. doi:10.1016/j.jaap.2021.105132
71. Jesionowski T, Zdarta J, Krajewska B. Enzyme immobilization by adsorption: A review. *Adsorption*. 2014;20:801–821. doi:10.1007/s10450-014-9623-y
72. Kilonzo P, Margaritis A, Bergougnou M. Effects of surface treatment and process parameters on immobilization of recombinant yeast cells by adsorption to fibrous matrices. *Bioresour Technol*. 2011;102:3662–3672. doi:10.1016/j.biortech.2010.11.055
73. Frankel ML, Bhuiyan TI, Veksha A, Demeter MA, Layzell DB, Helleur RJ, Hill JM, Turner RJ. Removal and biodegradation of naphthenic acids by biochar and attached environmental biofilms in the presence of co-contaminating metals. *Bioresour Technol*. 2016;216:352–361. doi:10.1016/j.biortech.2016.05.084
74. Lin B, Chen W. Sugarcane bagasse pyrolysis in a carbon dioxide atmosphere with conventional and microwave-assisted heating. *Front Energy Res*. 2015;3:1–9. doi:10.3389/fenrg.2015.00004

Section II

Water

7 An Analysis of Groundwater Movement and Realignment Factors for the Twin Mining System

A Modelling Approach

Abhay Kumar Soni, Paras Ranjan Pujari, and Ramesh Janipella

7.1 INTRODUCTION

Malanjkhand *Copper Project* (MCP) is an integrated copper mining project owned by Hindustan Copper Limited (HCL), India, which has a *twin mining system* in place meaning that at this mine, the copper ore excavation was continued by both surface mining and underground method in 2021. After 2022, the mining operation in the open pit was discontinued (decommissioning phase) and underground mining was scaled up (initial development stage). Thus, underground mining operations in the immediate vicinity (nearer) to an open-pit excavation are the typicality of this case study. In between the open mine and the underground mine, a natural rock barrier of 40 m, called a crown pillar, has been retained, and the development work of the underground mine is started from the upper benches (upper levels) of the open mine. Thus, mining at MCP is a classic example of mining complexity in contrast to the well-known Kennecott Copper Mine, Utah, USA (https://www.mining-technology.com/projects/bingham/).

For an operative *twin mining system,* disturbance in the groundwater, i.e. direction of groundwater flow as well as groundwater quantity, is quite conspicuous in the post-mining scenario. Hence, groundwater realignment chances, leading to water accumulation in the open mining pit, are fair enough. One of the most critical components in such assessment is that of the hydrological conditions of the area arising from the combination of both surface water and groundwater. Therefore, the research attempts should be based on the aquifer characteristics, water level fluctuations (WLFs) and behaviour of the water table, as done at MCP in this study. Site geology, on-going deep excavations and nearby water bodies interfere in groundwater

DOI: 10.1201/9781003442554-9

movement underground. Considering all these components, a simulative condition akin to the field has been created, and the mining consequences in terms of groundwater flow have been evaluated for the MCP pit located in the Indian peninsula. Besides this analysis, the research investigation has looked into yet another aspect which plays an important role in the mine life, i.e. on cessation of open-pit mining at MCP, particularly the excavation operation, is it feasible to convert the excavated open-pit area into a 'mining pit lake', as value addition?

Our review on research progress in groundwater of twin mining systems indicates that such cases are though present but limited. The research progress, particularly in the 'groundwater modelling for the mining areas' is available in plenty (Surinaidu et al., 2014; Fernández-Álvarez et al., 2016; Rapantova et al., 2007). This has been found and established as well that the modelling tool helps in the mine water management, both short-term and long-term, to mine developers/operators in particular (More et al., 2020). The surface water and groundwater management, quantitative estimation, prediction and assessment all are intended to benefit the mine planners because mining-influenced water is a valuable resource (More and Wolkersdorfer, 2022; Kurukulasuriya et al., 2022). Hence, modelling research, especially for the twin mining system, is a necessity to know the groundwater system – problem and progress in a better way. Practically, mining is a dynamic engineering operation, which has the potential to alter the groundwater system and facilitate hydraulic connections between surface water and groundwater (Kurukulasuriya et al., 2022); hence, extra emphasis should be laid on conducting such research.

7.2 STUDY AREA AND LOCATION DETAILS OF THE MINE SITE

MCP is a surface mine of copper ore owned by HCL (Estd. 1967). HCL is a well-known public-sector mining company in India and a recognized organization in the mineral industry. In this project, the production of copper ore commenced in 1982 from the surface mine. Subsequently, a matching ore concentrator plant, tailing disposal system and other auxiliary facilities were commissioned. The MCP premises prepare copper concentrates from the excavated Cu ore. This concentrate is dispatched to Khetri in Rajasthan state, India, for copper extraction. As indicated above, the surface mining at MCP was stopped by the end of 2021 and underground mining commenced. For this underground mine, the approach incline/decline is planned/located at the lower benches of the existing MCP open pit because the ore body is in continuation but at a greater depth than what was found in the open mine. From the already excavated mine benches, the copper ore body is closer; hence, the approach is easy and convenient.

Location details: Malanjkhand copper mine of HCL is located at Malanjkhand, Taluka: Baihar, District: Balaghat in the state of Madhya Pradesh (MP), India. The mine, 90 km from the district town of Balaghat, has no rail connectivity but is approachable from all-weather roads and connected with important towns of MP, namely Jabalpur, Gondia, Baihar, Lanji, Katangi Seoni and Balaghat. The location map of the study area is given in Figure 7.1a, and coordinate details are as follows:

- Latitude and longitude of mine site: 21°56′ to 22°05′N and 80°39′ to 80°46′E
- Survey of India, Topo sheet No.: 64 B/12, 64B/16, 64C/9 and 64C/13

FIGURE 7.1 (a) State map of Madhya Pradesh, India, showing the location of the Malanjkhand Copper Project (MCP). (b) MCP Pit in 2021 – a distant and closer view with pit water and developed bench faces.

Malanjkhand copper mine with 479.9 Ha of mining lease area and an annual copper ore production of 2 million tonnes (planned for open-pit mine) is an operative mechanized surface mine of the large category (Figure 7.1b). Till the year end of 2021, the Cu ore production was continued from the surface mining alone. Because of the higher stripping ratio, the extraction of near-surface ore has become uneconomical and slowly, underground mine planning has started to produce 05 MT per year (planned/targeted production) from the underground mine. The copper ore production from MCP underground was expected to begin shortly by the middle of 2022.

The first and initial phase of underground mining development (Figure 7.2) has been planned for up to 60 m depth (i.e. below 340 m RL), and planning for more depths will be done later. The underground mine of M/s HCL is expected to produce copper ore, mainly Chalcopyrite, with 1.31% copper.

7.3 MATERIALS AND METHOD

The water in the open-pit mine has a great concern for both production and productivity. The open-pit mining operation at Malanjkhand has reached 340 m RL (a reduced level from a measured reference). As such, the open-pit depth below the

FIGURE 7.2 A view of the Malanjkhand Copper Project underground workings [4] (A 1,500 m long drive at 210 m level connecting the North incline/decline and South incline/decline as an underground mine development activity).

ground level is 240 m, and the mine operation for ore excavation is done below the water table. Therefore, in an operative surface mine (a pit mine), the water accumulation and water flow path/direction assume great importance (Eqs. 7.1–7.3).

$$\text{Water In} - \text{Water Out} = \text{Change in Water Storage}(\Delta S). \tag{7.1}$$

If we discuss theoretically, the groundwater and its flow, in particular, are governed by Darcy's law and the conservation of mass, a general equation (Eq. 7.1) for conservation of mass (Karanth, 1993). Darcy's fundamental law (Eq. 7.2) states that the flow per unit area per unit time is directly proportional to the change in head and inversely proportional to the length of the flow path.

$$V = K(\partial h / \partial x) \tag{7.2}$$

where

V = flow per unit area per unit time
K = hydraulic conductivity
$\partial h/\partial x$ = hydraulic gradient

In steady-state conditions, the groundwater flow is expressed by the equation given below:

$$\frac{\partial}{\partial x}\left(\frac{K_x \partial h}{\partial x}\right) + \frac{\partial}{\partial y}\left(\frac{K_y \partial h}{\partial y}\right) + \frac{\partial}{\partial z}\left(\frac{K_z \partial h}{\partial z}\right) + W = 0 \tag{7.3}$$

where

K_x = hydraulic conductivity in the x-direction
K_y = hydraulic conductivity in the y-direction
K_z = hydraulic conductivity in the z-direction
h = hydraulic head and W is the source or sinks term.

The groundwater flow equation (Eq. 7.3) is solved subject to the initial and boundary conditions, some of which are assumed too (Section 3.1).

Having reviewed the theoretical aspects of water flow, numerical modelling is a better, more convenient and most modern tool to understand the complex mine-filled groundwater flow scenario. Hence, to predict and evaluate the groundwater forces as present in the mine, modelling is attempted for the typical hydrogeological setting that exists in MCP. In this field-orientated case study, Visual MODFLOW Flex (Version 7.0) has been used to simulate the groundwater conditions and assess the realignment of groundwater flow direction. The Visual MODFLOW-7.0 Flex is a cell-centred, 3D-Finite difference model and is the most widely used for the calculation of the steady-state or transient saturated groundwater flow (McDonald and Harbaugh, 1998). Obtained results could predict groundwater inflows (seepages) into the pit which vary over different periods. This groundwater together with the surface water, when occurs in the mining pit, is referred to as mine water. Thus, the modelling study provides the flow direction and the groundwater flux (magnitude of water), at the stipulated hydraulic head at the point of assessment. *Zone budgeting,* to know the water quantity of the mining pit has been applied for the developed model. The literature review, approach and the assumed conditions, to derive the modelling results, are described below.

7.4 LITERATURE REVIEW

Surinaidu et al. (2014) estimated groundwater inflows into the coal mines at different mine development stages using MODFLOW through hydrogeological and groundwater modelling. At various phases of mine development (surface mine), the study's findings can be utilized to determine the best places for groundwater dewatering and optimal groundwater pumping.

Rapantova et al. (2007) estimated dewatering in the Czech part of the Upper Silesian Coal Basin. The mining region where Poland and the Czech Republic traditionally extract deep hard coal is known as the Upper Silesian Coal Basin. The Ostrava-Karvina Coalfield, the main portion of the basin in the Czech Republic, is separated into three sub-basins; two of them are flooded, and groundwater levels are maintained at a certain level to avoid overflow to the third Karvina sub-basin, which is still being used for mining. Rapantova et al. (2007) were able to develop a reasonably accurate model prediction of 'detritus' dewatering as a result of water drainage into active or abandoned underground mines. The ensuing study shows extremely favourable results of model calibration on 28 calibration targets with respect to known regime of piezometric levels in monitoring boreholes. The predictions were extended till 2015 and are presented as flow vectors and contour maps of piezometric levels. Additionally, a simulation of water inrushes into the mine workings was also conducted.

An open-pit mine, when becomes deep, may in the future, be converted into an underground mine. In order to economize the cost of mining, such turning becomes essential because of the higher stripping ratio. There are lot of mixed mines, e.g. Rampura-Agucha, Lead-Zinc mine of Hindustan Zinc Limited in Rajasthan, India; Sukani gold mine, Egypt; Bingham Canyon copper mine, Utah, USA; Climax Molybdenum mine, Colarodo, USA; and Sudubury nickel ore body, Ontario, Canada. We do not know whether twin system is actually operational at these mines or not but can write with certainty that these are the sites where cent per cent possibility of twin system exists. Hence, the practical utility of this research is quite conspicuous, and its real field applicability will be as per the site-specific need.

Various mine case studies reveal that co-existence of surface and underground mining is limited world over. Records that narrate technical details, of both coal and metallic/non-metallic mineral extraction, show that similar field condition makes the natural groundwater system complex and its scientific evaluation becomes difficult. Hence, research input through modern modelling tool (modelling) solves such problems more accurately and rapidly. In this way, this research attempt enhances state of the art further and adds to the national /international contribution. The 'big hole' at Kimberly, South Africa, was converted to an underground mine with shafts to the surface and sub-terranean tunnels when the groundwater flooding, what was essentially a very deep well, could no longer be pumped out.

7.5 APPROACH AND ASSUMED CONDITIONS

In field, the geological and geo-hydrological factors impact groundwater system crucially which we have referred here as 'groundwater realignment'. This piece of the research work is thus aimed to assess the impact for an invariably and rare 'twin mining system' which is complex. Our approach to address the issue has been through the new analysis tool, i.e. groundwater modelling (MODFLOW software tool) because the analysis through computer tool is rather more accurate, precise and easy.

To build a *predictive model*, some assumptions are essential because these assumptions denote as well as represent the actual field conditions covering the risk and accuracy factors of a model in a wider perspective. For analyses and evaluation, it is assumed that the open pit may be filled with water, either fully or partially, and only the steady-state groundwater conditions exist and are considered for this analysis. The applied assumptions for the studied area include

i. The aquifer present is under water-table conditions meaning that the aquifer type is – 'An unconfined aquifer'.
ii. Hard rock formations are the principal formations of the study area.
iii. No consideration is given for first-order and second-order streams in surface drainage because there is no prominent perennial flowing water body lying in the vicinity. Mostly seasonal water streams mark the surface drainages all around.
iv. WLF trend is normal as per the seasons and water evaporation losses occur naturally.

MCP open-pit mine is a fully developed mine and reached the decommissioning phase, the last phase of the mine's life. At the mine site, undulating topography

TABLE 7.1
Location, Dimensions, RL, Depth and Water Condition Details of Underground Structures at MCP

S. No.	Name of Structure	Location		MRL (Below Ground Level)	Planned Depth or Length/ Excavation Reached up to	Actual Water Inflow Condition[a]
		Latitude Longitude	Dimension and Shape			
(a)	(b)	(c)	(d)	(e)	(f)	(g)
1.	South Ventilation Shaft	22°02′45″E 80°42′24″N	6.5 m dia; Circular	582.496 m RL	645 m 277.8 m	Heavy water seepage
2.	North Ventilation Shaft	22°02′3″E 80°43′3″N	6.5 m dia; Circular	573.358 m RL	633 m 185.9 m	Moderate seepage
3.	North Incline/ decline	22°01′41″E 80°42′55″N	5.5 m × 5.0 m Arc shaped	532 m RL (from the pit)	4610 m 2109.6 m	Moderate seepage
4.	South Incline/ decline	22°00′56″E 80°42′35″N	5.5 m × 5.0 m Arc shaped	436 m RL (from the pit)	3860 m 1320.7 m	Heavy water seepage
5.	Production shaft	22°01′27″E 80°42′41″N	7.0 m diameter Circular	583.602 m RL	695.0 m 685.9 m	Moderate seepage
6.	Service shaft	22°01′37″E 80°43′3″N	6.5 m diameter Circular	585.663 m RL	665.0 m 665.0 m	Moderate seepage

Note for inflow/seepage condition: In these underground structures the water inflow or seepage condition, as observed in the field, is moderate. Heavy seepage conditions are observed in the monsoon months from July–September. Here, low = <05 L/ft^2/min; moderate = >05 <10 L/ft^2/min and heavy = >20 L/ft^2/min

and temperate climatic conditions exist. The *watershed planning concept* has been considered in our approach to developing this simulative model for a mine area. The MCP open-pit mine encompasses significant water quantity in the rainy season (Figure 7.1b). During monsoon and post-monsoon seasons, groundwater seepage from the pit slopes (walls) remains visible. To commence the underground mining at MCP, several deep underground structures are under construction (Table 7.1), and as a result, alteration in the groundwater regime is bound to take place, though the mine is working below the water table. The phreatic surface has already been lowered as an impact of mining; hence, drawdown conditions will surely occur. With this background, our approach was to carry out groundwater modelling to know the realignment of groundwater forces for the existing twin mining system which is in place at Malanjkhand, uniquely. It is clear that both open and underground excavations and mining operations are responsible for groundwater disturbances.

Malanjkhand copper mine has three important areas concerning mine water, namely a lease area (479.9 Ha), an excavated area (pit area) and a catchment area

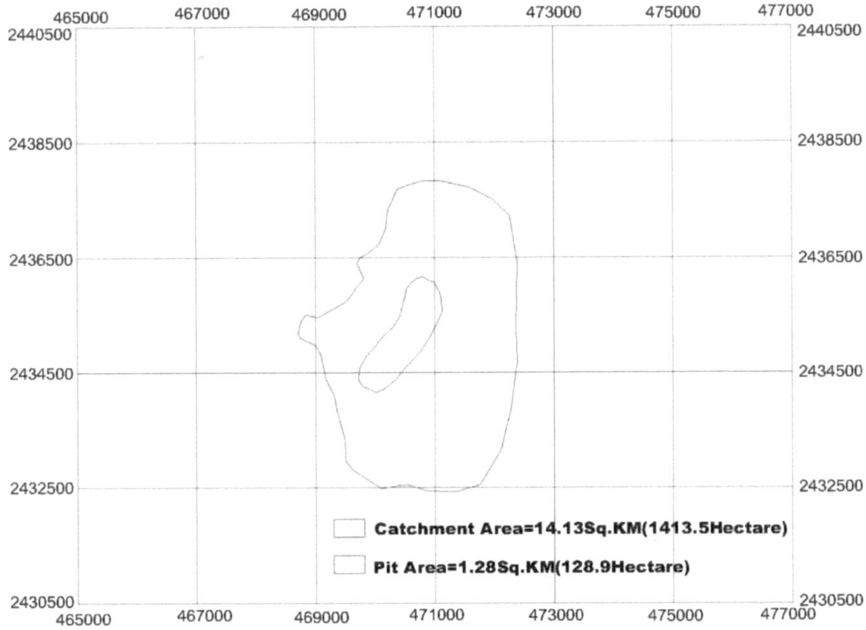

FIGURE 7.3 Three important mine water areas for the Malanjkhand Copper Project pit as drawn from GIS.

(1,413 Ha) besides the watershed area (Figure 7.3). The excavated pit area, measuring 128.9 Ha, is the most important area from the mine water angle because accumulated water is contained in this area forming the water source. Alternatively, the 'excavated area/pit area' or 'excavated pit area' is also termed as the 'mined-out area' or the 'dug-out area'. The pit area will always be smaller than the *catchment area*. All important features of the *twin mining system* have been given due emphasis in this study, e.g. an open pit and underground workings, when located closely, have a major chance of inundation, thereby endangering the mine safety. This is the reason why a 40 m rock barrier is considered in between the open pit and underground workings to avoid any mishaps (also left in the field). Secondly, the radius of influence, which is an area of influence in water science terminology, is important for seepage protection and safety; however, the available hydrostatic head is important for the water flow to occur and water pressure to develop (CSIR-CIMFR, 2018).

7.6 MODEL CONCEPTUALIZATION AND INPUT DATA

Grid design: The study area is delineated based on the watershed principle (ridge to valley) and is discretized into 40 rows and 20 columns (Figure 7.4), keeping the dimension of each cell as 90 m × 130 m. A single-layered groundwater system is conceptualized by taking into account the hydrogeology of the study area.

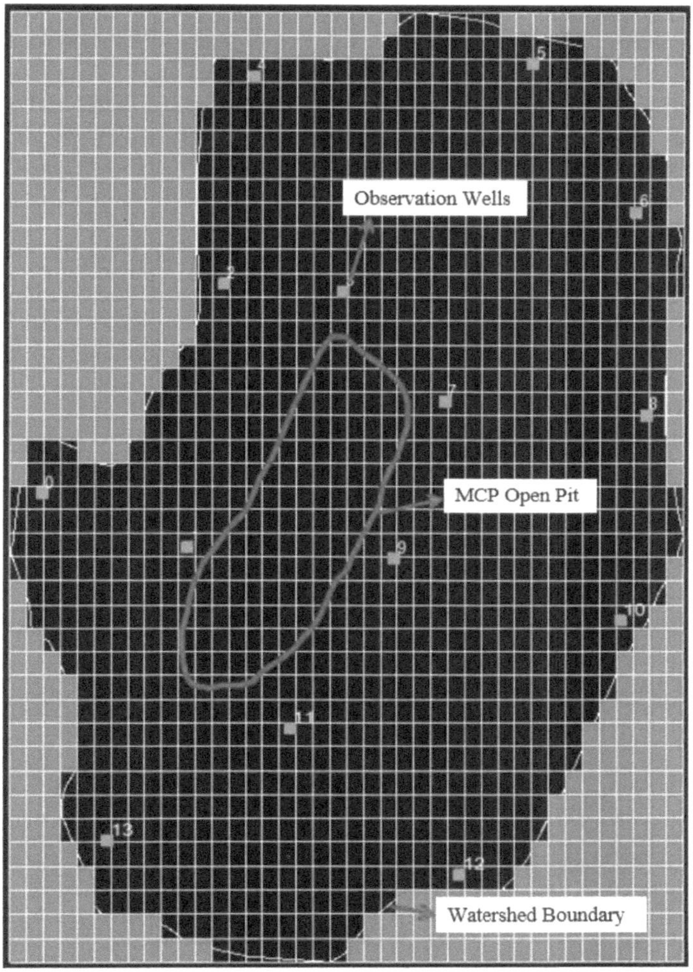

FIGURE 7.4 Malanjkhand Copper Project pit with watershed boundary and assigned observation wells to the model (Grid design of the study area).

Boundary conditions and input parameters: Boundary conditions represent locations in the model where water flows into or out of the model region due to external factors. The selection of boundary conditions is dependent on the hydrogeological conditions of the aquifer present in the study area and the objectives of the model. The following boundary conditions are assigned to the model evaluated in this study.

1. Specified head
   a. Constant Head (CHD)
2. Specified flux
   a. Recharge package (RCH)

Elevation: The elevation of the study area is assigned to the model based on the Digital Elevation Model of the study area. The extracted elevation data has been projected to the Universal Transverse Mercator projection and exported to the model. The topography of the watershed and the mine site has been considered and assigned to the model.

Observation wells: A network of 14 observation wells is selected for assigning the head to the model. These observation wells, as observed in the field, have different water levels, forming the basis of flow analysis (Figure 7.4).

Hydraulic conductivity: The hydraulic conductivity values, i.e. $K_x = 1.2\text{E-}5\,\text{m/s}$, $K_y = 1.2\text{E-}5\,\text{m/s}$ and $K_z = 1.2\text{E-}6\,\text{m/s}$ have been assigned to the model for the whole study area (CSIR-CIMFR, 2018).

Recharge: A recharge of 115 mm/year (10% of rainfall) has been assigned to the model according to the stipulations of groundwater estimation methodology as applied to Indian soils (GEC, 1997).

Constant head: The constant head (boundary condition constant head) has been assigned considering the actual water levels at the extreme northeast and southwest end of the study area.

The predictive model setup, as done in this case study is a conceptualized multi-layered model with different geological horizons as per the borehole litho log of the area having vertical hydraulic conductivity assigned to the model. Besides model conceptualization, other input data considered for modelling are pit depth = 240 m bgl (below the ground level); pit area (fixed) = 128.9 ha; average yearly rainfall = 1150.9 mm/year. In addition, all assumed conditions, mentioned in Section 3.1, namely hard rock type, unconfined aquifer, intercepted water table at the mining pit, normal seasonal WLFs and water evaporation losses by considering the rainfall infiltration factor (RIF) have been taken into account for the model development.

7.7 GROUNDWATER MODELLING: ANALYSIS OF RESULTS AND DISCUSSIONS

The groundwater modelling approach, best among others for advanced assessment prediction, is equally best for mines like MCP, which will be closed shortly. The end use of the open mine in terms of its value addition is yet another advantage of knowing the water conditions in advance. To simulate the water condition, existing MCP open-pit mine (Figure 7.1b) and large-sized underground structures (Table 7.1) that are likely to cause an impact on the groundwater movement have been modelled. Hence, for this modelling work, different scenarios and conditions are assumed. Principally, four different scenarios (Scenario 1–4) and two conditions (namely the dry pit and the water-filled pit) have been considered for assessment and evaluation. The dry pit condition and water-filled pit condition help in deciding the water flow direction, i.e. groundwater movement, which otherwise occurs in and around the pit mine naturally.

- Scenario (1)—Water-filled pit and no underground structure (Figure 7.5).
- Scenario (2)—Water-filled pit with all underground structures in place (Figure 7.6).

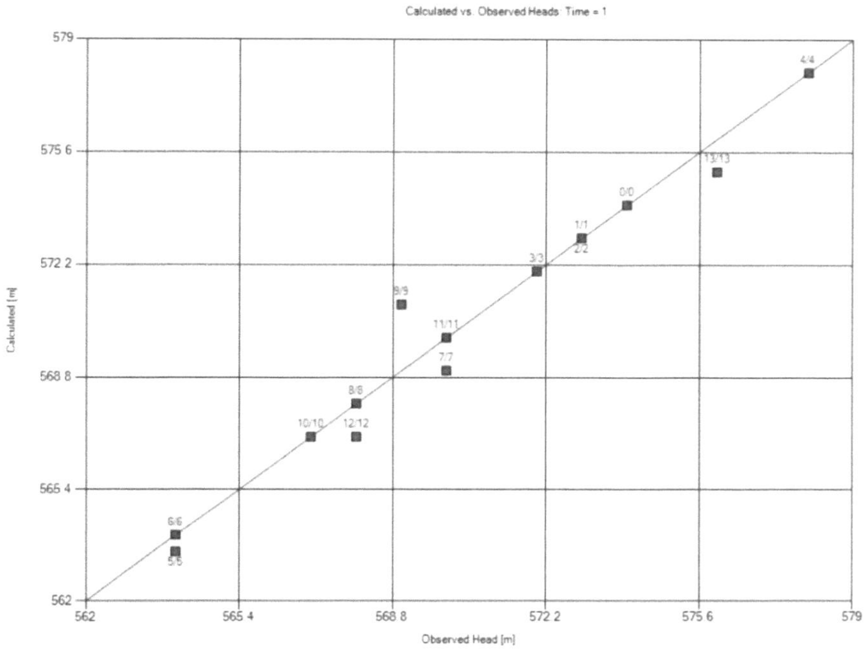

FIGURE 7.5 Calibration of the model [$R^2 = 0.91$ and NRMS = 12%].

- Scenario (3)—Dry pit (no water) and no underground structures (Figure 7.7).
- Scenario (4)—Dry pit (no water) with all underground structures in place (Figure 7.8).

In all scenarios mentioned above (1–4), the underground structures refer to those required for the development of an underground mine and include shafts, inclines and decline (Table 7.1, column b). These underground structures were fully developed in the year 2022 while the open mining pit is fully excavated /developed. The steady-state conditions, under four (04) different scenarios, have been analysed, and inferences about the groundwater flow direction and levels have been drawn. These are presented in Table 7.2, in column (iii). Figures 7.5–7.8 show the output results of the modelling.

These modelling results are further explained below point-wise.

a. The water level decline of 30 m, in a scenario (Figure 7.8) in the presence of underground structures, indicates that these underground structures are resulting in groundwater re-distribution or realignment of groundwater forces.
b. The groundwater recharge (flux) of 224.9 m^3/day is predicted based on the zone budgeting exercise.
c. The modelling results indicate that groundwater flow takes place from all directions. Groundwater flow direction is towards West to East when a

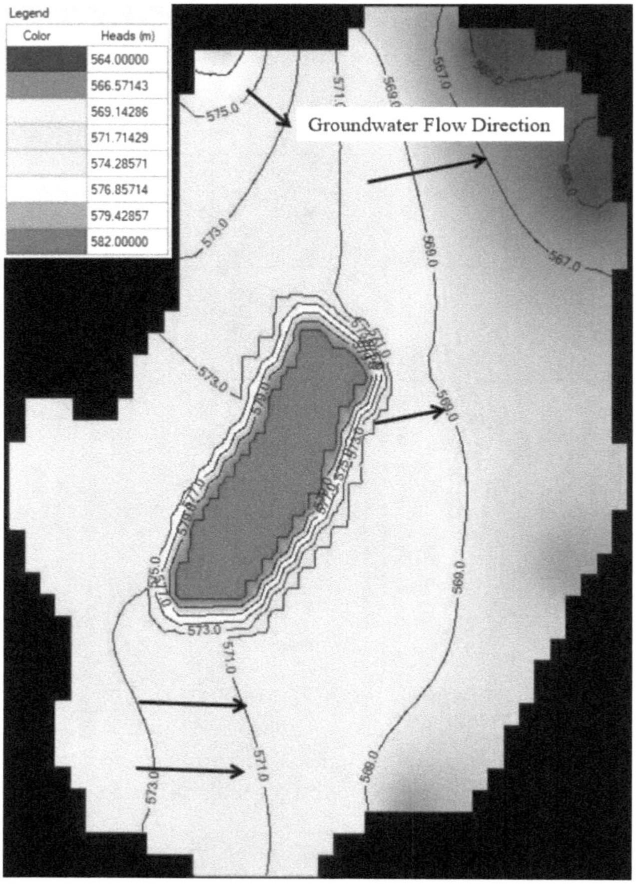

FIGURE 7.6 Model showing a water-filled pit with no underground structures.

water-filled pit has no underground structure in the vicinity (Figure 7.5). Multiple factors, both man-made (mining excavations) and natural, are affecting the flow directions.

d. The direction of the water flow towards the pit causes water accumulation.

The modelling results (Figures 7.5–7.8) confirm that the presence of underground structures in the vicinity of the open pit is leading to the realignment of groundwater flow. Such realignment is seen in all directions of the pit (N, S, E, W) also causing fluctuations in water levels in the watershed and near the pit. For the operative stage of the MCP mine and during the post-mining period, model calibration results are good and satisfactory, as evident from the best fit curve drawn (Figure 7.8). Results of the models in terms of 'water head' have been compared with the measured data (calculated values vs observed values) based on the mean absolute error and root

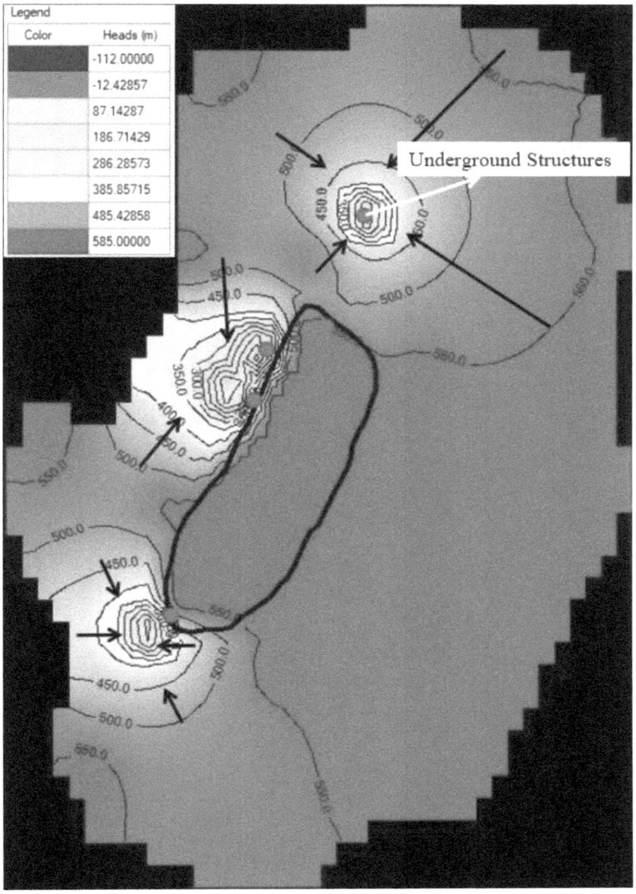

FIGURE 7.7 Model showing a water-filled pit with all underground structures.

mean square error. The developed groundwater flow model has been calibrated for steady-state conditions with the statistical values as – R^2 (Root square) = 0.91 and NRMS (normalized root-mean-square) = 12%. However, for non-steady conditions, the results may vary. In hydrogeology, NRMS deviation or root-mean-square deviation is used to evaluate the calibration of a groundwater model and is a measure of the differences between field values and the values observed, showing the error of the predicted model. It is an estimator that indicates how close you are to reality.

In modelling studies, zone budgeting (ZB) is useful for knowing the water quantity or magnitude of water likely to be available. Hence, a zone covering the pit has been delineated, considering that the lake formed by the pit is filled with water. Such a groundwater flow model with ZB indicated that the surrounding area of the pit lake will be recharged with an amount of 224.9 m^3/day. The specified flux (quantity/magnitude of groundwater, i.e. 224.9 m^3/day) under the assumed conditions shows the

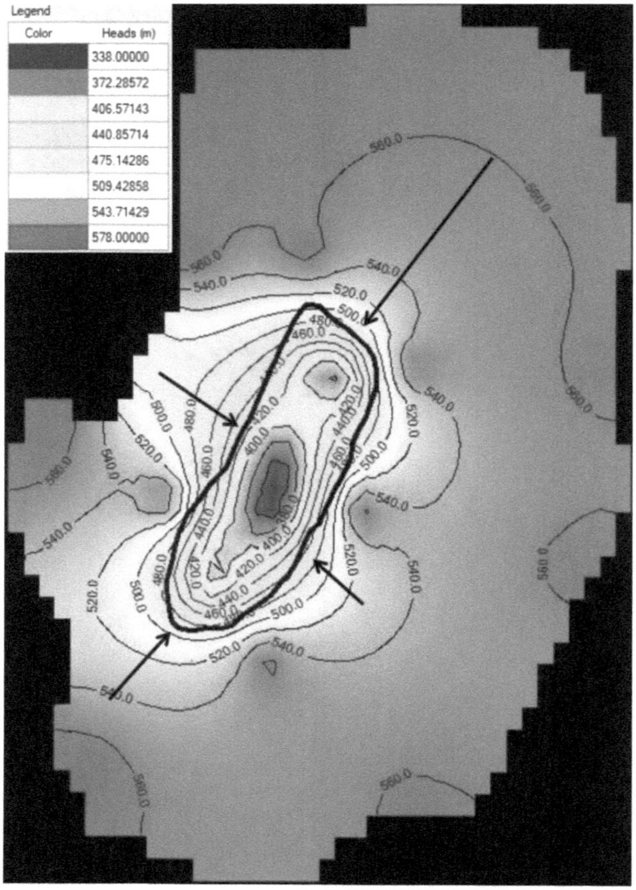

FIGURE 7.8 Model showing a fully excavated pit with no water and no underground structures.

water out principally and is helpful to assess the water availability in the mined-out pit of this open-pit copper mine.

When a mined-out pit is converted into a pit lake by water filling (Soni et al., 2014), the invariably favourable hydraulic gradient together with level differences (in general, the underground workings will be at comparatively lower levels) will guide the water flow. This may be either way, away from the pit or towards the pit, depending on the encountered field conditions. Other influencing factors, which have a causative effect on the groundwater movement and realignment, are discussed in the next sub-section. Thus, the groundwater modelling results and discussions as reported in this section, together with the influencing factors described later, will make the analysis complete in both qualitative and quantitative terms for the discussed problem. However, the scope for more detailed research will remain alive in the future.

TABLE 7.2
Summary Table: Results and Inferences from Groundwater Modelling

Figure Numbers	Assumed Condition	An Inference Drawn about Groundwater Flow Direction and Levels	
(i)	(ii)	(iii)	
Figure 7.5	Water-Filled Wet Pit (no underground structures in the vicinity).	Groundwater flow direction is towards West to East.	The water level at the pit lake is 582 m (MRL).
Figure 7.6	Water-Filled Wet Pit (underground structures in the vicinity).	Groundwater flow takes place from all directions wherever underground structures like shafts are assigned to the model. This causes water accumulation in the pit.	The water level at the pit lake remains at 582 m (MRL)
Figure 7.7	Fully Excavated Dry Pit (no underground structures in the vicinity).	Groundwater flow direction is towards the pit and from all directions.	The water level at the pit lake is 380 m (MRL).
Figure 7.8	Fully Excavated Dry Pit (all underground structures are in place in the vicinity)	The overall groundwater flow direction is towards the pit causing an accumulation of water.	The water level is reduced from 350 to 380 m (MRL) i.e. 30 m decline.

Note: The flux determined through zone budgeting for the MCP open-pit copper mine is 224.9 m^3/day.

7.8 INFLUENCING FACTORS

Geological and hydrological details are collected from the field for the model input. Since modelling input parameters are based on these real-field factors, their scrutiny became essential. When analysed, it is observed that *geological factors* and *hydrological factors* are intertwined and modelling results are correlated with. The influencing factors concerning the Malanjkhand mine are elaborated further in the following paragraphs.

7.8.1 Geological Factors

The local and regional geology of the MCP study area indicated that in the 10 km radius area, i.e. core zone (CZ) and buffer zone (BZ) area, Archaean rocks (*granite, conglomerate, arkosic grit and quartz with pelitic intercalations*) and basement rocks (*gneisses, schists and granitoid*) are dominant. In such rock types, either a high rate of recharge nor acute seepage conditions, from mine slopes or pit walls, have been observed in the past. As such, in hard rock formations, non-homogeneous conditions are witnessed, and the groundwater is contained in fractures and cracks. In general, the accumulated water in the mining pit is not larger, barring the rainy season of the monsoon. Water management is comparatively easier and not that tough.

'Aquifer' of any area is characterized by its groundwater quantity as well as groundwater flow direction to ensure its availability in the region. The aquifer of the MCP study area in which groundwater is contained includes Granitic rocks/Gneisses/Alluvium. Two aquifer types that are present here are:

- Shallow aquifer: Unconfined (areas with basaltic rock formations)
- Deep aquifer: Semi-confined (areas with granitic and archaic formations)

At MCP, the unconfined aquifer at shallow depth lies in immediate contact with irregular pit geometry. The MCP is located in the catchment boundaries of the *Banjar River* (a tributary of the Narmada river) and is a hard rock formation (basalt, granite and gneisses). Topographic depressions, the nature and extent of weathering and the presence of rock joints/fractures, which play an important role, are causing the movement of groundwater in the shallow aquifer. Similar groundwater flow in the deeper cracks is observed as well. Field investigation further indicated that the thickness of the aquifer at MCP ranges from 11 to 42 m (shallow to deep) and transmissivity (T) value from 62.12 to 237.63 m^2/day (average = 148.04 m^2/day) (CSIR-CIMFR, 2018). Accordingly, the hydraulic conductivity (K) value of 1.2E-5–1.2E-6 m/s (0.000012–0.0000012) has been assigned to the model. This K value of <1 m/d is good enough for the study area dominated by the Archaean rocks. However, basaltic rocks, present in many places, have higher K values (5–15 m/d). All these aquifer-related parameters, i.e. thickness, K and T values, guide us to know about the groundwater availability in the area (CSIR-CIMFR, 2018). The discontinuity and anisotropy in the MCP aquifer systems, induced fractures, caused either naturally or by the mine blasting process will play a major role because ore and waste-rock extraction in the mines comes across such common conditions.

7.8.2 Hydrological Factors

This influencing factor is related to the topography, drainage and existing water bodies in the area. The topography of the area is undulating where MCP is located and the surface drainage pattern of the mine area is dendritic. No backflow of water is allowed into the mining pit. As the mine workings are extended below the water table, the phreatic surface has attained the shape of topography (terrain) naturally. As far as water bodies near the mining excavations are concerned, the field studies indicated that the area has no important and prominent water body except a medium-sized pond called *Karamsara pond*, which is a stagnant surface water body mainly for local irrigational and pisciculture purposes. The perennial *Banjar River,* though falls in the buffer zone, flows more than 4 km (aerial distance) from the mine site) (CSIR-CIMFR, 2018). From these nearby water bodies, groundwater recharge and augmentation are taking place and groundwater movement, i.e. quantity and flow direction, is influenced.

Analysing further, the WLFs in the surface dug wells (open to the sky) of the area show that the seasonal fluctuation of groundwater level occurs, and it varies from 4.3

TABLE 7.3
Water Level Fluctuation in Pre-Monsoon and Post-Monsoon Season

S. No.	Name of Villages Core Zone = CZ Buffer Zone = BZ	Location Details			Depth of Water Table (DWT) and Water Level Fluctuation (WLF) in m		
		Latitude	Longitude	MSL (m)	Pre-Monsoon (June 2018)	Post-Monsoon (November 2017)	WLF (m)
1.	Bhimjori(CZ) *MethodistChurch*	22°02′37.150″N	80°43′39.020″E	535.87	7.5	4.94	2.56
2.	Khursipar(CZ)	22°01′07.270″N	80°43′59.237″E	506.20	7.67	5.4	2.27
3.	Borekheda(BZ)	21°59′40.787″N	80°43′29.151″E	491.61	10.27	8.48	1.79
4.	Birsa(BZ)	22°00′21.592″N	80°44′27.504″E	508.55	4.30	2.26	2.06
5.	Darbaritola(BZ)	21°59′53.484″N	80°44′19.587″E	524.9	8.4	3.95	4.45
6.	Nayatola(BZ)	21°59′21.810″N	80°44′25.130″E	518.6	10.49	3.67	6.87
7.	Kaingatola(BZ)	21°59′17.469″N	80°44′50.459″E	525.10	6.2	3.90	2.30
8.	Mararitola(BZ)	22°00′04.604″N	80°44′57.686″E	572.3	7.65	3.90	3.75
9.	Bhatlai(BZ)	22°01′12.752″N	80°44′35.705″E	526.0	11.7	6.55	5.15
10.	Dudhi(BZ)	22°00′49.120″N	80°44′35.705″E	511.46	10.3	4.34	5.96
11.	Sakha(BZ)	22°00′48.432″N	80°46′43.724″E	518.03	9.1	4.60	4.5
12.	Chinchgaon(BZ)	21°59′32.593″N	80°47′24.604″E	517.49	7.8	1.00	6.8
13.	Pipartola(BZ)	21°58′11.254″N	80°46′18.752″E	529.96	8.4	4.20	4.2
14.	Bhurruk(BZ)	21°57′42.340″N	80°40′36.626″E	488.72	5.1	3.7	1.4
15.	Mohgaon(BZ)	22°03′16.872″N	80°41′02.550″E	515.45	5.7	3.2	2.5
16.	Pauni(BZ)	22°02′48.646″N	80°41′37.491″E	504.57	11.8	7.9	3.9
17.	Chinditola(CZ)	21°59′46.107″N	80°41′26.455″E	522.10	13.8	6.90	6.9
18.	Suji(BZ)	22°00′28.516″N	80°40′51.021″E	519.4	7.0	2.0	5.0
19.	Karamsara(CZ)	22°1′12.190″N	80°41′49.911″E	527.15	9.20	3.4	5.8
		22°1′44.977″N	80°41′47.711″E	520.23	9.55	2.0	7.55
20.	Darjitola(BZ)	22°04′18.09″N	80°42′15.604″E	505.12	8.09	4.1	3.99
21.	Jagtatola(BZ)	22°0′26.041″N	80°43′50.299″E	519.4	9.8	5.54	4.26

Source: CSIR-CIMFR, 2018.
Note: Average WLF = 4.2709 m; range of WLF = 1.4–6.9 m.

to 13.8 m in pre-monsoon season and from 1.0 to 8.48 m in the post-monsoon season (Table 7.3). The average WLFs being 4.2709 m (maximum −6.9 m at *Chhinditola village* in the CZ and minimum −1.4 m at *Bhurruk village* in the BZ) have been observed (Table 7.3) (CSIR-CIMFR, 2018). This means that WLF and groundwater presence both are concomitant features, characterizing the MCP mine area. The project area receives rainfall from the S-W monsoon from June to September months. The annual average rainfall of the area is 1150.90 mm (based on the 17 years of data from 2001 to 2017), as recorded at the HCL observatory located about 2.5 km from

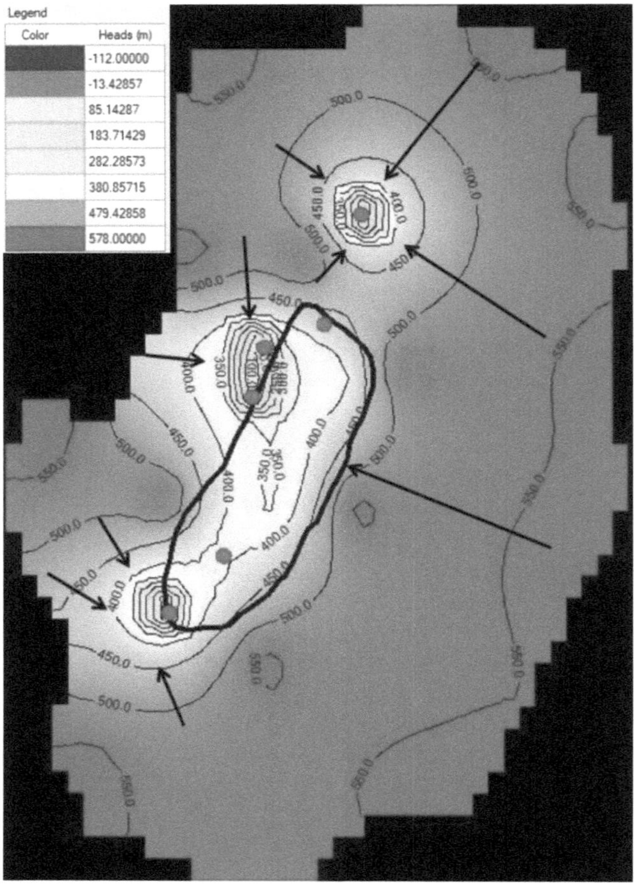

FIGURE 7.9 Model showing a fully excavated pit with no water and with all underground structures.

the mine site (CSIR-CIMFR, 2018). The study site/region, located in the central part of India, is characterized by temperate climate conditions. Central India's climate is neither very severe nor less. This climatic zone has an average daily temperature (30°C–40°C), humidity and wind speed, season-wise. All these prevailing climatic conditions make the area, semi-arid to humid causing normal water evaporation losses from the pit. Such conditions are, therefore, favourable for groundwater occurrences and availability. Considering all these factors, normal hydrological conditions have been considered for the model simulation input parameters. Notwithstanding the regional hydraulic gradient, in general, our analysis shows that the groundwater flow is towards the pit largely (Figures 7.6–7.8). In Scenario 4, when the fully excavated pit is dry and all the underground structures are completed (Figure 7.9), the groundwater direction towards the open pit indicates that water will be accumulated in the pit. In the water-filled pit, with no underground structures in place (Figure 7.5), the model shows that the flow direction is away from the pit. Thus, with and without the underground structures (Scenario: 1 and 4), it is evident that the direction of groundwater flow is impacted (Figure 7.5 vs Figure 7.8).

When water inflow conditions in the under-construction shafts, incline/decline at the MCP, are observed through field inspection, this fact (disturbance) has been noticed and verified. Broadly, low/moderate/heavy inflow conditions (low $= <05$ L/ft^2/min; moderate $= >05$ <10 L/ft^2/min and heavy $= >20$ L/ft^2/min) have been noticed. The groundwater seepage, i.e. inflow conditions, as grouped in the three mentioned categories, keeps on changing with time. During 2019–2022, these underground structures were at the construction stage and recorded observations were real.

Analysis of hydraulic gradient and transmissivity of strata confirms that the groundwater movement in future will occur and vary dynamically as the underground mine develops with time. The higher degree of uncertainty in the mining/geological domain greatly influences the dynamic groundwater movement process. Any groundwater system has a level of complexity under which it behaves deterministically. There is a critical level beyond which it becomes unstable, being able to change behaviour unexpectedly. The margin between the *operative level of complexity* and *critical complexity* has been predicted in this analysis through a modelling approach. The seepages to and fro, from the mine pit to underground workings, will take place and the realignment of groundwater forces will be a regular phenomenon. It is evident and inferred also that till a steady-state condition is attained, slow and gradual changes in groundwater movement are likely to be witnessed, requiring periodical monitoring. The seepage inflow of water, visible at the pit mine or in the underground workings, is of the medium order (neither more nor less), indicating that geological and hydrological factors both contribute to the groundwater movement and the realignment act. The aquifer encountered at MCP and the groundwater forces are impacted in the study area. Considering the dry-pit vs water-filled pit (Scenario 3 and 4; Scenario 1 and 2), the modelling results (output) show that groundwater flow directions are realigned and reoriented, whose main reason is the creation of underground structures in the vicinity (Figures 7.5–7.8).

To develop the open mine (pit) into a pit lake (end use of mine/value addition), the extent of the quarried area and mine depth are extremely important. Both these parameters continuously vary during mining activity. Now, these factors have been stabilized largely because the pit mine at MCP has reached the last phase with no extended ore extraction further. When the pit is converted into a pit lake (Soni et al., 2014), the need for the flow, i.e. its direction, will be essential for such conversion. However, water filling of a pit (either partial or full) could be done naturally or artificially depending on the surface drainage and groundwater inflow, the quarry floor levels (RLs and MRLs) and the surrounding area elevations.

7.9 CONCLUSIONS

Modelling of the groundwater flow regime for the twin mining system of MCP, as attempted in this study, concludes that groundwater disturbance, an act called groundwater realignment, has been observed at MCP. When large-sized and deep underground excavations, e.g. shafts/ incline /decline, are made near an open-pit mine, the groundwater movement, as happening naturally, gets disturbed. The predictive modelling results show that geo-hydrological factors are mainly influencing factors for the groundwater flow direction. The groundwater movement for different scenarios, when evaluated, indicates that the directions of groundwater and water levels are changed at the MCP. Such changes in the study area will be noticeable in the near future as well.

The groundwater quantity in the MCP mining pit is large, and the recharge occurring (224.9 m^3/day) is available for recycling and miscellaneous uses, thereby making the best value addition of the mine water. The research findings also indicate that it is not feasible to convert the Malanjkhand open-pit mine into a *mine-pit lake* as the end use of the mine because of the groundwater flow direction and the hydrostatic head/pressure that may act further as a cause of the water accumulation in the open pit (inundation). However, other end use of mined-out pits for miscellaneous applications may be tried, after a detailed feasibility assessment or scientific evaluation. A barrier pillar/crown pillar of 40 m in between, the open pit and an underground mine, is essentially required to be left intact. This barrier consists of natural rocks in situ and may have consequences on the groundwater flow because the rock barrier contains shale/clay/or other non-porous material exposed to the water underground. Overall, the groundwater movement at the studied site is complex and dynamic and keeps on altering directions. Therefore, the impact of mining on the groundwater regime is inevitable and needs to be regularly monitored for a deeper and better understanding of the groundwater scenario in the MCP area.

ACKNOWLEDGEMENTS

The authors thank the Director, CSIR-CIMFR (Central Institute of Mining & Fuel Research) and the Director, CSIR-NEERI (National Environmental Engineering Research Institute) for their support and encouragement. We are also thankful to Dr A K Raina, Chief Scientist, CSIR-CIMFR, Nagpur, for carefully editing and refining the manuscript. The help received from the project authorities particularly, Sri B.K. Gupta, Assistant General Manager (Geology), Hindustan Copper Limited HQ, Kolkata (W.B.), who was previously at Malanjkhand Copper Project (MCP), Balaghat, MP, India, for the help rendered during the fieldwork to collect and collate the data is duly acknowledged.

REFERENCES

A.K. Soni, B Mishra, S. Singh (2014), Pit lakes as an end-use of mining: A review. *Journal of Mining and Environment, Iran,* 5(2), pp. 99–111.

CSIR- CIMFR (2018), *Hydro-Geological Study of Malanjkhand Copper Project*. Project No. SSP/N/235/2017-18. Hindustan Copper Limited, District Balaghat, MP.

D. Kurukulasuriya, W. Howcroft E. Moon, K. Meredith, W. Timms (2022), Selecting environmental water tracers to understand groundwater around mines: opportunities and limitations. *Mine Water and the Environment*, 41(2), pp. 357–369. https://doi.org/10.1007/s10230-022-00845-y

GEC (1997), *Groundwater Estimation Methodology (GEC '97)*. Central Groundwater Board (CGWB), New Delhi, Government of India. https://cgwb.gov.in/ (accessed 8 July 2021).

HCL (2021), Tamralipi. *Journal of Hindustan Copper Limited*, 77, p. 11.

J.P. Fernández-Álvarez, L. Álvarez-Álvarez, R. Díaz-Noriega (2016), Groundwater numerical simulation in an open pit mine in a limestone formation using modflow. *Mine Water Environment,* 35, pp. 145–155. https://doi.org/10.1007/s10230-015-0334-8

K. R. Karanth (1993), *Hydrogeology.* McGraw Hill, New Delhi.

K. S. More, C. Wolkersdorfer, N. Kang, A. S. Elmaghraby (2020), Automated measurement systems in mine water management and mine workings: A review of potential methods. *Water Resources and Industry*, 24, 100136, https://doi.org/10.1016/j.wri.2020.100136

K. Samuel More, C. Wolkersdorfer (2022), *Predicting and Forecasting Mine Water Parameters Using a Hybrid Intelligent System, Water Resources Management.* Springer, New York. https://doi.org/10.1007/s11269-022-03177-2

L. Surinaidu, V.V.S. Gurunadha Rao, N. Srinivasa Rao, S. Srinu (2014), Hydrogeological and groundwater modelling studies to estimate the groundwater inflows into the coal Mines at different mine development stages using MODFLOW. *Andhra Pradesh, India, Water Resources and Industry*, 7–8, pp. 49–65. https://doi.org/10.1016/j.wri.2014.10.002

M. G. McDonald, A. W. Harbaugh (1988), *A Modular Three-Dimensional Finite-Difference Ground-Water Flow Model.* U.S. Geological Survey Techniques of Water-Resources Investigations.

N. Rapantova, A. Grmela, D. Vojtek, J. Halir, B. Michalek (2007), Groundwater flow modelling applications in mining hydrogeology. In: R. Cidu & F. Frau (Eds), *IMWA Symposium 2007: Water in Mining Environments*, 27–31 May 2007, Cagliari, Italy, pp. 349–353.

8 Metals Contamination of Water and Their Influences on Human and Aquatic Life

Mihaela Timofti

8.1 INTRODUCTION

Because the presence of the toxic metals (TMs) is a human risk assessment, it becomes more and more necessary to analyze the presence of these metals in environmental samples (Ali et al., 2023; Çamur et al., 2021; Shephard et al., n.d.; Vardhan et al., 2019). Some of the transition metals such as Zn, Cu, and Cr are vital for living organisms because of their participation in biological processes. Other transition metals such as Pb and Cd serve no biological purpose and pose significant risks, even if they are in very small quantities (Çamur et al., 2021; González-Muñoz et al., 2008; Li et al., 2020; Moody et al., 2020).

For the survival and proper functioning of living organisms, several metals are crucial as they play indispensable roles in various biological processes. They serve as cofactors in enzymes, facilitate electron transfers, and play crucial roles in various biochemical pathways. Proper intake and regulation of these metals are important for maintaining good health and preventing deficiencies or toxicities. Some of the necessary metals in living organisms include:

Chromium (Cr) because it supports glucose metabolism and insulin function.
Cobalt (Co) because it is a component of vitamin B12, and it is essential for red blood cell formation and nervous system function.
Copper (Cu) because of its involvement in electron transfer reactions, enzyme activity, and iron metabolism.
Iron (Fe) because of its essential role in transporting oxygen in hemoglobin and its catalytic role in various enzymatic reactions.
Manganese (Mn) because of its requirement in enzyme activation and its role as antioxidant defence.
Molybdenum (Mo) because of its participation in enzymatic reactions involved in nitrogen metabolism.
Nickel (Ni) because of its certain enzymatic activities.
Zinc (Zn) because of its role in enzyme function, gene regulation, and immune system support.

DOI: 10.1201/9781003442554-10

TMs with no biological purpose are elements that have no known essential function in living organisms and can be harmful or toxic even at low levels. These metals can accumulate in the body over time, leading to various health issues and pose significant health risks, and exposure to them should be minimized to protect human health and the environment. Regulatory measures and awareness campaigns are in place to reduce exposure to these hazardous substances. Some of the TMs with no biological purpose include:

Aluminum (Al): Although is a common element, excessive exposure through sources such as antacids or certain cooking materials can lead to neurotoxicity and has been associated with Alzheimer's disease, although the link is still not entirely clear (Crisponi et al., 2012a; Zhao et al., 2020).

Arsenic (As): Found in groundwater in some parts of some countries and certain foods, long-term exposure to arsenic can lead to skin, lung, and bladder cancer, as well as cardiovascular and neurological issues (Ahsan et al., 2006; Chen et al., 2009; Chung et al., 2014).

Beryllium (Be): Primarily used in certain industries, beryllium can cause chronic lung disease and skin allergies when inhaled or contacted (Omeed & Raja, 2017; Saltini & Amicosante, 2001).

Cadmium (Cd): Often present in industrial processes, batteries, and certain foods, cadmium can damage the kidneys, bones, and respiratory system (Hernández-Cruz et al., 2022; Satarug, 2018).

Lead (Pb): Found in certain industrial materials, paints, and contaminated water, lead exposure can cause developmental and neurological damage, especially in children (Zhang et al., 2023).

Mercury (Hg): Released into the environment through industrial activities, especially in coal-fired power plants, mercury can contaminate fish and other seafood. Mercury exposure can harm the nervous system, particularly in foetuses and young children (Joorabian Shooshtari et al., 2023; Sakamoto et al., 2017).

8.2 TOXIC METALS WITH NO BIOLOGICAL PURPOSE

8.2.1 Aluminum (Al)

Aluminum toxicity refers to the harmful effects that excessive exposure to this metal can have on living organisms, including humans and aquatic life (Crisponi et al., 2012b; Shuhaimi-Othman et al., 2013a). This metal is a naturally occurring element, and its presence in the environment is widespread (Liu et al., 2024). However, human activities, such as industrial processes, mining, and the use of aluminum-containing products, can lead to its increased levels in certain areas (Brough & Jouhara, 2020; Donoghue et al., 2014; Matúš et al., 2004).

Aluminum toxicity has some potential health effects in both humans and aquatic life. In humans, it can have: breast cancer risk, neurological effects such as memory loss, cognitive impairment, motor dysfunction (Crisponi et al., 2012b; Hao et al., 2023; Liu et al., 2022; Moussaron et al., 2023), bone and skeletal issues due to the interference with bone mineralization (Maya et al., 2016), respiratory issues such as coughing and shortness of breath (Crisponi et al., 2012b), gastrointestinal problems including nausea, vomiting, and stomach pain, and impaired kidney function (Iqbal et al., 2022).

In aquatic life, aluminum toxicity can also have adverse effects when present in elevated concentrations in water bodies. While some aquatic organisms can tolerate low levels of aluminum, excessive exposure can be harmful to various species (Shuhaimi-Othman et al., 2013b). The impact of aluminum toxicity on aquatic life is referring to: impaired growth and reproduction including fish, invertebrates, and algae (Liu et al., 2022), damaged gills and respiration leading to impaired respiration and oxygen uptake (Gokul et al., 2023), changes in water quality affecting nutrient cycling and ecosystem dynamics (Yan et al., 2022), and the accumulation within the cell wall and intercellular gaps of plant leaves (Rzayev et al., 2022).

8.2.2 Arsenic (As)

Arsenic toxicity refers to the harmful effects of excessive exposure to arsenic, a naturally occurring element that can be found in the environment. Arsenic is widely distributed in the earth's crust and can be released into the environment through both natural processes such as dissolution of minerals containing arsenic or volcanic emissions, and human activities, such as mining, industrial processes, and the use of certain agricultural chemicals (Marchant et al., 2017; Mishra et al., 2022; Pandey et al., 2023; Punshon et al., 2017; Simionov et al., 2021a). Various routes of exposure to arsenic include ingestion of contaminated water, food, or air, as well as dermal contact with arsenic-containing substances (Chung et al., 2014). Chronic exposure to low levels of arsenic is a significant concern as it can accumulate in the body over time and can cause human health issues (Dutta et al., 2015; Rodríguez et al., 2010; Yang et al., 2021).

Arsenic toxicity in humans refers to: skin issues including hyperpigmentation, lesions, and a condition known as "arsenic keratosis" (Ahsan et al., 2006; Chung et al., 2014), respiratory problems such as coughing and shortness of breath (Chen et al., 2009; Clancy et al., 2012), gastrointestinal issues including nausea, vomiting, abdominal pain, and diarrhoea (Bobak & Guerrant, 2015), cardiovascular effects including heart disease and hypertension (Kaufman et al., 2021; Kuo et al., 2022), neurological effects due to the fact that arsenic can cross the blood–brain barrier and affect the central nervous system leading to neurological symptoms such as peripheral neuropathy and cognitive impairments (Shayan et al., 2023), increased cancer risk including skin, lung, bladder, and liver cancer (Clancy et al., 2012; Kuo et al., 2022; Matthews et al., 2019).

It's important to note that arsenic toxicity in humans is a significant public health concern, especially in areas with contaminated water sources (Kumar et al., 2020; Nguyen & Mulligan, 2023; Zhang et al., 2011).

Aquatic organisms can be exposed to arsenic through direct uptake from water or through the food chain (Córdoba-Tovar et al., 2022; Zhang et al., 2022). The presence of arsenic can disturb the growth of plankton, crustaceans, and molluscs, while also reducing photosynthetic processes, which in turn affects the ecological dynamics and productivity of aquatic ecosystems (Córdoba-Tovar et al., 2022).

8.2.3 Cadmium (Cd)

Cadmium is not a naturally occurring element but is a by-product of other metal mining and refining processes. Cadmium toxicity refers to the harmful effects of excessive exposure to cadmium, a toxic heavy metal that is released into the environment through various human activities and the use of cadmium-containing products such as jewellery (Nowak & Chmielnicka, 2000; Salles et al., 2021; Shephard et al., n.d.; Shuhaimi-Othman et al., 2013).

Due to its toxicity, cadmium can cause a range of health issues in humans through the inhalation of cadmium dust and fumes, consumption of cadmium-contaminated food and water, or skin contact with cadmium-containing materials. (Ngure & Kinuthia, 2020; Salles et al., 2021). The respiratory system can become irritated through the inhalation of cadmium dust or fumes, potentially causing symptoms such as coughing and shortness of breath (Shafaghat et al., 2016). The presence of cadmium in human body can cause gastrointestinal disturbances, including nausea, vomiting, abdominal pain, and diarrhea (Salles et al., 2021; Schaefer et al., 2022). Chronic exposure to elevated cadmium levels may lead to kidney damage because of its strong affinity to this organ and can interfere with bone metabolism and affect bone density (osteoporosis) when long-term exposure it is made (Cirovic et al., 2022; Minzala et al., 2022). Some studies suggest that chronic cadmium exposure may have adverse effects on reproductive health in both men and women (Ramezanifar et al., 2023). Also, prolonged exposure to cadmium has been linked to a higher risk of developing certain cancers, such as lung and prostate cancer. (Ngure & Kinuthia, 2020; Salles et al., 2021; Shafaghat et al., 2016).

In terms of aquatic life, cadmium is a major environmental concern. Cadmium can enter water bodies through industrial discharges, urban runoff, and agricultural activities (Kasbaji et al., 2022; Rocha et al., 2023). Once released into water, cadmium can be absorbed by aquatic organisms, and it tends to accumulate in the food chain, becoming more concentrated at higher trophic levels (Das et al., 2023; Lazăr et al., 2024; Simionov et al., 2021b). High cadmium concentrations in water can negatively affect the growth and reproductive success of aquatic organisms, including fish, invertebrates, and algae. Cadmium toxicity can damage the gills and respiratory organs of aquatic organisms, leading to impaired respiration and oxygen uptake (Das et al., 2023; Lee et al., 2023) and can alter the structure and dynamics of aquatic ecosystems, affecting biodiversity and ecosystem functioning (Cheng et al., 2023).

8.3 METALS THAT ARE NEEDED IN LIVING ORGANISMS

8.3.1 Chromium (Cr)

The mining operations carried out in Sukinda Valley produce a variety of pollutants containing various highly toxic and carcinogenic forms of chromium, as established by reference (Mohanty et al., 2023a; Rathinam et al., 2007). Significant amounts of chromium are released into the natural environment during the production process, including groundwater, soil, sediments, air, and waste discharges, which negatively impact agricultural fields, as noted in the reference (Mishra et al., 1970; Mohanty et al., 2023a).

Contaminated groundwater containing Cr(VI) has been identified as a substantial concern for human health in Finland, France, Germany, Italy, Switzerland, and the United Kingdom (Tiwari & De Maio, 2017; Coetzee et al., 2020a; Gunkel-Grillon et al., 2014; Tiwari et al., 2019)

Regulations governing the acceptable levels of chromium in both work environments and drinking water are in place to safeguard human health against excessive exposure. Therefore, in 2010, the Occupational Safety and Health Administration has implemented a significant reduction in the permissible exposure limit for Cr(VI) in the workplace to an exposure limit of 5 µg Cr(VI)/m³ of air (National Emphasis Program – Hexavalent Chromium (osha.gov) accessed at 22 August 2023), and the National Institute for Occupational Safety and Health (NIOSH) of the USA has recommended a 10-hour TWA exposure limit for all Cr (VI) compounds of 1 µg Cr (VI)/m³ (Chromium (Cr) Toxicity|ATSDR—CSEM (cdc.gov) accessed at 22 August 2023). Also, the Scientific Committee on Occupational Exposure Limits, which was set up in 1995 by the European Commission, assesses the possible health consequences of being exposed to chemicals in the workplace and their recommendations have served as the foundation for regulatory actions concerning occupational exposure limits. These limits are established at the Union level in accordance with the Chemical Agents Directive (Directive 98/24/EC) and the Carcinogens and Mutagens Directive (Directive 2004/37/EC) to safeguard workers from chemical hazards. Industries employing chromium should prioritize occupational safety practices, including adequate ventilation, the use of protective gear, and adherence to specific handling procedures. Furthermore, individuals can lower their risk of chromium toxicity by refraining from consuming water or food that may be contaminated. Should concerns arise regarding potential chromium exposure, it is imperative to seek medical attention and adhere to safety guidelines diligently in order to mitigate associated risks.

A prevalent element in the earth's crust and seawater, chromium exists in nine distinct oxidation states in our environment (Monga et al., 2022), primarily as metallic (Cr^0), trivalent (Cr^{+3}), and hexavalent (Cr^{+6}) chromium.

The hexavalent form, predominantly produced through the oxidation of the more common trivalent chromium, is highly water-soluble across the entire pH range and profoundly toxic (Eid & Ibrahim, 2021; Shokri & Fard, 2022). Its ability to permeate biological membranes in the human body leads to a range of mutagenic and carcinogenic effects. (Mohanty et al., 2023a) On the other hand, trivalent chromium readily forms an insoluble precipitate, known as [$Cr(OH)_3$], in both alkaline and neutral conditions being present in most foods and nutrient supplements (Cefalu & Hu, 2004; Shokri & Fard, 2022). It is an essential nutrient with minimal toxicity (Cefalu & Hu, 2004; Eid & Ibrahim, 2021) and plays a role in regulating glucose, lipid, and protein metabolism (Mohanty et al., 2023a). The environmental concentrations of Cr(III) are on the rise due to the increased manufacturing of leather, textiles, and steel (Mortada et al., 2023).

Chromium toxicity can result from both occupational and environmental exposures, and its toxicity refers to the harmful effects that can occur when an individual is exposed to excessive amounts of chromium, particularly in its hexavalent (Cr(VI)) and sometimes trivalent (Cr(III)) forms (Boşgelmez & Güvendik, 2017; Coetzee et al., 2020b). Hexavalent chromium is often produced during industrial processes, such as chrome plating, stainless steel production, and leather tanning (Ariram et al., 2022; Guan et al., 2023; Liu et al., 2022; Mekicha et al., 2020; Pradeep et al., 2021). It can also contaminate water sources due to improper disposal of industrial waste (Mohanty et al., 2023b).

Inhaling hexavalent chromium compounds can lead to respiratory system irritation, manifesting as symptoms such as coughing, wheezing, and breathlessness. Extended exposure may escalate to more serious conditions such as asthma and bronchitis (Monga et al., 2022). Direct skin contact with hexavalent chromium can result in skin irritation and allergic reactions, with symptoms such as redness, itching, and dermatitis (Bregnbak et al., 2015). Sustained exposure to elevated levels of hexavalent chromium has been linked to a higher risk of developing lung cancer (Matthews et al., 2019). Consuming significant amounts of hexavalent chromium can lead to gastrointestinal problems, including stomach ulcers, as well as harm to the liver and kidneys (Yan et al., 2020). Some research indicates that exposure to hexavalent chromium might have detrimental effects on reproductive health and potentially increase the likelihood of birth defects (Chakraborty et al., 2022; Xia et al., 2015).

Chromium toxicity can exert significant impacts on aquatic ecosystems when it is present, particularly in the hexavalent chromium (Cr(VI)) form (Mohanty et al., 2023a). The introduction of hexavalent chromium into aquatic environments can occur through multiple avenues, including industrial discharges, runoff from contaminated sites, and improper disposal of industrial waste (Kerur et al., 2020; Monga et al., 2022). Once within water bodies, chromium can have detrimental effects on aquatic life. Hexavalent chromium (Cr(VI)) has been observed to contribute to the reduction of dissolved oxygen in water, a vital element for the survival of aquatic organisms (Mortada et al., 2023). Diminished oxygen levels can lead to hypoxia, potentially suffocating fish and other aquatic species. Chromium exposure can disrupt the reproductive processes of aquatic organisms, resulting in reduced fertility rates and hindrance in the development of eggs and embryos. Due to its toxicity, chromium can alter the behavior of aquatic organisms, affecting their feeding, mating, and predator avoidance patterns (Mortada et al., 2023). Chromium can bioaccumulate in the tissues of aquatic organisms, particularly in filter-feeding species such as mussels and clams (Aharchaou et al., 2022). As predators consume these contaminated organisms, chromium can biomagnify up the food chain, leading to higher concentrations in top predators. Fish and other aquatic organisms can experience damage to their gills, liver, and other organs when exposed to elevated levels of hexavalent chromium (Jiang et al., 2021). This damage can impair their overall health and survival. The toxicity of chromium can disrupt aquatic food chains, affecting the abundance and distribution of various species within the ecosystem (Han et al., 2023).

In summary, hexavalent chromium in aquatic environments can lead to a range of adverse effects on aquatic life, from oxygen depletion and reproductive disruption to behavioral changes and damage to vital organs, ultimately affecting the balance and dynamics of aquatic ecosystems.

8.3.2 Cobalt (Co)

Cobalt compounds predominantly exist in two valence states: cobaltous (Co^{2+}) and cobaltic (Co^{3+}). The cobaltous state is more commonly available and widespread in both commercial and environmental settings. (Leyssens et al., 2017). In small quantities, cobalt is a vital trace element needed by living organisms to support various biochemical processes. It is a component of vitamin B12 (cobalamin), which plays a crucial role in red blood cell formation, nervous system function, and DNA synthesis (Arbo et al., 2022; Biuu et al., 1964; Hypotkm et al., 1992; Santos et al., 2023). In appropriate quantities, cobalt is not toxic and is necessary for good health. However, cobalt can become toxic when exposure exceeds the body's ability to handle it or when individuals are exposed to high levels of certain cobalt compounds (Leyssens et al., 2017). Cobalt toxicity is more commonly associated with occupational exposure such as professional motorcyclists or workers from industry of manufacturing of certain alloys, batteries, and pigments (Arbo et al., 2022; Broding et al., 2009; Carvalho et al., 2018; Nordberg, 1994).

Cobalt toxicity may lead to some potential health effects such as: respiratory issues, allergic reactions, cardiotoxicity, thyroid dysfunction, and neurological effects. For example, the inhalation of cobalt by diamond polishers through dust, fumes, or aerosols has been shown to irritate the respiratory system, leading to symptoms such as coughing, wheezing, and shortness of breath. (Demedts et al., 1984; Leyssens et al., 2017). Some individuals may develop allergic reactions to cobalt, especially when in contact with cobalt-containing metals (Broding et al., 2009). Prolonged exposure to high levels of cobalt can lead to cardiomyopathy, a condition where the heart muscle becomes weakened and less efficient in pumping blood (Jarvis, 1992) or may affect the thyroid function, affecting hormone production and metabolism (Lantin et al., 2011; Li et al., 2023). In extreme cases, cobalt toxicity can lead to neurological symptoms, including difficulty walking, tremors, and impaired coordination due to cobalt's capability to traverse the highly selective blood–brain barrier and accumulate within the brain (Leyssens et al., 2017).

It's important to note that cobalt toxicity is rare in everyday life and is more likely to occur in industrial settings where exposure levels are significantly higher. Regulatory measures and workplace safety guidelines are in place to limit cobalt exposure and protect workers from potential toxicity. The Cobalt Institute (CI) acknowledges the revised harmonized classification regulations for cobalt metal within the European Union, as outlined in the 14th Adaptation to Technical Progress (ATP) to the Classification, Labelling, and Packaging (CLP) Regulation. These regulations were officially published in the EU Official Journal on February 18, 2020, introducing a temporary Generic Concentration Limit (GCL) set at $\geq 0.1\%$. These updated rules came into effect on March 9, 2020, and are slated to be enforced starting from October 1, 2021 (Registry of CLH intentions until outcome – ECHA (europa.eu), accessed at 29 August 2023).

The toxicity of cobalt can impact aquatic life when present in elevated concentrations in water bodies. Cobalt is naturally found in aquatic environments in trace amounts, but anthropogenic activities, such as industrial discharges and mining activities, can lead to an increase in cobalt levels in water, soil, and sediment (Leyssens et al., 2017; Popa et al., 2023; Simionov et al., 2023; Vasquez Ugaz et al., 2023).

Elevated cobalt levels can alter the chemical composition of water, potentially affecting the overall water quality and leading to ecological imbalances. Fish exposed to high concentrations of cobalt can experience various health issues because cobalt can accumulate in their tissues and lead to adverse effects on their organs, including the liver and kidneys (Ali et al., 2023b; Lall & Kaushik, 2021). This accumulation can also disrupt normal physiological functions and lead to reduced growth, impaired reproduction, and increased mortality in fish populations (Alves et al., 2022). In the case of molluscs and crustaceans, cobalt toxicity lead to reduced feeding rates, impaired reproduction, and changes in behavior when they are exposed to elevated cobalt levels (Zaynab et al., 2022). The presence of cobalt in water among the other trace metals with toxic potential may have a direct impact on carbon cycle or nutrient cycle; thus, the aquatic systems will be directly impacted ((Yan et al., 2022). Also, cobalt toxicity can disrupt aquatic food chains due to the established fact that if certain species at lower trophic levels are affected, it can have cascading effects on other organisms that depend on them for food (Zaynab et al., 2022).

8.3.3 Copper (Cu)

In everyday life, most people get adequate copper through their diet without experiencing toxicity. Copper from food sources is usually regulated well by the body, and any excess is excreted. However, excessive consumption of copper supplements such as B12 vitamin or accidental ingestion of copper-containing materials can lead to toxicity (Rebelo et al., 2023). Observational data have established a connection between elevated copper levels in the blood and an increased susceptibility to cardiovascular diseases (Jäger et al., 2022).

In small quantities, copper is a vital trace element necessary for living organisms to support various biological functions, including enzyme activity and lipid metabolism. (Jäger et al., 2022). In proper quantities, copper is necessary for good health and is involved in various physiological processes. However, copper can become toxic when there is an excessive buildup in the body or when individuals are exposed to high levels of copper through various sources (Jaishankar et al., 2014; Loredo et al., 2008). Copper toxicity can occur due to various reasons, including occupational exposure, excessive consumption of copper supplements, or ingestion of contaminated food or water (Copper—Consumer (nih.gov), accessed on 29 August 2023) (Haase et al., 2021; Jaishankar et al., 2014; Simionov et al., 2022).

Copper toxicity has some potential health effects such as: gastrointestinal issues (nausea, vomiting, and stomach cramps), liver damage (liver inflammation (hepatitis) and impaired liver function), haemolytic anemia (destruction of red blood cells), neurological symptoms (confusion, disorientation, and seizures), and renal dysfunction (impair kidney function or even kidney damage) (Borak et al., 2000).

It is important to note that copper toxicity is relatively uncommon and is more likely to affect specific vulnerable groups, such as individuals with Wilson's disease (a genetic disorder affecting copper metabolism) or those exposed to occupational hazards involving copper. (Haase et al., 2021; Jaishankar et al., 2014; Kasztelan-Szczerbinska & Cichoz-Lach, 2021). If someone suspects that they have been exposed to high levels of copper or experiences symptoms related to copper toxicity, they should seek medical attention promptly. Healthcare professionals can assess the situation, provide appropriate treatment, and advise on further preventive measures.

When copper is present in elevated concentrations in water bodies, its toxicity can have significant adverse effects on aquatic life. Copper is a common metal that can enter aquatic ecosystems through various sources, including mining activities, industrial discharges, agricultural runoff, and natural weathering of rocks and soil (Haase et al., 2021; Jaishankar et al., 2014). Copper can bind to sediment particles, accumulating in the bottom of water bodies. This can result in long-term exposure to bottom-dwelling organisms and pose risks to benthic communities (Loredo et al., 2008). Fish are particularly sensitive to copper toxicity because elevated copper levels in water can impair their gill function, leading to difficulty in obtaining oxygen, and disrupt their ion regulation (Anderson & Spear, 1980) This can result in reduced growth, impaired reproduction, and increased mortality in fish populations or other aquatic organisms (Malhotra et al., 2020; Zaynab et al., 2022). Copper can inhibit photosynthesis in aquatic plants and algae, which are crucial for maintaining the oxygen levels and overall health of aquatic ecosystems (Yong et al., 2021). Copper toxicity can also disrupt the aquatic food chain. If certain species at lower trophic levels are affected, it can have cascading effects on other organisms that depend on them for food (Kumar et al., 2020).

8.4 CONCLUSIONS AND RECOMMENDATIONS

To protect both human health and aquatic ecosystems from the harmful effects of aluminum, arsenic, cadmium, chromium, cobalt, and copper toxicity, it is crucial to manage and control sources of this TM pollution.

Regulatory measures and proper waste management are essential to minimize Al, As, Cd, Cr, Co and Cu discharges into the environment (EU Restriction of Hazardous Substances Directive (RoHS): An Essential Guide (compliancegate.com), accessed on 31 August 2023, REACH Annex XVII Substances List: An Overview (compliancegate.com), accessed on 31 August 2023, List of Products Covered by the EU Battery Directive (compliancegate.com), accessed on 31 August 2023, (EC Regulation No 1223, 2009). Additionally, sustainable practices and awareness of the potential impacts of these TMs are essential to preserve the health and balance of ecosystems.

Efforts are made to control industrial discharges and minimize TM contamination in the environment. Environmental monitoring programs are essential to assess Al, As, Cd, Cr, Co, and Cu levels in water bodies and ensure that they remain within safe limits for aquatic organisms.

Additionally, sustainable practices and awareness of the potential impacts of Al, As, Cd, Cr, Co, and Cu on aquatic ecosystems are crucial to safeguarding the health and balance of these environments.

To manage and control the sources of TM pollution, such as aluminum (Al), arsenic (As), cadmium (Cd), chromium (Cr), cobalt (Co), and copper (Cu), it requires a comprehensive approach. Regular assessment and adaptation of strategies based on new research findings and technologies are essential for effective long-term management of TM pollution. In the following, some recommendations will be presented for reducing the amount of the TMs arriving into the environment especially in water bodies.

Waste management practices implemented with effectiveness will reduce the quantity of the TMs into the environment. This can be done by encouraging recycling and proper disposal of electronic waste, batteries, and other products containing these metals and by establishing and enforcing regulations for industries to manage and treat their waste responsibly to prevent the leaching of TMs into soil and water.

Investing in the water treatment facilities and technologies for removing TMs from wastewater before it is released into water bodies by including in it advanced filtration systems and/or new coagulation and precipitation methods are other methods that will reduce the amount of the TMs released into the aquatic systems. Also, implementing regular monitoring programs will assess the quality of water sources and identify potential contamination because early detection allows for timely intervention to prevent further pollution.

By *providing incentives for companies* that invest in green practices, it will encourage industries to adopt cleaner technologies and production processes that minimize the use and release of TMs into the environment.

Development of land use plans that consider the potential impact of industrial, agricultural, and residential activities on soil quality and *elaboration of zoning regulations* can help in preventing the concentration of TMs in specific areas.

Implementing soil remediation programs in areas with existing contamination will may involve the removal and treatment of contaminated soil.

It is a well-known fact that informed and engaged communities can play a crucial role in holding industries accountable and advocating for stronger environmental protection measures. So, *increasing public awareness* about the sources and risks of TMs through educational campaigns and educating communities about the proper disposal of products containing these metals and the importance of reducing personal exposure will encourage citizen participation in monitoring and reporting potential sources of pollution.

Keep in mind that the *implementing these recommendations requires collaboration between government agencies, industries, communities, and environmental organizations.*

REFERENCES

Ahamad, A., & Kumar, J. (2023). Pyrethroid pesticides: An overview on classification, toxicological assessment and monitoring. *Journal of Hazardous Materials Advances*, *10*, 100284. https://doi.org/10.1016/j.hazadv.2023.100284

Aharchaou, I., Maul, A., Pons, M. N., Pauly, D., Poirot, H., Flayac, J., Rodius, F., Rousselle, P., Beuret, M., Battaglia, E., & Vignati, D. A. L. (2022). Effects and bioaccumulation of Cr(III), Cr(VI) and their mixture in the freshwater mussel Corbicula fluminea. *Chemosphere*, *297*. https://doi.org/10.1016/j.chemosphere.2022.134090

Ahsan, H., Chen, Y., Parvez, F., Zablotska, L., Argos, M., Hussain, I., Momotaj, H., Levy, D., Cheng, Z., Slavkovich, V., Van Geen, A., Howe, G. R., & Graziano, J. H. (2006). Arsenic exposure from drinking water and risk of premalignant skin lesions in Bangladesh: Baseline results from the health effects of arsenic longitudinal study. *American Journal of Epidemiology*, *163*(12), 1138–1148. https://doi.org/10.1093/aje/kwj154

Ali, M. U., Wang, C., Li, Y., Li, R., Yang, S., Ding, L., Feng, L., Wang, B., Li, P., & Wong, M. H. (2023). Heavy metals in fish, rice, and human hair and health risk assessment in Wuhan city, central China. *Environmental Pollution, 328*. https://doi.org/10.1016/j.envpol.2023.121604

Alves, A. C. F., Saiki, P. T. O., da Silva Brito, R., Scalize, P. S., & Rocha, T. L. (2022). How much are metals for next-generation clean technologies harmful to aquatic animal health? A study with cobalt and nickel effects in zebrafish (Danio rerio). *Journal of Hazardous Materials Advances, 8*, 100160. https://doi.org/10.1016/j.hazadv.2022.100160

Anderson, P. D., & Spear, P. A. (1980). Copper pharmacokinetics in fish gills ii body size relationships for accumulation and tolerance. *Water Research, 14*. https://doi.org/10.1016/0043-1354(80)90160-8

Arbo, M. D., Garcia, S. C., Sarpa, M., Da Silva Junior, F. M. R., Nascimento, S. N., Garcia, A. L. H., & Da Silva, J. (2022). Brazilian workers occupationally exposed to different toxic agents: A systematic review on DNA damage. *Mutation Research - Genetic Toxicology and Environmental Mutagenesis*, 879–880. https://doi.org/10.1016/j.mrgentox.2022.503519

Ariram, N., Pradeep, S., Sundaramoorthy, S., & Madhan, B. (2022). Single pot low float chromium tanning: Cleaner pathway approach to environment friendly leather manufacturing. *Process Safety and Environmental Protection, 167*, 434–442. https://doi.org/10.1016/j.psep.2022.09.024

Ashwani Kumar, T., & Marina, D. M. (2017). Assessment of risk to human health due to intake of chromium in the groundwater of the Aosta Valley region, Italy. *Human and Ecological Risk Assessment, 23*(5), 1153–1163.

Bakshe, P., & Jugade, R. (2023). Phytostabilization and rhizofiltration of toxic heavy metals by heavy metal accumulator plants for sustainable management of contaminated industrial sites: A comprehensive review. *Journal of Hazardous Materials Advances, 10*, 100293. https://doi.org/10.1016/j.hazadv.2023.100293

Biuu, B., Cooperuin, A., & Lowenstein, L. (1964). Relative folate deficiency of erythrocytes in pernicious anemia and its correction with cyanocobalamin. *Blood, 24*(5), 502–521.

Bobak, D. A., & Guerrant, R. L. (2015). Nausea, vomiting, and noninflammatory diarrhea. In J. E. Bennett, R. Dolin, & M. J. Blase (Eds.), *Principles and Practice of Infectious Diseases* (Vol. Part II, 8th edn). Churchill Livingstone, New York.

Borak, J., Cohen, H., & Hethmon, T. A. (2000). Copper exposure and metal fume fever: Lack of evidence for a causal relationship. *AIHA Journal, 61*(6), 832–836. https://doi.org/10.1080/15298660008984594

Boşgelmez, I. İ., & Güvendik, G. (2017). N-acetyl-l-cysteine protects liver and kidney against chromium(VI)-induced oxidative stress in mice. *Biological Trace Element Research, 178*(1), 44–53. https://doi.org/10.1007/S12011-016-0901-2

Bregnbak, D., Johansen, J. D., Jellesen, M. S., Zachariae, C., Menné, T., & Thyssen, J. P. (2015). Chromium allergy and dermatitis: Prevalence and main findings. *Contact Dermatitis, 73*(5), 261–280. https://doi.org/10.1111/COD.12436

Broding, H. C., Michalke, B., Göen, T., & Drexler, H. (2009). Comparison between exhaled breath condensate analysis as a marker for cobalt and tungsten exposure and biomonitoring in workers of a hard metal alloy processing plant. *International Archives of Occupational and Environmental Health, 82*(5), 565–573. https://doi.org/10.1007/s00420-008-0390-5

Brough, D., & Jouhara, H. (2020). The aluminium industry: A review on state-of-the-art technologies, environmental impacts and possibilities for waste heat recovery. *International Journal of Thermofluids*, 1–2. https://doi.org/10.1016/j.ijft.2019.100007

Çamur, D., Topbaş, M., İlter, H., Albay, M., Ayoğlu, F. N., Can, M., Altın, A., Demirtaş, Y., Somuncu, B. P., Aydın, F., & Açıkgöz, B. (2021). Heavy metals and trace elements in whole-blood samples of the fishermen in turkey: The fish/ermen heavy metal study (FHMS). *Environmental Management, 67*(3), 553–562. https://doi.org/10.1007/s00267-020-01398-y

Carvalho, R. B., Carneiro, M. F. H., Barbosa, F., Batista, B. L., Simonetti, J., Amantéa, S. L., & Rhoden, C. R. (2018). The impact of occupational exposure to traffic-related air pollution among professional motorcyclists from Porto Alegre, Brazil, and its association with genetic and oxidative damage. *Environmental Science and Pollution Research, 25*(19), 18620–18631. https://doi.org/10.1007/s11356-018-2007-1

Cefalu, W. T., & Hu, F. B. (2004). *Role of Chromium in Human Health and in Diabetes.* https://diabetesjournals.org/care/article-pdf/27/11/2741/562350/zdc01104002741.pdf

Chakraborty, R., Renu, K., Eladl, M. A., El-Sherbiny, M., Elsherbini, D. M. A., Mirza, A. K., Vellingiri, B., Iyer, M., Dey, A., & Valsala Gopalakrishnan, A. (2022). Mechanism of chromium-induced toxicity in lungs, liver, and kidney and their ameliorative agents. *Biomedicine and Pharmacotherapy, 151.* https://doi.org/10.1016/j.biopha.2022.113119

Chen, Y., Parvez, F., Gamble, M., Islam, T., Ahmed, A., Argos, M., Graziano, J. H., & Ahsan, H. (2009). Arsenic exposure at low-to-moderate levels and skin lesions, arsenic metabolism, neurological functions, and biomarkers for respiratory and cardiovascular diseases: Review of recent findings from the Health Effects of Arsenic Longitudinal Study (HEALS) in Bangladesh. *Toxicology and Applied Pharmacology, 239*(2), 184–192. https://doi.org/10.1016/j.taap.2009.01.010

Cheng, C. H., Ma, H. L., Liu, G. X., Fan, S. G., Deng, Y. Q., Jiang, J. J., Feng, J., & Guo, Z. X. (2023). Toxic effects of cadmium exposure on intestinal histology, oxidative stress, microbial community, and transcriptome change in the mud crab (Scylla paramamosain). *Chemosphere, 326.* https://doi.org/10.1016/j.chemosphere.2023.138464

Chung, J. Y., Yu, S. Do, & Hong, Y. S. (2014). Environmental source of arsenic exposure. *Journal of Preventive Medicine and Public Health, 47*(5), 253–257. https://doi.org/10.3961/jpmph.14.036

Cirovic, A., Denic, A., Clarke, B. L., Vassallo, R., Cirovic, A., & Landry, G. M. (2022). A hypoxia-driven occurrence of chronic kidney disease and osteoporosis in COPD individuals: New insights into environmental cadmium exposure. *Toxicology, 482.* https://doi.org/10.1016/j.tox.2022.153355

Clancy, H. A., Sun, H., Passantino, L., Kluz, T., Muñoz, A., Zavadil, J., & Costa, M. (2012). Gene expression changes in human lung cells exposed to arsenic, chromium, nickel or vanadium indicate the first steps in cancer. *Metallomics, 4*(8), 784–793. https://doi.org/10.1039/C2MT20074K

Coetzee, J. J., Bansal, N., & Chirwa, E. M. N. (2020a). Chromium in environment, its toxic effect from chromite-mining and ferrochrome industries, and its possible bioremediation. *Exposure and Health, 12*(1), 51–62. https://doi.org/10.1007/s12403-018-0284-z

Coetzee, J. J., Bansal, N., & Chirwa, E. M. N. (2020b). Chromium in environment, its toxic effect from chromite-mining and ferrochrome industries, and its possible bioremediation. *Exposure and Health, 12*(1), 51–62. https://doi.org/10.1007/S12403-018-0284-Z

Córdoba-Tovar, L., Marrugo-Negrete, J., Barón, P. R., & Díez, S. (2022). Drivers of biomagnification of Hg, As and Se in aquatic food webs: A review. *Environmental Research, 204.* https://doi.org/10.1016/j.envres.2021.112226

Crisponi, G., Nurchi, V. M., Bertolasi, V., Remelli, M., & Faa, G. (2012a). Chelating agents for human diseases related to aluminium overload. *Coordination Chemistry Reviews, 256*(1–2), 89–104. https://doi.org/10.1016/j.ccr.2011.06.013

Crisponi, G., Nurchi, V. M., Bertolasi, V., Remelli, M., & Faa, G. (2012b). Chelating agents for human diseases related to aluminium overload. *Coordination Chemistry Reviews, 256*(1–2), 89–104. https://doi.org/10.1016/j.ccr.2011.06.013

Das, S., Kar, I., & Patra, A. K. (2023). Cadmium induced bioaccumulation, histopathology, gene regulation in fish and its amelioration: A review. *Journal of Trace Elements in Medicine and Biology, 79.* https://doi.org/10.1016/j.jtemb.2023.127202

Demedts, M., Gheysens, B., Nagels, J., Verbeken, E., Lauweryns, J., van den Eeckhout A., Lahaye, D., & Gyselen, A. (1984). Cobalt lung in diamond polishers. *American Review of Respiratory Disease, 130*(1), 130–135.

Donoghue, A. M., Frisch, N., & Olney, D. (2014). Bauxite mining and alumina refining: Process description and occupational health risks. *Journal of Occupational and Environmental Medicine*, *56*(5). https://doi.org/10.1097/JOM.0000000000000001

Dutta, K., Prasad, P., & Sinha, D. (2015). Chronic low level arsenic exposure evokes inflammatory responses and DNA damage. *International Journal of Hygiene and Environmental Health*, *218*(6), 564–574. https://doi.org/10.1016/j.ijheh.2015.06.003

EC Directive no. 37 (2004). *On the Protection of Workers from the Risks Related to Exposure to Carcinogens or Mutagens at Work*. European Commision.

EC Regulation No 1223. (2009). *Regulation (EC) No 1223/2009 of the European Parliament and of the Council of 30* November 2009 *on Cosmetic Products*. European Commision.

Eid, B. M., & Ibrahim, N. A. (2021). Recent developments in sustainable finishing of cellulosic textiles employing biotechnology. *Journal of Cleaner Production*, *284*. https://doi.org/10.1016/j.jclepro.2020.124701

Gokul, T., Kumar, K. R., Veeramanikandan, V., Arun, A., Balaji, P., & Faggio, C. (2023). Impact of particulate pollution on aquatic invertebrates. *Environmental Toxicology and Pharmacology*, *100*. https://doi.org/10.1016/j.etap.2023.104146

González-Muñoz, M. J., Peña, A., & Meseguer, I. (2008). Monitoring heavy metal contents in food and hair in a sample of young Spanish subjects. *Food and Chemical Toxicology*, *46*(9), 3048–3052. https://doi.org/10.1016/J.FCT.2008.06.004

Guan, X., Zhang, B., Liu, S., An, M., Han, Q., Li, D., & Rao, P. (2023). Facile degradation of chitosan-sodium alginate-chromium (III) gel in relation to leather re-tanning and filling. *International Journal of Biological Macromolecules*, *240*. https://doi.org/10.1016/j.ijbiomac.2023.124437

Gunkel-Grillon, P., Laporte-Magoni, C., Le Mestre, M., Bazire, N., & Lemestre, M. (2014). Toxic chromium release from nickel mining sediments in surface waters, New Caledonia Toxic chromium release from nickel mining sediments in surface waters Toxic chromium release from nickel mining sediments in surface waters, New Caledonia. *Environmental Chemistry Letters*, 10. https://doi.org/10.1007/s10311-014-0475-1

Haase, L. M., Birk, T., Bachand, A. M., & Mundt, K. A. (2021). A health surveillance study of workers employed at a copper smelter: Effects of long-term exposure to copper on lung function using spirometric data. *Journal of Occupational and Environmental Medicine*, *63*(8), E480–E489. https://doi.org/10.1097/JOM.0000000000002252

Han, L., Gu, H., Lu, W., Li, H., Peng, W. xi, Ling Ma, N., Lam, S. S., & Sonne, C. (2023). Progress in phytoremediation of chromium from the environment. *Chemosphere*, *344*. https://doi.org/10.1016/j.chemosphere.2023.140307

Hao, W., Zhu, X., Liu, Z., Song, Y., Wu, S., Lu, X., Yang, J., & Jin, C. (2023). Aluminum exposure induces central nervous system impairment via activating NLRP3-medicated pyroptosis pathway. *Ecotoxicology and Environmental Safety*, *264*, 115401. https://doi.org/10.1016/j.ecoenv.2023.115401

Hernández-Cruz, E. Y., Amador-Martínez, I., Aranda-Rivera, A. K., Cruz-Gregorio, A., & Pedraza Chaverri, J. (2022). Renal damage induced by cadmium and its possible therapy by mitochondrial transplantation. *Chemico-Biological Interactions*, *361*. https://doi.org/10.1016/j.cbi.2022.109961

Hypotkm, M., Regland', B., & Gottfrles+, C. G. (1992). Slowed synthesis of DNA and methionine is a pathogenetic mechanism common to dementia in Down's syndrome, Al DS and Alzheimer's disease? *Medical Hypotheses*, *38*, 11–19.

Iqbal, I., Sharma, S., & Pathania, A. R. (2022). An insight into application and harmful effects of aluminium, aluminium complexes and aluminium nano-particles. *Materials Today: Proceedings*, *62*, 4365–4369. https://doi.org/10.1016/j.matpr.2022.04.870

Jäger, S., Cabral, M., Kopp, J. F., Hoffmann, P., Ng, E., Whitfield, J. B., Morris, A. P., Lind, L., Schwerdtle, T., & Schulze, M. B. (2022). Blood copper and risk of cardiometabolic diseases: A Mendelian randomization study. *Human Molecular Genetics*, *31*(5), 783–791. https://doi.org/10.1093/hmg/ddab275

Jaishankar, M., Tseten, T., Anbalagan, N., Mathew, B. B., & Beeregowda, K. N. (2014). Toxicity, mechanism and health effects of some heavy metals. *Interdisciplinary Toxicology*, *7*(2), 60–72). https://doi.org/10.2478/intox-2014-0009

Jarvis. (1992). Cobalt cardiomyopathy: A report of two cases from mineral assay laboratories and a review of the literature. *Journal of Occupational Medicine*, *34*(6), 620.

Jiang, D., Hu, X., Jin, X., Ma, A., & Yin, D. (2021). Oxidized nanoscale zero-valent iron changed the bioaccumulation and distribution of chromium in zebrafish. *Chemosphere*, *263*. https://doi.org/10.1016/j.chemosphere.2020.128001

Joorabian Shooshtari, S., Abdollahzadeh, E., Esmaili-Sari, A., & Ghasempouri, S. M. (2023). A review of mercury contamination in representative flora and fauna of Iran: Seafood consumption advisories. *Journal of Hazardous Materials Advances*, *10*, 100291. https://doi.org/10.1016/j.hazadv.2023.100291

Kasbaji, M., Mennani, M., Grimi, N., Barba, F. J., Oubenali, M., Simirgiotis, M. J., Mbarki, M., & Moubarik, A. (2022). Implementation and physico-chemical characterization of new alkali-modified bio-sorbents for cadmium removal from industrial discharges: Adsorption isotherms and kinetic approaches. *Process Biochemistry*, *120*, 213–226. https://doi.org/10.1016/j.procbio.2022.06.010

Kasztelan-Szczerbinska, B., & Cichoz-Lach, H. (2021). Wilson's disease: An update on the diagnostic workup and management. *Journal of Clinical Medicine*, *10*(21). https://doi.org/10.3390/jcm10215097

Kaufman, J. A., Mattison, C., Fretts, A. M., Umans, J. G., Cole, S. A., Voruganti, V. S., Goessler, W., Best, L. G., Zhang, Y., Tellez-Plaza, M., Navas-Acien, A., & Gribble, M. O. (2021). Arsenic, blood pressure, and hypertension in the strong heart family study. *Environmental Research*, *195*. https://doi.org/10.1016/j.envres.2021.110864

Kerur, S. S., Bandekar, S., Hanagadakar, M. S., Nandi, S. S., Ratnamala, G. M., & Hegde, P. G. (2020). Removal of hexavalent chromium-industry treated water and wastewater: A review. *Materials Today: Proceedings*, *42*, 1112–1121. https://doi.org/10.1016/j.matpr.2020.12.492

Kumar, M., Gupta, N., Ratn, A., Awasthi, Y., Prasad, R., Trivedi, A., & Trivedi, S. P. (2020). Biomonitoring of heavy metals in river ganga water, sediments, plant, and fishes of different trophic levels. *Biological Trace Element Research*, *193*(2), 536–547. https://doi.org/10.1007/s12011-019-01736-0

Kuo, C. C., Balakrishnan, P., Gribble, M. O., Best, L. G., Goessler, W., Umans, J. G., & Navas-Acien, A. (2022). The association of arsenic exposure and arsenic metabolism with all-cause, cardiovascular and cancer mortality in the Strong Heart Study. *Environment International*, *159*. https://doi.org/10.1016/j.envint.2021.107029

Lall, S. P., & Kaushik, S. J. (2021). Nutrition and metabolism of minerals in fish. *Animals*, *11*(9). https://doi.org/10.3390/ani11092711

Lantin, A. C., Mallants, A., Vermeulen, J., Speybroeck, N., Hoet, P., & Lison, D. (2011). Absence of adverse effect on thyroid function and red blood cells in a population of workers exposed to cobalt compounds. *Toxicology Letters*, *201*(1), 42–46. https://doi.org/10.1016/j.toxlet.2010.12.003

Lazăr, N. N., Simionov, I. A., Petrea, Ştefan M., Iticescu, C., Georgescu, P. L., Dima, F., & Antache, A. (2024). The influence of climate changes on heavy metals accumulation in Alosa immaculata from the Danube River Basin. *Marine Pollution Bulletin*, *200*. https://doi.org/10.1016/j.marpolbul.2024.116145

Lee, J.-W., Jo, A.-H., Lee, D.-C., Choi, C. Y., Kang, J.-C., & Kim, J.-H. (2023). Review of cadmium toxicity effects on fish: Oxidative stress and immune responses. *Environmental Research*, *236*, 116600. https://doi.org/10.1016/j.envres.2023.116600

Leyssens, L., Vinck, B., Van Der Straeten, C., Wuyts, F., & Maes, L. (2017). Cobalt toxicity in humans: A review of the potential sources and systemic health effects. *Toxicology*, *387*, 43–56. https://doi.org/10.1016/j.tox.2017.05.015

Li, A., Zhou, Q., Mei, Y., Zhao, J., Zhao, M., Xu, J., Ge, X., Li, Y., Li, K., Yang, M., & Xu, Q. (2023). Thyroid disrupting effects of multiple metals exposure: Comprehensive investigation from the thyroid parenchyma to hormonal function in a prospective cohort study. *Journal of Hazardous Materials*, *459*. https://doi.org/10.1016/j.jhazmat.2023.132115

Li, Y., Yu, Y., Zheng, N., Hou, S., Song, X., & Dong, W. (2020). Metallic elements in human hair from residents in smelting districts in northeast China: Environmental factors and differences in ingestion media. *Environmental Research*, *182*. https://doi.org/10.1016/j.envres.2019.108914

Liu, M., Zhu, H., Fang, Y., Liu, C., Li, X., Zhang, X., Ma, L., Wang, K., Yu, M., Sheng, W., & Zhu, B. (2024). An ultra-sensitive fluorescent probe for recognition of aluminum ions and its application in environment, food, and living organisms. *Spectrochimica Acta - Part A: Molecular and Biomolecular Spectroscopy*, *307*. https://doi.org/10.1016/j.saa.2023.123578

Liu, P., Guo, C., Cui, Y., Zhang, X., Xiao, B., Liu, M., Song, M., & Li, Y. (2022). Activation of PINK1/Parkin-mediated mitophagy protects against apoptosis in kidney damage caused by aluminum. *Journal of Inorganic Biochemistry*, *230*. https://doi.org/10.1016/j.jinorgbio.2022.111765

Liu, P., Liu, Z., Chu, M., Yan, R., Li, F., Tang, J., & Feng, J. (2022). Detoxification and comprehensive recovery of stainless steel dust and chromium containing slag: Synergistic reduction mechanism and process parameter optimization. *Process Safety and Environmental Protection*, *164*, 678–695. https://doi.org/10.1016/j.psep.2022.06.034

Loredo, J., Álvarez, R., Ordóñez, A., & Bros, T. (2008). Mineralogy and geochemistry of the Texeo Cu-Co mine site (NW Spain): Screening tools for environmental assessment. *Environmental Geology*, *55*(6), 1299–1310. https://doi.org/10.1007/s00254-007-1078-y

Malhotra, N., Ger, T. R., Uapipatanakul, B., Huang, J. C., Chen, K. H. C., & Hsiao, C. Der. (2020). Review of copper and copper nanoparticle toxicity in fish. *Nanomaterials*, *10*(6), 1–28. https://doi.org/10.3390/nano10061126

Marchant, B. P., Saby, N. P. A., & Arrouays, D. (2017). A survey of topsoil arsenic and mercury concentrations across France. *Chemosphere*, *181*, 635–644. https://doi.org/10.1016/j.chemosphere.2017.04.106

Matthews, N. H., Fitch, K., Li, W. Q., Morris, J. S., Christiani, D. C., Qureshi, A. A., & Cho, E. (2019). Exposure to trace elements and risk of skin cancer: A systematic review of epidemiologic studies. *Cancer Epidemiology Biomarkers and Prevention*, *28*(1), 3–21. https://doi.org/10.1158/1055-9965.EPI-18-0286

Matúš, P., Kubová, J., Bujdoš, M., Streško, V., & Medved', J. (2004). Chemical partitioning of aluminium in rocks, soils, and sediments acidified by mining activity. *Analytical and Bioanalytical Chemistry*, *379*(1), 96–103. https://doi.org/10.1007/s00216-004-2562-9

Maya, S., Prakash, T., Madhu, K. Das, & Goli, D. (2016). Multifaceted effects of aluminium in neurodegenerative diseases: A review. *Biomedicine and Pharmacotherapy*, *83*, 746–754. https://doi.org/10.1016/j.biopha.2016.07.035

Mekicha, M. A., de Rooij, M. B., Matthews, D. T. A., Pelletier, C., Jacobs, L., & Schipper, D. J. (2020). The effect of hard chrome plating on iron fines formation. *Tribology International*, *142*. https://doi.org/10.1016/j.triboint.2019.106003

Minzala, D.-N., Simionov, I.-A., & Petrea, Ştefan -M. (2022). Evaluation of potentially toxic elements in black sea fishery resources: A review. *Scientific Papers. Series E. Land Reclamation, Earth Observation & Surveying, Environmental Engineering*, *XI*, 136–147.

Mishra, A., Oliinyk, P., Lysiuk, R., Lenchyk, L., Rathod, S. S. S., Antonyak, H., Darmohray, R., Dub, N., Antoniv, O., Tsal, O., & Upyr, T. (2022). Flavonoids and stilbenoids as a promising arsenal for the management of chronic arsenic toxicity. *Environmental Toxicology and Pharmacology, 95*. https://doi.org/10.1016/j.etap.2022.103970

Mishra, V., Samantaray, D. P., Dash, S. K., Mishra, B. B., & Swain, R. K. (1970). Study on hexavalent chromium reduction by chromium resistant bacterial isolates of sukinda mining area. *Our Nature, 8*(1), 63–71. https://doi.org/10.3126/on.v8i1.4313

Mohanty, S., Benya, A., Hota, S., Kumar, M. S., & Singh, S. (2023a). Eco-toxicity of hexavalent chromium and its adverse impact on environment and human health in Sukinda Valley of India: A review on pollution and prevention strategies. *Environmental Chemistry and Ecotoxicology, 5*, 46–54. https://doi.org/10.1016/j.enceco.2023.01.002

Mohanty, S., Benya, A., Hota, S., Kumar, M. S., & Singh, S. (2023b). Eco-toxicity of hexavalent chromium and its adverse impact on environment and human health in Sukinda Valley of India: A review on pollution and prevention strategies. *Environmental Chemistry and Ecotoxicology, 5*, 46–54. https://doi.org/10.1016/j.enceco.2023.01.002

Monga, A., Fulke, A. B., & Dasgupta, D. (2022). Recent developments in essentiality of trivalent chromium and toxicity of hexavalent chromium: Implications on human health and remediation strategies. *Journal of Hazardous Materials Advances, 7*, 100113. https://doi.org/10.1016/J.HAZADV.2022.100113

Moody, K. H., Hasan, K. M., Aljic, S., Blakeman, V. M., Hicks, L. P., Loving, D. C., Moore, M. E., Hammett, B. S., Silva-González, M., Seney, C. S., & Kiefer, A. M. (2020). Mercury emissions from Peruvian gold shops: Potential ramifications for Minamata compliance in artisanal and small-scale gold mining communities. *Environmental Research, 182*. https://doi.org/10.1016/j.envres.2019.109042

Mortada, W. I., El-Naggar, A., Mosa, A., Palansooriya, K. N., Yousaf, B., Tang, R., Wang, S., Cai, Y., & Chang, S. X. (2023). Biogeochemical behaviour and toxicology of chromium in the soil-water-human nexus: A review. *Chemosphere, 331*. https://doi.org/10.1016/j.chemosphere.2023.138804

Moussaron, A., Alexandre, J., Chenard, M.-P., Mathelin, C., & Reix, N. (2023). Correlation between daily life aluminium exposure and breast cancer risk: A systematic review. *Journal of Trace Elements in Medicine and Biology, 79*, 127247. https://doi.org/10.1016/j.jtemb.2023.127247

Ngure, V., & Kinuthia, G. (2020). Health risk implications of lead, cadmium, zinc, and nickel for consumers of food items in Migori Gold mines, Kenya. *Journal of Geochemical Exploration, 209*. https://doi.org/10.1016/j.gexplo.2019.106430

Nguyen, P. M., & Mulligan, C. N. (2023). Study on the influence of water composition on iron nail corrosion and arsenic removal performance of the Kanchan arsenic filter (KAF). *Journal of Hazardous Materials Advances, 10*, 100285. https://doi.org/10.1016/j.hazadv.2023.100285

Nordberg, G. (1994). The science of the total environment assessment of risks in occupational cobalt exposures. *The Science of the Total Environment, 150*, 201–207.

Nowak, B., & Chmielnicka, J. (2000). Relationship of lead and cadmium to essential elements in hair, teeth, and nails of environmentally exposed people. *Ecotoxicology and Environmental Safety, 46*(3), 265–274. https://doi.org/10.1006/eesa.2000.1921

Omeed, S., & Raja, T. (2017). Berylliosis (2021st ed.). https://europepmc.org/article/NBK/nbk470364#free-full-text

Pandey, V., Yadav, R., Singh, A., Mishra, D., Shanker, K., Singh, S., & Khare, P. (2023). Differential behaviour of four genotypes of Andrographis paniculata (Burm.f.) Nees toward combined toxicity of As, Cd, and Pb: An ionomics and metabolic interpretation. *Journal of Hazardous Materials Advances*, 100274. https://doi.org/10.1016/j.hazadv.2023.100274

Popa, M. D., Simionov, I. A., Petrea, S. M., & Dima, F. M. (2023). Heavy metals accumulation in the tissues of the common reed (phragmites australis). *Scientific Papers. Series E.*

Land Reclamation, Earth Observation & Surveying, Environmental Engineering, 86–91. https://www.researchgate.net/publication/376522408

Pradeep, S., Sundaramoorthy, S., Sathish, M., Jayakumar, G. C., Rathinam, A., Madhan, B., Saravanan, P., & Rao, J. R. (2021). Chromium-free and waterless vegetable-aluminium tanning system for sustainable leather manufacture. *Chemical Engineering Journal Advances*, *7*. https://doi.org/10.1016/j.ceja.2021.100108

Punshon, T., Jackson, B. P., Meharg, A. A., Warczack, T., Scheckel, K., & Guerinot, M. Lou. (2017). Understanding arsenic dynamics in agronomic systems to predict and prevent uptake by crop plants. *Science of the Total Environment*, *581–582*, 209–220. https://doi.org/10.1016/j.scitotenv.2016.12.111

Ramezanifar, S., Beyrami, S., Mehrifar, Y., Ramezanifar, E., Soltanpour, Z., Namdari, M., & Gharari, N. (2023). Occupational exposure to physical and chemical risk factors: A systematic review of reproductive pathophysiological effects in women and men. *Safety and Health at Work*, *14*(1), 17–30. https://doi.org/10.1016/j.shaw.2022.10.005

Rathinam, A., Kalarical Janardhanan, S., Jonnalagadda Raghava, R., & Balachandran Unni, N. (2007). Biological removal of carcinogenic chromium(VI) using mixed Pseudomonas strains. *The Journal of General and Applied Microbiology*, *53*, 71–79.

Rebelo, A., Duarte, B., Freitas, A. R., Almeida, A., Azevedo, R., Pinto, E., Peixe, L., Antunes, P., & Novais, C. (2023). Uncovering the effects of copper feed supplementation on the selection of copper-tolerant and antibiotic-resistant Enterococcus in poultry production for sustainable environmental practices. *The Science of the Total Environment*, *900*, 165769. https://doi.org/10.1016/j.scitotenv.2023.165769

Rocha, G. S., de Palma Lopes, L. F., de Medeiros, J. F., Montagner, C. C., & Gaeta Espíndola, E. L. (2023). Environmental concentrations of cadmium and fipronil, isolated and combined, impair the survival and reproduction of a Neotropical freshwater copepod. *Environmental Pollution*, *336*. https://doi.org/10.1016/j.envpol.2023.122415

Rodríguez, V. M., Limón-Pacheco, J. H., Carrizales, L., Mendoza-Trejo, M. S., & Giordano, M. (2010). Chronic exposure to low levels of inorganic arsenic causes alterations in locomotor activity and in the expression of dopaminergic and antioxidant systems in the albino rat. *Neurotoxicology and Teratology*, *32*(6), 640–647. https://doi.org/10.1016/j.ntt.2010.07.005

Rzayev, F. H., Gasimov, E. K., Agayeva, N. J., Manafov, A. A., Mamedov, C. A., Ahmadov, I. S., Khusro, A., Valan Arasu, M., Sahibzada, M. U. K., Al-Dhabi, N. A., & Choi, K. C. (2022). Microscopic characterization of bioaccumulated aluminium nanoparticles in simplified food chain of aquatic ecosystem. *Journal of King Saud University - Science*, *34*(1). https://doi.org/10.1016/j.jksus.2021.101666

Sakamoto, M., Kakita, A., Domingo, J. L., Yamazaki, H., Oliveira, R. B., Sarrazin, S. L. F., Eto, K., & Murata, K. (2017). Stable and episodic/bolus patterns of methylmercury exposure on mercury accumulation and histopathologic alterations in the nervous system. *Environmental Research*, *152*, 446–453. https://doi.org/10.1016/j.envres.2016.06.034

Salles, F. J., Tavares, D. J. B., Freire, B. M., Ferreira, A. P. S. da S., Handakas, E., Batista, B. L., & Olympio, K. P. K. (2021). Home-based informal jewelry production increases exposure of working families to cadmium. *Science of the Total Environment*, *785*. https://doi.org/10.1016/j.scitotenv.2021.147297

Saltini, C., & Amicosante, M. (2001). Beryllium disease. *The American Journal of the Medical Sciences*, *321*(1), 89–98. https://doi.org/10.1097/00000441-200101000-00013

Santos, A. J. M., Khemiri, S., Simões, S., Prista, C., Sousa, I., & Raymundo, A. (2023). The importance, prevalence and determination of vitamins B6 and B12 in food matrices: A review. *Food Chemistry*, *426*. https://doi.org/10.1016/j.foodchem.2023.136606

Satarug, S. (2018). Dietary cadmium intake and its effects on kidneys. *Toxics*, *6*(1). https://doi.org/10.3390/toxics6010015

Schaefer, H. R., Flannery, B. M., Crosby, L., Jones-Dominic, O. E., Punzalan, C., & Middleton, K. (2022). A systematic review of adverse health effects associated with oral cadmium exposure. *Regulatory Toxicology and Pharmacology*, *134*. https://doi.org/10.1016/j.yrtph.2022.105243

Shafaghat, A., Keyvanfar, A., Manteghi, G., & Lamit, H. Bin. (2016). Environmental-conscious factors affecting street microclimate and individuals' respiratory health in tropical coastal cities. *Sustainable Cities and Society*, *21*, 35–50. https://doi.org/10.1016/j.scs.2015.11.001

Shayan, M., Barangi, S., Hosseinzadeh, H., & Mehri, S. (2023). The protective effect of natural or chemical compounds against arsenic-induced neurotoxicity: Cellular and molecular mechanisms. *Food and Chemical Toxicology*, *175*. https://doi.org/10.1016/j.fct.2023.113691

Shephard, B. K., Mcintosh, A. W., Atchison, G. J., & Nelson1, D. W. (n.d.). Aspects of the aquatic chemistry of cadmium and zinc in a heavy metal contaminated lake. *Water Research*, *14*, 1061–1066.

Shokri, A., & Fard, M. S. (2022). A critical review in Fenton-like approach for the removal of pollutants in the aqueous environment. *Environmental Challenges*, *7*. https://doi.org/10.1016/j.envc.2022.100534

Shuhaimi-Othman, M., Nadzifah, Y., Nur-Amalina, R., & Umirah, N. S. (2013a). Deriving freshwater quality criteria for copper, cadmium, aluminum and manganese for protection of aquatic life in Malaysia. *Chemosphere*, *90*(11), 2631–2636. https://doi.org/10.1016/j.chemosphere.2012.11.030

Shuhaimi-Othman, M., Nadzifah, Y., Nur-Amalina, R., & Umirah, N. S. (2013b). Deriving freshwater quality criteria for copper, cadmium, aluminum and manganese for protection of aquatic life in Malaysia. *Chemosphere*, *90*(11), 2631–2636. https://doi.org/10.1016/j.chemosphere.2012.11.030

Simionov, I. A., Cristea, D. S., Petrea, Ştefan, M., Mogodan, A., Jijie, R., Ciornea, E., Nicoară, M., Rahoveanu, M. M. T., & Cristea, V. (2021a). Predictive innovative methods for aquatic heavy metals pollution based on bioindicators in support of blue economy in the Danube river basin. *Sustainability (Switzerland)*, *13*(16). https://doi.org/10.3390/su13168936

Simionov, I. A., Cristea, D. S., Petrea, S. M., Mogodan, A., Nicoara, M., Plavan, G., Baltag, E. S., Jijie, R., & Strungaru, S. A. (2021b). Preliminary investigation of lower Danube pollution caused by potentially toxic metals. *Chemosphere*, *264*. https://doi.org/10.1016/j.chemosphere.2020.128496

Simionov, I. A., Călmuc, M., Antache, A., Călmuc, V., Petrea, Ştefan -M., Nica, A., Cristea, V., & Neculiţă, M. (2022). The use of pectinatella magnifica as bioindicator for heavy metals pollution in Danube delta. *Scientific Papers. Series E. Land Reclamation, Earth Observation & Surveying, Environmental Engineering*, *XI*, 181–186.

Simionov, I. A., Calmuc, V., Petrea, S. M., Antache Alina, Nica Aurelia, Iticescu Catalina, Georgescu Puiu Lucian, & Cristea Victor. (2023). Evaluation of heavy metals concentrations in the black sea turbot and elements correlation analysis. *Scientific Papers. Series E. Land Reclamation, Earth Observation & Surveying, Environmental Engineering*, 296–302. https://www.researchgate.net/publication/376522530

Tiwari, A. K., & De Maio, M. (2017). Assessment of risk to human health due to intake of chromium in the groundwater of the Aosta Valley region, Italy. *Human and Ecological Risk Assessment: An International Journal*, *23*(5), 1153–1163.

Tiwari, A. K., Orioli, S., & De Maio, M. (2019). Assessment of groundwater geochemistry and diffusion of hexavalent chromium contamination in an industrial town of Italy. *Journal of Contaminant Hydrology*, *225*. https://doi.org/10.1016/j.jconhyd.2019.103503

Vardhan, K. H., Kumar, P. S., & Panda, R. C. (2019). A review on heavy metal pollution, toxicity and remedial measures: Current trends and future perspectives. *Journal of Molecular Liquids*, *290*. https://doi.org/10.1016/j.molliq.2019.111197

Vasquez Ugaz, C. A., León-Roque, N., Nuñez-León, J. L., Hidalgo-Chávez, D. W., & Oblitas, J. (2023). Geochemical and environmental assessment of potential effects of trace elements in soils, water, and sediments around abandoned mining sites in the northern Iberian Peninsula (NW Spain). *Heliyon*, *9*(3). https://doi.org/10.1016/j.heliyon.2023.e14659

Xia, W., Hu, J., Zhang, B., Li, Y., Pierce Wise Sr, J., Bassig, B. A., Zhou, A., Savitz, D. A., Xiong, C., Zhao, J., du, X., Zhou, Y., Pan, X., Yang, J., Wu, C., Jiang, M., Peng, Y., Qian, Z., Zheng, T., & Xu, S. (2015). A case-control study of maternal exposure to chromium and infant low birth weight in China. *Chemosphere*, *144*, 1484–1489. https://doi.org/10.1016/j.chemosphere.2015.10.006

Yan, J., Huang, H., Liu, Z., Shen, J., Ni, J., Han, J., Wang, R., Lin, D., Hu, B., & Jin, L. (2020). Hedgehog signaling pathway regulates hexavalent chromium-induced liver fibrosis by activation of hepatic stellate cells. *Toxicology Letters*, *320*, 1–8. https://doi.org/10.1016/j.toxlet.2019.11.017

Yan, J., Sharma, N., Flynn, E. D., Giammar, D. E., Schwartz, G. E., Brooks, S. C., Weisenhorn, P., Kemner, K. M., O'Loughlin, E. J., Kaplan, D. I., & Catalano, J. G. (2022). Consistent controls on trace metal micronutrient speciation in wetland soils and stream sediments. *Geochimica et Cosmochimica Acta*, *317*, 234–254. https://doi.org/10.1016/j.gca.2021.10.017

Yang, K., Chen, C., Brockman, J., Shikany, J. M., & He, K. (2021). Low- and moderate-levels of arsenic exposure in young adulthood and incidence of chronic kidney disease: Findings from the CARDIA Trace Element Study. *Journal of Trace Elements in Medicine and Biology*, *63*. https://doi.org/10.1016/j.jtemb.2020.126657

Yong, W. K., Sim, K. S., Poong, S. W., Wei, D., Phang, S. M., & Lim, P. E. (2021). Interactive effects of warming and copper toxicity on a tropical freshwater green microalga Chloromonas augustae (Chlorophyceae). *Journal of Applied Phycology*, *33*(1), 67–77. https://doi.org/10.1007/s10811-020-02087-3

Zaynab, M., Al-Yahyai, R., Ameen, A., Sharif, Y., Ali, L., Fatima, M., Khan, K. A., & Li, S. (2022). Health and environmental effects of heavy metals. *Journal of King Saud University – Science*, *34*(1). https://doi.org/10.1016/j.jksus.2021.101653

Zhang, Q., Zheng, Q., & Sun, G. (2011). Arsenic-contaminated cold-spring water in mountainous areas of Hui County, Northwest China: A new source of arsenic exposure. *Science of the Total Environment*, *409*(24), 5513–5516. https://doi.org/10.1016/j.scitotenv.2011.08.055

Zhang, W., Miao, A. J., Wang, N. X., Li, C., Sha, J., Jia, J., Alessi, D. S., Yan, B., & Ok, Y. S. (2022). Arsenic bioaccumulation and biotransformation in aquatic organisms. *Environment International*, *163*. https://doi.org/10.1016/j.envint.2022.107221

Zhang, Z., Li, J., Jiang, S., Xu, M., Ma, T., Sun, Z., & Zhang, J. (2023). Lactobacillus fermentum HNU312 alleviated oxidative damage and behavioural abnormalities during brain development in early life induced by chronic lead exposure. *Ecotoxicology and Environmental Safety*, *251*. https://doi.org/10.1016/j.ecoenv.2023.114543

Zhao, Y., Dang, M., Zhang, W., Lei, Y., Ramesh, T., Priya Veeraraghavan, V., & Hou, X. (2020). Neuroprotective effects of Syringic acid against aluminium chloride induced oxidative stress mediated neuroinflammation in rat model of Alzheimer's disease. *Journal of Functional Foods*, *71*. https://doi.org/10.1016/j.jff.2020.104009

Copper — Consumer. nih.gov (accessed on 29 August 2023).

EU — EU Restriction of Hazardous Substances Directive (RoHS): An Essential Guide. compliancegate.com (accessed 31 August 2023).

EU — List of Products Covered by the EU Battery Directive. compliancegate.com (accessed 31 August 2023).
National Emphasis Program — Hexavalent chromium. osha.gov (accessed 22 August 2023).
National Emphasis Program — Chromium (Cr) Toxicity|ATSDR — CSEM. cdc.gov (accessed 22 August 2023).
REACH Annex XVII Substances List: An Overview. compliancegate.com (accessed 31 August 2023).
Registry of CLH intentions until outcome — ECHA. europa.eu (accessed 29 August 2023).

9 Surface Water Chemistry of Sohagpur Coalfield, Madhya Pradesh, India

Ashwani Kumar Tiwari, Pramod Kumar Singh, Abhay Kumar Singh, and T.N. Singh

9.1 INTRODUCTION

Water resources (i.e., surface and groundwater) are essential for mankind's survival, environment, and development of a nation. Contamination of water resources is a serious concern for any nation and their society. Demand for freshwater resources as well as contamination of water resources is growing day by day mainly due to the increasing population. Several anthropogenic activities (i.e., agricultural, mining, sewage discharge, industrial effluent, solid waste, and others) are responsible for surface water resources pollution (Rösner, 1998; Karn and Harada, 2001; Subrahmanyam and Yadaiah, 2001; Lee and Bukaveckas, 2002; Srinivasa and Govil, 2008; Pastor and Hernández, 2012; Obiadi et al., 2016). Moreover, geogenic processes are also responsible for contamination of surface water resources (Schneider et al., 2017).

Deterioration of surface water quality in mining regions is a common problem. Researchers have reported that mining and related activities impacted the surface water quality (Huang et al., 2010; Ning et al. 2011; Tiwari et al., 2015). Particularly, surface water contamination with major and trace elements in the mining area is a general issue (Wu et al., 2009; Tiwari and De Maio, 2018). Hence, surface water quality monitoring is essential in the mining region for sustainable use and management. Therefore, Indian researchers have reported the surface water quality status of coalfields, such as Jharia (Abhishek and Sinha, 2006), Talcher (Sahoo et al., 2016), West Bokaro (Tiwari et al., 2016), IB Valley (Sahoo et al., 2021), East Bokaro (Mahato et al., 2021), Raniganj (Kumar and Singh, 2022), Singrauli (Varshney et al., 2021), and others. However, such information on the surface water quality of the Sohagpur coalfield is lacking. Thus, the key aim of this study is to assess the surface water and chemistry for suitability for drinking and domestic and irrigation uses in the coalfield.

9.2 STUDY AREA

The Sohagpur coalfield is located in the district of Shahdol, Madhya Pradesh (Figure 9.1). In the district, from December to February it is winter, and from March to mid-June it is hot summer (CGWB 2013). The district received an average rainfall of 1131.4 mm (CGWB 2013). The Son River and its small tributaries pass through

DOI: 10.1201/9781003442554-11

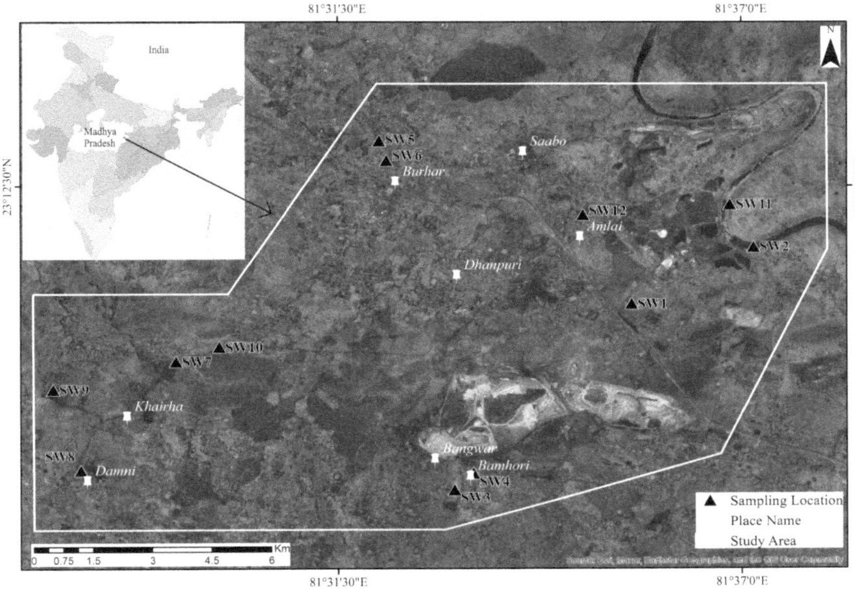

FIGURE 9.1 Surface water sampling location map.

the study area. The study area has agriculture, mining, paper industry, and small business. The irrigation practices in the area depend on canals and tanks/ponds. Moreover, groundwater resources are also used for irrigation (CGWB 2013). The Sohagpur coalfield geology comprises the Barakar, Tachir, Lameta, Deacan trap, Pali, and metamorphic Formations. The details of these formations are described elsewhere (Pareek 1987, Agnihotri et al. 2016, Mondal et al. 2020).

9.3 SAMPLING AND ANALYSIS

Twelve surface water samples were collected from various locations, such as Bemhauri, Budhar, Navaganv, Almai, Kandhoha, and Karkati, in the Sohagpur coalfield in March 2022:nine samples from Ponds, two samples (SW1 and SW11) from the Son River, and one sample (SW9) from Nala (channel), respectively (Figure 9.1). For the collection of samples, 500 ml prewashed polyethylene bottles were used. Electrical conductivity (EC) and pH values were measured on-site using a portable pH and conductivity meter (Hanna HI98130). The Thermo Orion turbidity meter (AQ3010) was used for turbidity value measurement in the lab. To analyze major ions, surface water samples were filtered through 0.45 μm millipore membrane filters. Bicarbonate (HCO_3^-) was determined using the titration method (APHA 1998). The Ion Chromatography (IC) was used for the analysis of major anions (F^-, Cl^-, NO_3^-, and SO_4^{2-}) (Metrohm 883, Basic Ion Chromatography Plus with MetrosepaSupp 5 - 250/4.0 column) and major cations (Ca^{2+}, Mg^{2+}, Na^+, and K^+) (Metrohm 930 Compact IC Flex with Metrosep C4-150/4.0 column).

9.4 QUALITY CONTROL

During the sample collection and analysis of surface water, proper quality control was exercised to avoid impurities. For pH and EC, some of the surface samples were determined using a pH and conductivity meter (Thermo Scientific VSATAR 52) in addition to a handy pH and conductivity meter. Also, Na^+ and K^+ were analyzed in some of the samples with a Flame Photometer in addition to IC. The results were compared, and a deviation of <5% was observed. A charge balance error was calculated and was <5% for samples except for one sample.

9.5 WATER QUALITY INDICES

Water quality indices are essential for determining the water quality for irrigation and drinking. In this study, WQI (water quality index) for surface water samples of the coalfield was estimated considering 11 significant parameters, such as turbidity, TDS, pH, SO_4^{2-}, NO_3^-, Cl^-, F^-, TH, Mg^{2+}, Na^+, and Ca^{2+}, to assess the appropriateness of the samples for drinking purpose. Every parameter was assigned a weightage factor of 2–5 grounded on their influence on water quality and health (Vasanthavigar et al. 2010, Tiwari et al. 2017). For instance, a weight of 5 was assigned to TDS, F^-, and NO_3^- while 4 to turbidity, pH, Cl^-, and SO_4^{2-}. Besides, three weights were assigned to total hardness and two to Na^+, Mg^{2+}, and Ca^{2+}. Table 9.1 shows the equations that were applied to estimate the WQI (Vasanthavigar et al. 2010, Tiwari et al. 2017). Water quality is classified into five classes based on the value of WQI, such as WQI <50 (Excellent water), WQI 50–100 (Good water), WQI 100–200 (Poor water), WQI 200–300 (Very poor water), and WQI >300 (Unfit for drinking use) (Vasanthavigar et al. 2010, Tiwari et al. 2017).

TABLE 9.1
Equation to Estimate the WQI

$$W_i = w_i / \sum_{i=1}^{n} w_i \tag{9.1}$$

where n = number of parameters and w_i = weightage of each water parameter.

$$q_i = \left(\frac{C_i}{S_i}\right) \times 100 \tag{9.2}$$

q_i (quality rating) = calculated for each parameter which is done by dividing its estimated value in each water sample (C_i) by the corresponding drinking water quality standards (S_i) (BIS 2012, WHO 2017). Then the result is multiplied by 100. C_i and S_i are given in mg/L except for pH and turbidity.

$$\mathrm{SI}_i = W_i \times q_i \tag{9.3}$$

SI (Sub Index) = estimated, which is the product of the relative weight (W_i) and q_i.

$$\mathrm{WQI} = \sum_{i=1}^{n} \mathrm{SI}_i \tag{9.4}$$

SI_i = sub-index of ith parameter, n = number of considered parameters.

TABLE 9.2
Parameters Indicating the Suitability of Water for Irrigation

SAR	=	$Na/[(Ca + Mg)/2]^{0.5}$	(9.5)
Na%	=	$Na + K/(Ca + Mg + Na + K) \times 100$	(9.6)
RSC	=	$(CO_3 + HCO_3) - (Ca + Mg)$	(9.7)
KI	=	$Na/(Ca + Mg)$	(9.8)
MH	=	$Mg/(Ca + Mg) \times 100$	(9.9)

Source: After Shainberg and Oster (1976) and Todd (1980).
Note: All concentrations are in meq/L.

9.6 SUITABILITY FOR IRRIGATION

The sodium adsorption ratio (SAR), percent sodium (%Na), Kelley index (KI), residual sodium carbonate (RSC), and magnesium hazard (MH) were calculated to determine whether water was suitable for irrigation use. The estimation formula is presented in Table 9.2.

9.7 RESULTS AND DISCUSSION

The statistically analyzed parameters for Sohagpur coalfield's surface water samples ($n = 12$), such as minimum, maximum, mean, and standard error, are provided in Table 9.3.

9.7.1 pH, Turbidity, EC, and TDS

pH of the samples in the Sohagpur coalfield ranges from 5.9 to 8.8, with an average value of 7.6. The study area's surface water is acidic to alkaline in nature. Turbidity in the surface water samples varies from 0.93 NTU to 50.1 NTU, with an average value of 11 NTU in the study area (Table 9.3). EC in the surface water samples varies from 62.9 to 2,130 µS/cm with a mean value of 668 µS/cm in the coalfield (Table 9.3). Total dissolved solids (TDS) values of the surface water samples range between 51.1 and 1,636 mg/L, with a mean value of 464 mg/L in the study area (Table 9.3). Freeze and Cherry (1979) have classified water into four important classes based on the value of TDS in water, such as freshwater (TDS <1,000 mg/L), brackish water (>1,000 mg/L), saline water (>10,000 mg/L), and brine water (100,000 mg/L), respectively. Based on the Freeze and Cherry (1979) classification, around 92% of the samples belong to the freshwater category, and one sample (SW6) belongs to the brackish water class in the Sohagpur coalfield.

9.7.2 Major Ion Chemistry

In this study, HCO_3^-, Cl^-, and Ca^{2+} are leading dissolved ions in the surface water of the coalfield and contribute around 33%, 28%, and 18% to TDS concentration along with

TABLE 9.3
Surface Water Samples Exceeding the Desirable Limits Prescribed by the Bureau of Indian Standards (BIS, 2012) for Drinking Purposes

	BIS 2012 (IS: 10500)		Groundwater ($n=12$)				
Water Quality Parameters	Maximum Permissible Limits	Maximum Acceptable Limits	Minimum	Maximum	Mean	Standard Error	The Number of Samples and % of Samples Exceeded Acceptable Limits
pH	–	6.5–8.5	5.9	8.8	7.6	0.2	2[a] and 17[a]
EC	–	–	62.9	2130	668	167	–
Turbidity	–	<5.0	0.93	50.1	11	3.97	07 and 58
HCO_3^-	–	–	31.1	286.4	155.7	24.1	–
F^-	1.5	1.0	0.24	0.64	0.37	0.04	–
Cl^-	1,000	250	3.5	760	132	65.3	2 and 17
NO_3^-	–	45	0.10	27.12	9.13	2.86	–
SO_4^{2-}	400	200	0.98	123.3	32.2	12.25	–
Na^+	–	–	2.7	72.3	26.9	7.1	–
Ca^{2+}	200	75	5.2	385	86.2	30.0	04 and 33
Mg^{2+}	100	30	2.5	67.2	20.1	5.3	03 and 25
K^+	–	–	1.6	17.2	6.8	1.3	–
TDS	2,000	500	51.1	1636	464	122	04 and 33
TH	600	200	23.5	1238	298	94	05 and 42

Note: Unit: Concentration in mg/L, except EC (μS/cm), Turbidity (NTU) and pH.
BDL, below detection limit; EC, electrical conductivity; TDS, total dissolved solids; TH, total hardness.

[a] Not within the acceptable range.

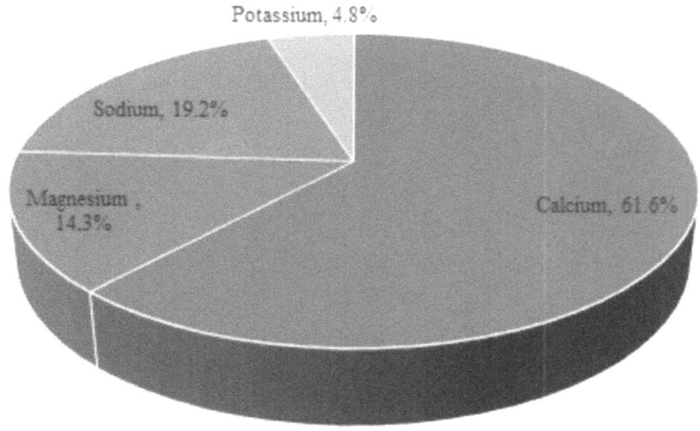

FIGURE 9.2 Major cation chemistry of the Sohagpur coalfield surface water.

SO_4^{2-} (7%), Na^+ (6%), and Mg^{2+} (4%), respectively. Moreover, NO_3^-, K^+, and F^- contribute very little (<2%) to TDS concentration in the study area. In the case of cation chemistry, calcium and sodium are the primary dominant ions to surface water cation chemistry, followed by magnesium and potassium ions in the coalfield, respectively. The order of cation abundance in the surface water samples is $Ca^{2+}>Na^+>Mg^{2+}>K^+$ in the coalfield. Calcium concentration in the surface water samples varies from 5.2 to 385 mg/L with a mean value of 86.5 mg/L in the coalfield (Table 9.3) and contributes 61.6% to surface water cation chemistry in the area (Figure 9.2). Magnesium concentration ranges between 2.5 and 67.2 mg/L with an average concentration of 20.1 mg/L in the samples and contributes 14.3% to surface water cation chemistry (Figure 9.2). However, sodium and potassium vary from 2.7 to 72.3 mg/L (avg. 26.9 mg/L) and from 1.6 to 17.2 mg/L (mean 6.8 mg/L) (Table 9.3) and contribute 19.2% and 4.8% surface water cation chemistry in the area, respectively (Figure 9.2).

In the case of anion chemistry, bicarbonate and chloride are the chief dominant ions to surface water anion chemistry, followed by sulfate, nitrate, and fluoride ions in the coalfield, respectively. The order of anions abundance is $HCO_3^->Cl^->SO_4^{2-}>NO_3^->F^-$ in the surface water samples of the coalfield. Concentration of bicarbonate varies from 31.1 to 286.4 mg/L with a mean value of 155.7 mg/L (Table 9.3) and contributes 47.3% to the surface water anion chemistry of the study area (Figure 9.3). Concentrations of chloride and nitrate range between 3.5 and 760 mg/L (avg. 132 mg/L) and between 0.98 and 123.3 mg/L (mean 32.2 mg/L) (Table 9.3) and contribute 40% and 9.8% the surface water anion chemistry, respectively (Figure 9.3). However, nitrate and fluoride vary from 1 to 27.12 mg/L (mean 9.13 mg/L) and 0.24 to 0.64 mg/L (avg. 0.37) (Table 9.3) and contribute 2.8% and 0.1% to the surface water anion chemistry, respectively (Figure 9.3).

9.7.3 Water Type

The Piper (1944) diagram helps to understand the associations of different dissolved elements in water and classify the water types based on their chemical characteristics.

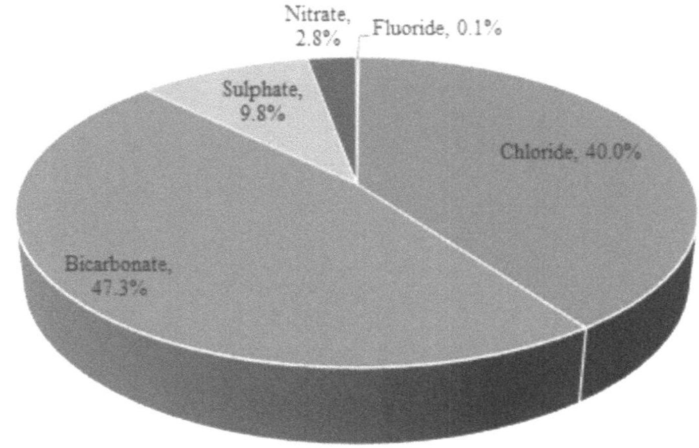

FIGURE 9.3 Major anion chemistry of the Sohagpur coalfield surface water.

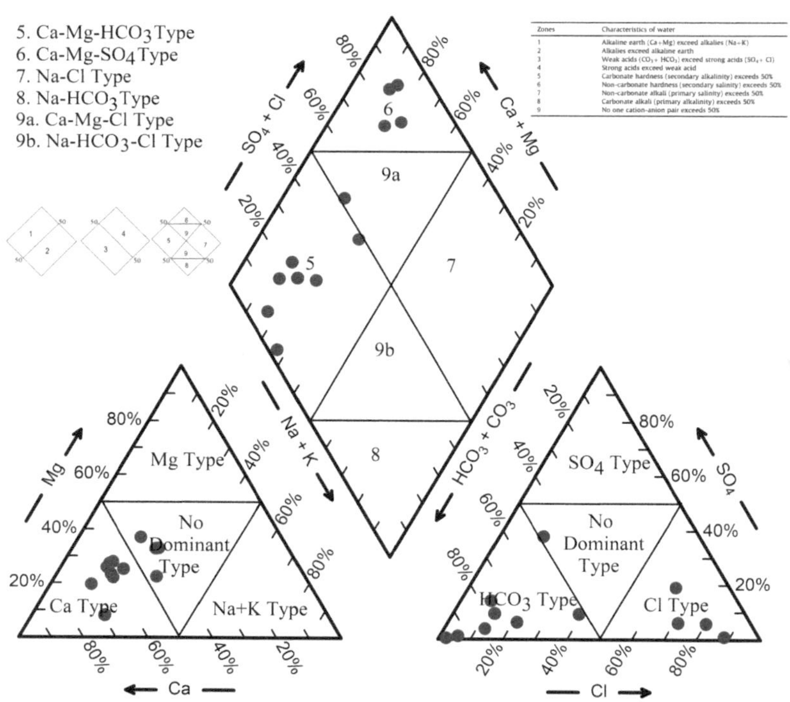

FIGURE 9.4 Piper's trilinear diagram showing the relationship between dissolved ions and water type.

The triangular anionic zone of the Piper diagram shows that about 58% of the surface water samples fall into the bicarbonate zone, while around 33% of the samples (SW1, SW4, SW5, and SW6) fall into the chloride zone (Figure 9.4). Only one sample

(SW9) falls into a no-dominant zone in the study area. It is suggested that weak acid (HCO_3^-) dominates over strong acid ($SO_4^{2-}+Cl^-$) in the coalfield surface water samples. However, the triangular cationic zone of the Piper diagram suggests that 67% of the samples fall into the calcium-dominated zone, while the rest of the 33% samples (SW8, SW9, SW10 and SW12) fall into a no-dominant zone in the study area (Figure 9.4). It indicates that $Ca^{2+}+Mg^{2+}$ (alkaline earth metals) exceed Na^++K^+ (alkali metal cations) in the coalfield surface water samples.

The diamond shape of the Piper diagram reveals that 58% of the samples fall into zone 5 and 33% of the samples in zone 6, respectively, while one of the samples falls into zone 9a in the study area (Figure 9.4). The Piper diagram reveals that the six surface water samples (SW2, SW3, SW7, SW8, SW10, and SW11) have Ca^{2+}–Mg^{2+}–HCO_3^- water type, while the rest of the samples have Ca^{2+}–Mg^{2+}–$SO4_2^-$–Cl^- water type in the Sohagpur coalfield (Figure 9.4).

9.7.4 Suitability for Drinking and Domestic Uses

The analyzed parameters of surface water are compared with the Bureau of Indian Standards (BIS 2012) guideline for drinking and health of the public (Table 9.3). About 83% of the samples are within the BIS (2012) acceptable range of 6.5–8.5 in the study area. However, the pH of one sample (SW4) is below the range of 6.5, and one sample (SW2) is above the range of 8.5 of the BIS (2012), respectively (Table 9.3), suggesting that both samples are not appropriate for direct consumption. Drinking water that is excessively turbid shows a health risk because it can shield pathogenic microorganisms from the effects of disinfectants and encourage bacterial growth during storage. (Singh et al. 2013; Tiwari and Singh 2014). In this study, around 58% of the samples exceed the turbidity limit of 5 NTU (BIS 2012) and are unsuitable for direct drinking use. However, the remaining samples are well within the recommended BIS (2012) turbidity limit in the study area. TDS concentration exceeds the BIS (2012) acceptable limit of 500 mg/L in 33% of surface water samples, while all the samples are within the maximum permissible limit of 2,000 mg/L in the study area (Table 9.3). Water can be classified into four hardness classes based on the value of hardness, such as soft water (75 mg/L), moderately hard water (75–150 mg/L), hard water (150–300 mg/L), and very hard water (>300 mg/L) (Sawyer and McCarty 1967). In the study area, one surface water sample belongs to the soft and hard water classes, while five samples belong to moderately hard to very hard water classes. About 42% of the samples exceed the BIS total hardness acceptable limit of 200 mg/L in the study area (Table 9.3). The concentrations of F^-, Cl^-, and SO_4^{2-} are within the BIS acceptable limit of 1.0 mg/L, 250 mg/L, and 200 mg/L in the surface water samples (Table 9.3). However, around 17% of the surface samples exceed the nitrate acceptable limit of 45 mg/L as per the BIS guidelines in the study area. About 33% and 25% of the samples are above the acceptable limit of calcium (75 mg/L) and magnesium (30 mg/L) in the study area.

The calculated WQI shows that 42% and 42% of the surface water samples belong to the very good and good water category and are for use for domestic purposes (Table 9.4). However, one sample (SW4) is in the poor water class, and one sample (SW6) is in the very poor water class. Both samples are unsuitable for drinking use and require appropriate treatment before use.

TABLE 9.4
Calculated Water Quality Index

Location	WQI	Class
SW1	85	Good
SW2	35	Very Good
SW3	48	Very Good
SW4	132	Poor
SW5	84	Good
SW6	228	Very Poor
SW7	32	Very Good
SW8	58	Good
SW9	82	Good
SW10	26	Very Good
SW11	37	Very Good
SW12	83	Good

9.7.5 Suitability for Irrigation Uses

Sodium (Na) is an important parameter in understanding the suitability of water for irrigation uses. If water has a high sodium value, it may lead to the formation of alkaline soil. Similarly, the EC concentration plays a vital role in classifying irrigation water types. If water has a high EC value, it may lead to the formation of saline soil. The calculated SAR value varies from 0.13 to 1.5, with an average value of 0.63 in the study area. All the tested samples belong to a low class (SAR <6) (Figure 9.5) and suggest that the Sohagpur coalfield surface samples can be used for irrigation. However, the EC of the surface water samples varies between 62.9 and 2,130 μS/cm with an average of 668 μS/cm in the coalfield. About 67% of the samples belong to medium-salinity classes (250–750 μS/cm). In comparison, the rest of the 33% surface water samples belong to high-salinity class (750–2,250 μS/cm) (USSL 1954) in the study area (Figure 9.5). Samples SW1, SW5, and SW6 have high EC concentrations and can be unsuitable for irrigation of normal crops (Figure 9.5).

Percent sodium (%Na) is another significant parameter to assess the suitability of water for irrigation uses (Singh et al. 2008). The %Na values in the surface water samples range from 12.7% to 31.5% with an average value of 19.8%, and all the tested samples are below the proposed Indian Standard's highest value of 60% (Singh et al. 2008). The Wilcox diagram (1955) shows that about 67% of the samples fall into the excellent to good category, while around 25% of the samples fall into the good to permissible class. These samples can be used for irrigation without any risk in the study area, except for one sample (Figure 9.6).

The RSC (Residual Sodium Carbonate) is also an important calculation to understand water suitability for irrigational uses. In the present study, the values of RSC

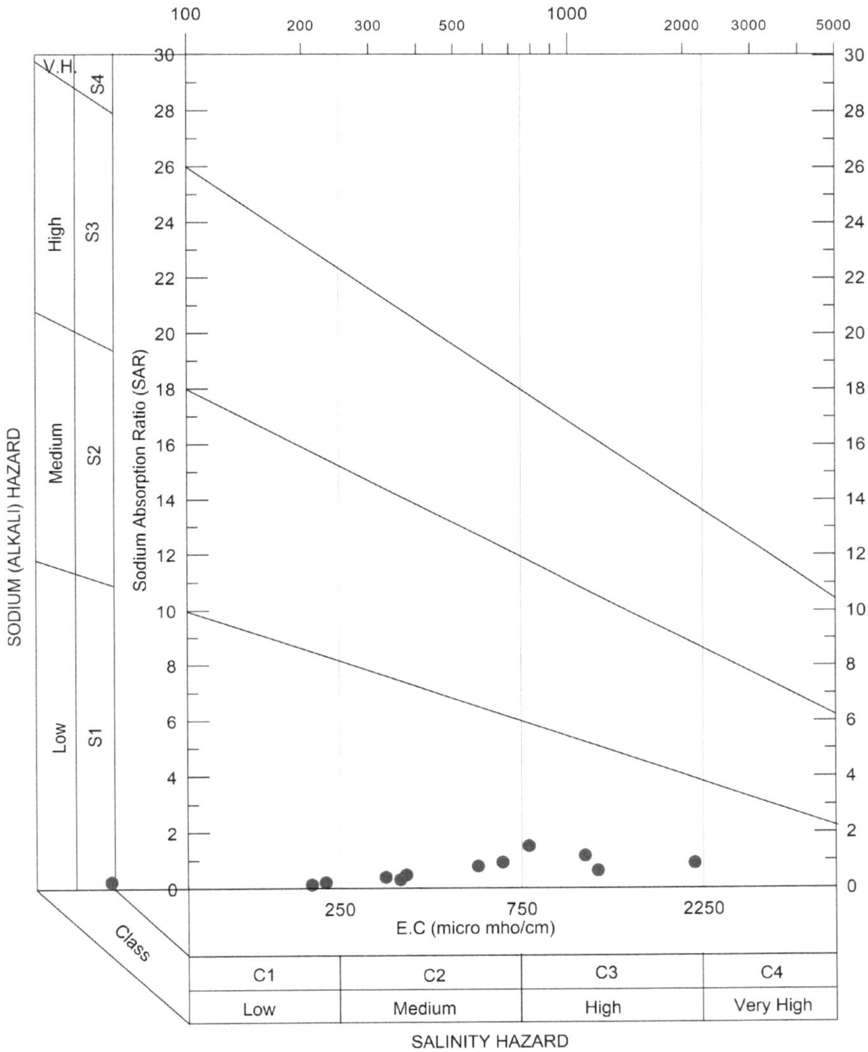

FIGURE 9.5 US salinity diagram (USSL) for classification of irrigation water.

in the samples vary from −20.7 to 0.80, with an average of −3.40, suggesting that samples are suitable for irrigation purposes. The Kelley index (KI) and magnesium hazard (MH) are also noteworthy in categorizing water for irrigation uses. Water with a value of KI > 1.0 and MH > 50% is inappropriate for irrigation uses (Paliwal 1967, Kelley 1947). In this study, surface water samples are below the recommended values of KI (>1.0) and MH (>50%), suggesting that surface water is suitable for irrigation uses in the area.

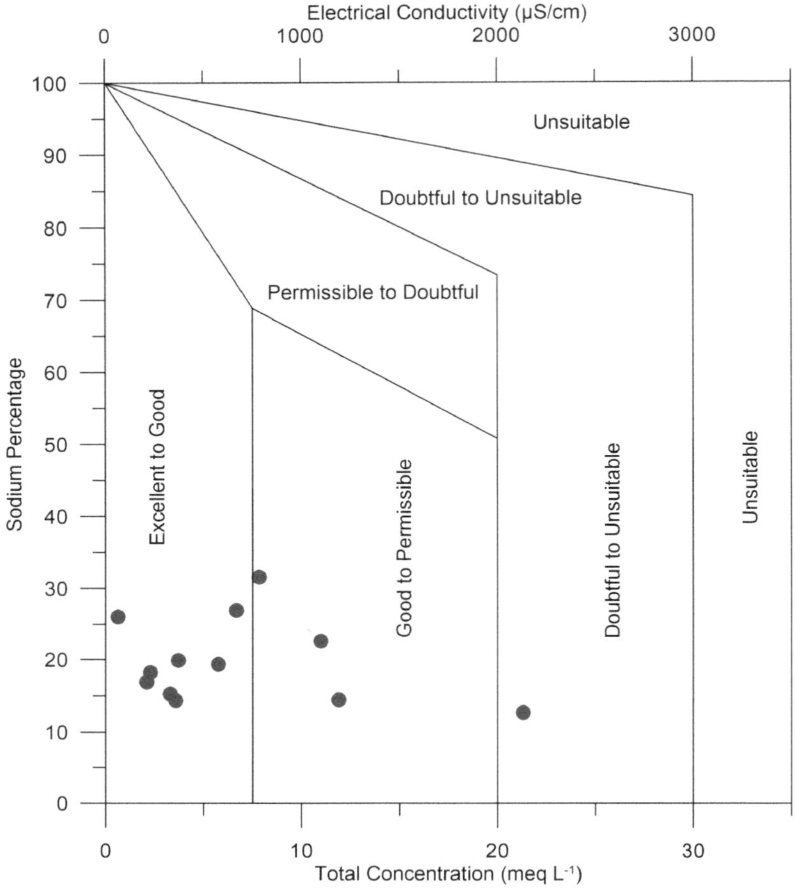

FIGURE 9.6 Plot of sodium percent versus electrical conductivity (after Wilcox 1955) of the groundwater in the study area.

9.8 CONCLUSIONS

1. The coalfield surface water is acidic to alkaline in nature and has an average EC of 668 μS/cm.
2. HCO_3^-, Cl^-, and Ca^{2+} are leading dissolved ions in the surface water of the coalfield and contribute around 33%, 28%, and 18% to TDS concentration along with SO_4^{2-} (7%), Na^+ (6%), and Mg^{2+} (4%), respectively.
3. The study surface water samples are Ca^{2+}–Mg^{2+}–HCO_3^- and Ca^{2+}–Mg^{2+}–SO_4^{2-}–Cl^- water types.
4. Some samples with high turbidity, chloride, TDS, TH, and WQI values are unsuitable for drinking uses and require appropriate treatment before utilization.
5. Irrigation indices suggest that the surface water samples are appropriate for irrigation uses in the study area, except for salinity risk.

ACKNOWLEDGMENT

The authors thank the Director of CSIR-CIMFR, Dhanbad, and the Vice-chancellor of JNU, New Delhi, for research facilities. One author (AKT) is grateful to the University Grants Commission (UGC) for the financial support through the BSR-UGC start-up grant (No.F. 30-553/2021(BSR)). The authors thank Mr. Saurabh Kumar Singh, School of Environmental Science, JNU for his help.

REFERENCES

Abhishek, T. R., & Sinha, S. K. (2006). Status of surface and ground water quality in coal mining and industrial areas of Jharia coalfield. *Indian Journal of Environmental Protection*, 26(10), 905–910.

Agnihotri, D., Tewari, R., Pillai, S. S. K., Jasper, A., & Uhl, D. (2016). Early Permian Glossopteris from the Sharda Open Cast mine, Sohagpur Coalfield, Shahdol District, Madhya Pradesh. *Paleobotony*, 65, 97–107

APHA (1998). *Standard Methods for the Examination of Water and Wastewater*, 20th edn. American Public Health Association, Washington DC.

BIS (2012). *Drinking Water Specifcations,* IS:10500, 2nd revision. Bureau of Indian Standards, New Delhi. ftp://law.resource.org/in/bis/S06/is.10500.2012.pdf

CGWB (2013). Report on District at a Glance, Shahdol district, Madhya Pradesh, North Central Region, Bhopal

Freeze, R. A., & Cherry, J. A. (1979). *Groundwater.* Prentice-Hall, London.

Huang, X., Sillanpää, M., Gjessing, E. T., Peräniemi, S., & Vogt, R. D. (2010). Environmental impact of mining activities on the surface water quality in Tibet: Gyama valley. *Science of the Total Environment*, 408(19), 4177–4184.

Karn, S. K., & Harada, H. (2001). Surface water pollution in three urban territories of Nepal, India, and Bangladesh. *Environmental Management*, 28, 483–496.

Kelley, W. P. (1946). Permissible composition and concentration of irrigation waters. *Proceeding American Society of Civil Engineering,* 106. https://doi.org/10.1061/TACEAT.0005384.

Kumar, P., & Singh, A. K. (2022). Hydrogeochemistry and quality assessment of surface and sub-surface water resources in Raniganj coalfield area, Damodar Valley, India. *International Journal of Environmental Analytical Chemistry*, 102(19), 8346–8369.

Lee, A. A., & Bukaveckas, P. A. (2002). Surface water nutrient concentrations and litter decomposition rates in wetlands impacted by agriculture and mining activities. *Aquatic Botany*, 74(4), 273–285.

Mahato, M. K., Singh, A. K., Singh, P. K., & Singh, G. (2021). Evaluation of factors influencing surface water quality in a coalfield area of Damodar valley, India: A sustainable uses. *International Journal of Environmental Analytical Chemistry*, 1–23. doi:10.1080/03067319.2021.1946685

Mondal, D., Roy, P. N. S., & Kumar, M. (2020). Monitoring the strata behavior in the destressed zone of a shallow Indian longwall panel with hard sandstone cover using mine-microseismicity and borehole televiewer data. *Engineering Geology,* 271(105593), 1–25.

Ning, L., Liyuan, Y., Jirui, D., & Xugui, P. (2011). Heavy metal pollution in surface water of Linglong gold mining area, China. *Procedia Environmental Sciences*, 10, 914–917.

Obiadi, I. I., Obiadi, C. M., Akudinobi, B. E. B., Maduewesi, U. V., & Ezim, E. O. (2016). Effects of coal mining on the water resources in the communities hosting the Iva Valley and Okpara Coal Mines in Enugu State, Southeast Nigeria. *Sustainable Water Resources Management*, 2, 207–216.

Paliwal, K. V. (1967). Effects of gypsum application on the quality of irrigation water; The Madras. *Agriculture Journal,* 59, 646–647

Pareek, H. S. (1987). Petrographic, chemical and trace–elemental composition of the coal of SohagpurCoalfeld, Madhya Pradesh, India. *International Journal of Coal Geology,* 9, 187–207

Pastor, J., & Hernández, A. J. (2012). Heavy metals, salts and organic residues in old solid urban waste landfills and surface waters in their discharge areas: Determinants for restoring their impact. *Journal of Environmental Management*, 95, S42–S49.

Piper, A. M. (1944). A graphical procedure in the geochemical interpretation of water analysis. *Transactions of the American Geophysical Union,* 25, 914–928

Rösner, U. (1998). Effects of historical mining activities on surface water and groundwater-an example from northwest Arizona. *Environmental Geology*, 33, 224–230.

Sahoo, B. P., Sahu, H. B., & Pradhan, D. S. (2021). Hydrogeochemistry and surface water quality assessment of IB valley coalfield area, India. *Applied Water Science*, 11(9), 153.

Sahoo, M., Mahananda, M. R., & Seth, P. (2016). Physico-chemical analysis of surface and groundwater around Talcher coal field, district Angul, Odisha, India. *Journal of Geoscience and Environment Protection*, 4(2), 26.

Sawyer, C. N., & McCarty, P. L. (1967). *Chemistry of Sanitary Engineers*, 2nd edn. McGraw Hill, New York

Schneider, P., Nilius, U., Gottschalk, N., Süß, A., Schaffrath, M., Löser, R., & Lange, T. (2017). Determination of the geogenic metal background in surface water: Benchmarking methodology for the rivers of Saxony-Anhalt, Germany. *Water*, 9(2), 75.

Shainberg, I., & Oster, J. D. (1976). *Quality of Irrigation Water*. International Irrigation Information Center.

Singh, A. K., Raj, B., Tiwari, A. K., & Mahato, M. K. (2013). Evaluation of hydrogeochemical processes and groundwater quality in the Jhansi district of Bundelkhand region, India. *Environmental Earth Science,* 70(3):1225–1247

Singh, A. K., Mondal, G. C., Kumar, S., Singh, T. B., Tewary, B. K., & Sinha, A. (2008). Major ion chemistry, weathering processes and water quality assessment in upper catchment of Damodar River basin, India. *Environmental Geology*, 54, 745–758.

Srinivasa, S., & Govil, P. K. (2008). Distribution of heavy metals in surface water of Ranipet industrial area in Tamil Nadu, India. *Environmental Monitoring and Assessment*, 136, 197–207.

Subrahmanyam, K., & Yadaiah, P. (2001). Assessment of the impact of industrial effluents on water quality in Patancheru and environs, Medak district, Andhra Pradesh, India. *Hydrogeology Journal*, 9, 297–312.

Tiwari, A. K.,& Singh, A. K. (2014). Hydrogeochemical investigation and groundwater quality assessment of Pratapgarh district, Uttar Pradesh. *Journal of Geological Society of India,* 83(3), 329–343

Tiwari, A. K., & De Maio, M. (2018). Assessment of sulphate and iron contamination and seasonal variations in the water resources of a Damodar Valley Coalfield, India: A case study. *Bulletin of Environmental Contamination and Toxicology*, 100, 271–279.

Tiwari, A. K., De Maio, M., Singh, P. K., & Mahato, M. K. (2015). Evaluation of surface water quality by using GIS and a heavy metal pollution index (HPI) model in a coal mining area, India. *Bulletin of Environmental Contamination and Toxicology*, 95, 304–310.

Tiwari, A. K., Singh, P. K., & Mahato, M. K. (2016). Hydrogeochemical investigation and qualitative assessment of surface water resources in West Bokaro coalfield, India. *Journal of the Geological Society of India*, 87, 85–96.

Tiwari, A. K., Singh, A. K., Singh, A. K., & Singh, M. P. (2017). Hydrogeochemical analysis and evaluation of surface water quality of Pratapgarh district, Uttar Pradesh, India. *Applied Water Science*, 7, 1609–1623.

Todd, D. (1980). *Ground Water Hydrology*, 2nd edn, Wiley, New York.

USSL. (1954). *Diagnosis and Improvement of Saline and Alkali Soils*. US Department of Agriculture Hand Book No 60 . US Salinity Laboratory.

Varshney, R., Modi, P., Sonkar, A. K., Singh, P., & Jamal, A. (2022). Assessment of surface water quality in and around Singrauli coalfield, India and its remediation: An integrated approach of GIS, water quality index, multivariate statistics and phytoremediation. *Arabian Journal of Geosciences*, 15(18), 1530.

Vasanthavigar, M., Srinivasamoorthy, K., Vijayaragavan, K., Ganthi, R. R., Chidambaram, S., Anandhan, P., & Vasudevan, S. (2010). Application of water quality index for groundwater quality assessment: Thirumanimuttar sub-basin, Tamilnadu, India. *Environmental Monitoring and Assessment,* 171, 595–609

WHO (2017). *Guidelines for Drinking-Water Quality*, 4th Edition, Incorporating the First Addendum. World Health Organization, Geneva. https://apps.who.int/iris/bitstream/handle/10665/254637/9789241549950-eng.pdf?sequence=1

Wilcox, L. V. (1955). *Classification and Use of Irrigation Waters.* US Department of Agriculture Cir 969, US Department of Agriculture, Washington, DC

Wu, P., Tang, C., Liu, C., Zhu, L., Pei, T., & Feng, L. (2009). Geochemical distribution and removal of As, Fe, Mn and Al in a surface water system affected by acid mine drainage at a coalfield in Southwestern China. *Environmental Geology*, 57, 1457–1467.

10 Water Scarcity and Desertification of Mining Landscapes

Anita Punia and Saurabh Kumar Singh

10.1 INTRODUCTION

Soil a productive layer of the earth crust accomplished by the variety of organic and inorganic components needed for the survival of life. The soil is storehouse of nutrients required for the growth of vegetation. The desertification is a land degradation process, which converts fertile land to unproductive land caused by lack of water and poor vegetation cover. The good water-holding capacity of soil favors the dissolution of minerals and uptake of nutrients supporting the growth and development of vegetation (Bergkamp 1995; Peng et al. 2020). The growth of plants and stoppage of erosion of topsoil improve the water holding capacity and nutrient content of soil (Wang et al. 2012). In a simple word, the desertification is mainly caused by the lack of adequate water. The inadequate water is attributed to a lack of precipitation and overexploitation of water resources.

Desertification is mainly reported in arid and semiarid regions due to natural (dry climate and high wind speed) and anthropogenic activities (overgrazing, overcultivation, exploitation of groundwater and fuelwood) (Tao 2014) creating economic, social, and environmental problems. The decline in freshwater resources is observed in the Asian region as well as at the global level suggesting the seriousness of the issue (Foster et al. 2013; Huang et al. 2021). The decline of freshwater resources leads to food scarcity and also disturbs the functioning of an ecosystem. The mines exploit water resources excessively and leading to rapid decline in availability of freshwater (Custodio et al. 2017). The semiarid and arid regions face major problems in accessing the freshwater resources compared to wet regions. The decrease in availability of water resources due to mining in the arid and semiarid regions further contributes to rise in intensity of desertification. The number of mines are continuously increasing in the drylands across the globe (Shen et al. 2021a).

The land use and land cover studies confirm the depletion of water resources and vegetation cover in the mining landscapes (Punia et al. 2021). The decrease in water resource and loss of vegetation could result in desertification of the region (Albaladejo et al. 1998; Amin 2004). In recent times, many anthropogenic activities such as excessive exploitation of water resources for agriculture, drinking purpose, and other development activities further aggravate the situation. It is very difficult to justify the mining as only reason for the desertification; however, it can be stated that mines intensify the process of desertification in the landscape.

DOI: 10.1201/9781003442554-12

Environmental pollution such as water, soil, air and vegetation, and degradation of natural resources due to mines are well reported in the literature (Tepanosyan et al. 2018; Keovilignavong 2019; Vendrell-Puigmitja et al. 2021). Mining activities during ore excavation such as dredging, digging, drilling and blasting, transportation, and mineral processing lead to dispersal of fine particles of particulate matter resulting in environmental contamination (Petavratzi and Lowndes 2005; Saarikoski et al. 2018). The unsafe execution of ore excavation processes could result in faults and fractures in aquifer leading to leaching of elements into the groundwater (Hebblewhite 2020; Song and Liang 2021). In contrast, during the ore beneficiation process, one of the main causes of elemental pollution is use of different chemicals (e.g. mercury, arsenic, and cyanide) to concentrate the minerals (Wu et al. 2017; Niane et al. 2019; Zhou et al. 2021; Larrabure and Rodríguez-Reyes 2021). Mercury (Hg) is being used to concentrate gold through amalgamation process. The gold mines are known for Hg pollution and impacting the organisms of the area (Salomons 1995). Recent studies show rise in contamination of other metals such as Zn, Cr, Cd, Mn, Pb, and As in the mining environment due to gold mines (Barcelos et al. 2020). Artisanal or small-scale gold mines active in Southern Ecuadorian Amazon are polluting the Nangaritza River basin with other metals specifically Mn in addition to Hg (González-Merizalde et al. 2016).

Mines attract migration of people and also create employment opportunities resulting in rise in population. Industrialization and urbanization are one of the main causes for the overexploitation of water resources leading to depletion of aquifers (Calderhead et al. 2012; Zhang et al. 2019). Thus, it can be stated that mines directly or indirectly play an important role in desertification of the region. The main objectives of the study are to understand the factors leading to desertification of mining region and mitigation methods for its prevention.

10.2 IMPACTS OF MINES ON WATER RESOURCES

The water resources are essential for vegetation growth and survival. The mining activities adversely impact the water resources in terms of both quality and quantity, and it results in the decline of vegetation cover. In the following section, the adverse impacts of mines on water resources are discussed.

10.2.1 Quality

Mining activities deteriorate the quality of water, which is not safe for drinking purposes without treatment (Mhlongo et al. 2018; Gbedzi et al. 2022; Punia and Siddaiah 2017). Impacts of mines on the surface and groundwater depend on bedrock geology controlled majorly by rock–water interaction (Zhang et al. 2021; Arkoc et al. 2016). The rise of elemental concentration specifically As, Pb, Hg, Cd, and Se in the drinking water of Abakaliki, southeast Nigeria, due to lead-zinc mining, impacts health of the people residing in the area (Obasi and Akudinobi 2020). The contamination of water due to mines also impacts biodiversity of aquatic ecosystem (Simonin et al. 2021). The concentration of metals is observed higher in the bottom feeders and organisms at higher trophic level suggesting that biomagnification of metals from prey to predator in the acid mine drainage impacted the areas of southern China (Chan et al. 2021). The water quality in Brajrajnagar coal mining area of India is not suitable for irrigation purpose (Sahoo and Khaoash 2020). The leached elements from mine spoils neutralized on reacting with aluminosilicate and carbonate dissolution leading to decrease in elemental concentration (Clark et al. 2018). The recovery rate of water chemistry deteriorated because mining is very slow and takes decades to reach its natural geochemical conditions (Cianciolo et al. 2020).

The concentration of metals is mostly observed high near the waste dumps (Punia et al. 2017; Neiva et al. 2016). The remediation of abandoned mine or waste dumps is not always enough to the prevent the adverse impacts. The metal concentrations (As, Al, Mn, and Ni) were observed above the acceptable limits in the surface water close to open-pit mine and waste dumps at Murçós mining complex, NE Portugal, even after the remediation of sites (Antunes et al. 2016). Most of water sources in the region contain As in As (III) form, which is more toxic and mobile in nature.

The sulfide-enriched waste from Pb, Zn, Fe, and Cu mines generates Acid Mine Drainage (AMD) and mobilizes metals, i.e., Cd, Cr, Pb, Zn, Cu, Co, etc., into the surroundings (Nieva et al. 2018; Cánovas et al. 2021; Helser et al. 2022). It is difficult to name particular metal contamination in the sulfidic mining areas because sulfides form acid on reacting with water and oxygen, which lower the pH and mobilize elements from the parent rock. The red mud generated from bauxite mining is enriched in iron and aluminum (Al) oxides and major source of Al contamination of nearby environment. It results in loss of soil fertility, adversely impacts the survival of microorganisms, and poses high risk to public health (Pradhan et al. 2023). The soil and water bodies close to chromite ore processing waste dumping sites are highly contaminated with Cr (Pattnaik and Equeenuddin 2016; Tiwary et al. 2017). The calculated cancer risk for sites close to chromite ore processing residues at Rania, Uttar Pradesh, India, is observed to be 1–2 higher than the safe limit, indicating the human health risk (Guleria et al. 2022). In the uranium mining regions across the world, the major concerns for policymakers are health hazards related to radiations and elevated level of radioactive elements specifically U (Carvalho et al. 2016; Semenova et al. 2020; de Souza Pereira et al. 2020).

10.2.2 Quantity

In India, the guidelines for water quantification during the operation of open and underground mines are lacking (Soni 2020). The national average water resources depletion rate due to coal mining is 1.32 m^3 per ton of coal in China (Alun and Li 2017). In south China, the depletion rate of water resources is three times more than the national average, which is attributed to higher rate of pollution and consumption due to coal mining. In the mining region of Tim Mersoï basin, northwestern Niger, groundwater is being used for mining activities. Groundwater shows drop in the water table in the mining areas, indicating that it is one of the main reasons for its depletion (Dobi et al. 2021; Sonkar et al. 2023). In the Jinci spring area of Shanxi, China, the overexploitation of water for mining is one of the main reasons for the decline in quantity of groundwater (Yang et al. 2018).

Decrease in freshwater resources after the establishment of mines leads to conflicts in the local community. The conflict among mining authorities and local communities over the rights of water and land resources is expected to rise in future also. The authorities play with the laws and rules to meet their demand of natural resources (Stoltenborg and Boelens 2016). The variability in magnitude of water consumption by an individual mine in different region varies depending on the ore processing and climatic conditions. Large-scale mining influences water stress conditions in some parts of the world when consumption exceeds the natural availability of water (Meißner 2021). The corporate reporting of water accounting would improve water management practices in mining sector at global level (Northey et al. 2019).

Mines result in decline of groundwater level (Yang et al. 2018; Soni 2019), and the land use land cover (LULC) studies also confirm loss of water resources from the mining landscapes (Dube et al. 2024; Gyawali et al. 2022; Owolabi et al. 2021; Prosper et al. 2015). Additionally, the change in LULC due to mines influences the water balance (Awotwi et al. 2019), and the depletion of groundwater levels is attributed to decrease in recharge rate and increase in withdrawal in mining area (Chaulya 2004). Some of mining regions where decline in groundwater is reported are as follows: sand mining of Papagani catchment in Karnataka (MChandrasekhara 2006), copper mines at Kherti, Rajasthan (Punia et al. 2021), Singrauli coalfield (Sonkar et al. 2023), and Alwar district of Rajasthan due to illegal mines (Jitendra 2019).

10.3 PROCESSES AFFECTING QUANTITY OF THE FRESHWATER RESOURCES

A huge quantity of groundwater is withdrawn from aquifers during mining, and overexploitation of resources is one of the reasons of water scarcity in the landscape. The mining activities that result in decline in water resources quantity are discussed in the following section.

10.3.1 Ore Beneficiation Process

The ore beneficiation is both dry (crushing) and wet (grinding, milling, and froth floatation) operations needed water in significant quantity for dust settling or to dissolve the material. Specifically, the froth floatation process requires huge quantity of water and also generates a significant amount of wastewater along with tailings. The higher proportion of water content into the tailings is mainly evaporated into the atmosphere. Recently, with the decrease in availability of freshwater resources, wastewater is being reused in the mineral beneficiation process. The reuse of wastewater from froth floatation process not only saves freshwater but also reduces the demand of chemical reagents (Lin et al. 2020a). The presence of residual reagents in the wastewater increases the efficiency of mineral recovery, decreasing the economic cost of production.

The wastewater discharge is observed more than the withdrawal in most of underground mines compared to opencast mines (Lin et al. 2020b). The reuse of wastewater generated from mining in the water stress regions reduces the withdrawal of freshwater. The implementation of dewatering technologies for the thickening of tailings before its discharge into dam would reduce the generation of wastewater. The thickening and low content of water prevent the formation of acid mine drainage in the tailings and sustainable approach to mitigate its adverse impacts on environment (Jones and Boger 2012). Dewatering of tailings through sedimentation, filtration, and thermal drying produces thick and dry concentrate with moisture content of 5–10% (Haldar 2013).

Reducing the gangue minerals before the grinding ore and forth floatation process reduces the burden on further beneficiation process and the requirement of water. The use of dry sand fluidized-bed separator and heap leaching methods for early rejection of gangue is economical and also saves energy and water as it reduces the mass flow rate to grinding and floating (Franks et al. 2015). The dry ore beneficiation methods are tested in the semiarid and arid regions to decrease the water requirement. The dry ore beneficiation reduces the economic and environmental cost of ore beneficiation process (Baawuah et al. 2020). Additionally, dry ore beneficiation reduces water consumption and generates no tailings during the process (Pudasainee et al. 2020).

10.3.2 In-Rushing of Groundwater and Dewatering of Open-Pit Mines

The surface or underground mine working got filled with groundwater (Dong et al. 2021a) and needs dewatering to continue the ore excavation process. After dewatering stopped, the water rebound in underground mine working leading to dissolution of AMD and increase in metal concentration (Wright et al. 2018). The formation and dissolution of AMD due to dewatering and rebound of water result in deterioration of its quality, and excessive withdrawal or pumping out of rebounded water from mine workings impacts its quantity. Degradation of quality and quantity of surface or groundwater resources due to open-mine pit lakes threatens the ecology and local residents living in the downstream (Koch et 2005).

Predication and prevention of water inrush should be planned considering the hydrogeology and geochemistry of the area. Predication of water inrushing is not 100% certain and a dynamic process (Dong et al. 2021b). Measures should be taken to avoid water in-rush through faults, blockage of channels connecting mines with aquifers, backfilling of mining sites with waste, and recharging of aquifers with processed mine water (Yin et al. 2016).

10.4 DESERTIFICATION OF MINING LANDSCAPES

The change in LULC with respect to expansion of mining is significant at regional level. The analysis of NDVI, surface temperature, and albedo extracted from time series of satellite images provides information about the process and the extent of desertification (Scrivani et al. 2014). The conversion of vegetation covered by cropland and forest into mining activity leads to land degradation and desertification (Awotwi et al. 2018). The loss of croplands and wetlands in the Holingol region of Inner Mongolia from 1978 to 2011 is attributed to surface coal mining (Qian et al. 2014). The spatiotemporal change analysis using Landstat images of Lingbei mining areas, Ganzhou, China, shows rise in desertification during the initial phase of mining, and desertified land coverage slowly declines and stabilizes during the ecological restoration and reclamation phase (Li et al. 2021). The process of ecological restoration is slow and takes decades to reach premining stage.

Mines pose high ecological risks to water and sediment quality (Kusin et al. 2017) impacting the terrestrial and aquatic ecosystems (Beck et al. 2020). Accumulation of metals than the normal concentration adversely impacts plants growth, metabolism, and water uptake efficiency. Metal accumulation and survival capacity of plants vary from species to species. The soils of Pb-Zn mining area of Guangxi, Southwest China, are contaminated with Cd, Pb, and Zn. Metals are toxic for the plants, and *C. crepidioides* is found to be hyperaccumulator of Cd (Zhu et al. 2018), suggesting its probable use for remediation. Metal concentrations are observed higher in trees compared to vegetables and cultivated plants in the Morila mining area of Mali (Bokar et al. 2020).

The farmland soil surrounding the active or abandoned Yaoposhan polymetallic mine is highly contaminated with toxic metalloid(s). The high concentration of toxic metalloid(s) in the vegetables and rice grown on severely contaminated soil poses carcinogenic and noncarcinogenic health threats to local community living in Guangdong province of China (Sun et al. 2020). The impacts of mining waste on the vegetation cover remain for decades or centuries. The plants grown on waste heaps of historical mines accumulate significant amount of metals posing threat to environment (Stefanowicz et al. 2016). Mining disturbs the vegetation cover in nearby locations (Huang et al. 2014). The contamination reduces the growth of plants converting the landscape into a desert-like look.

The overexploitation of water for mining decreases its availability in water scarce regions. In China, Provinces of Shanxi and Shaanxi are facing huge water stress condition due to coal production (Zhang et al. 2021). The proper management and allocation of water resources considering the water supply and demand lead to sustainable development of the region. The shallow water resources are more prone to depletion and quality degradation due to mining activities. The fractures created due to mines lead to leaking of shallow water resources (Qiao et al. 2016) and systematic mining considering the geology of the region is a sustainable approach. The management of phreatic water resources consists of proper utilization of water resources, adoption of suitable mining methods with minimal adverse impact, and reconstruction of water resistant layer to prevent the contact of mines and water resources (Liu et al. 2019). The adverse impact on the phreatic water resources is more due to longwall mining compared to special mining methods such as strip mining, filling mining, and room pillar mining.

Groundwater storage is critical for the growth of vegetation in the arid and semiarid regions. The depletion of groundwater due to extensive exploitation during mining adversely impacts groundwater-dependent vegetation cover (Shen et al. 2021b). The decline in vegetation due to mining-induced dewatering results in the conversion of mining landscapes into drylands (Shen et al. 2021a). The semiarid and arid regions are deficit in water resources, and the exploitation of water for the mining further reduces its availability for plant growth. The semiarid regions are covered with scares vegetation compared to wet region. The deforestation is less noticeable in arid and semiarid region as the area is covered by scarce vegetation.

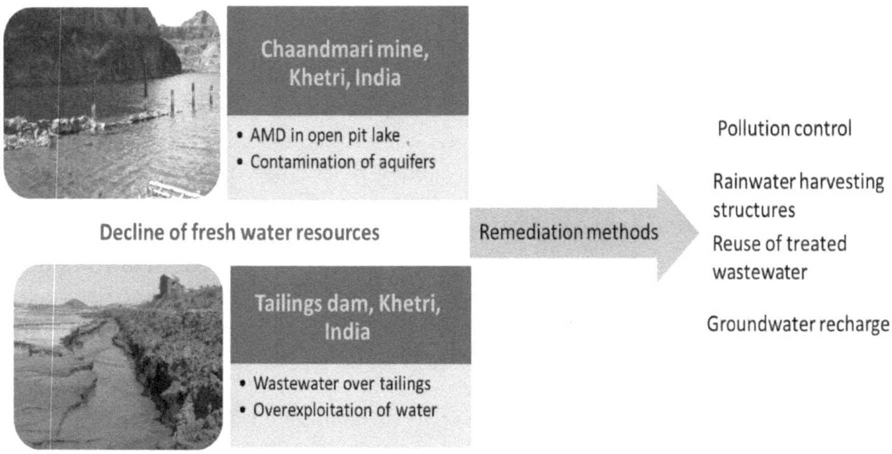

FIGURE 10.1 Shows factors affecting the water resources and remediation measures. (Photographs are by author.)

10.5 RECOMMENDATIONS TO MITIGATE DESERTIFICATION

Some of the recommendations are given to address the problem of degradation in quality and quantity of water due to mines (Figure 10.1). The proper management of water resources prevents the desertification of mining landscapes and an approach to sustainable mining.

10.5.1 Pollution Control

Mining activities also lead to mixing of different aquifers resulting in higher concentration of ions in uncontaminated groundwater (Liu et al. 2017). The reasons for mixing of aquifers are hydrogeological structures or anthropogenic factors such as fractures and faults developed via mining. Measures should be adopted to prevent the mixing of contaminated mine water with uncontaminated aquifers. Increase in desertification is reported in the rare earth mining region of Jiangxi Province, southern China, and the government initiatives enhanced the process of ecological restoration. It is difficult to curb the moderate and light desertification due to mining and needs continuous monitoring (Li et al. 2023). The remediation of contaminated sites or water bodies would reduce the damage to vegetation cover.

The application of sulfate-reducing bacteria decreases the acid generation potential of tailings from Ingaldahl copper mines, Karnataka, India (Natarajan et al. 2006), suggesting the feasibility of biotechnological aspects to control mining pollution. The Sukhinda chromite mine of Orissa, India, is well known for toxic Cr (VI) contamination and its health risk in the community (Naz et al. 2016). The use of water hyacinth, i.e., *Eichhornia crassipes,* is found 99.5% efficient in removal of Cr (VI) from wastewater released from chromite mine within 15 days (Saha et al. 2017).

10.5.2 Land Restoration

The methods should be adopted to increase organic matter content to restore the quality of degraded mine land. The tree plantation on mine spoils increases the organic matter and nutrients in spoils heaps, accelerating the recovery of natural habitat. The application of *B. thuringiensis* enhances the weathering of carbonate-enriched parent rocks and increases the Ca and Mg content of soil. It promotes the plant growth in the rocky desertification region (Wu et al. 2017). In semiarid and arid regions, the restoration of degraded and waste lands is challenging due to lack of water resources. The humidity condensation or air moisture condensation facilitates the thriving of vegetation cover on the mining waste dumps (Juřička et al. 2016). The cold surface of waste dumps and canavas attributed to their specific shape and design enable the condensation of warm air.

In the mining region, deforestation and expansion of coverage area of overburden rocks are major changes in land use pattern (Thakur et al. 2022a). The revegetation of deforested land of Kotma Coalmines, Anuppur, Madhya Pradesh, is not sufficient to prevent the declining trend of sparse and dense vegetated land (Thakur et al. 2022b). The acidic pH and poor soil characteristics of mined land are major hindrances in achieving the revegetation. The plantation of herbaceous monocots (Dowarah et al. 2009) and native species (Gairola et al. 2023) accelerates the ecological restoration of degraded mined land. In India, the regrassing of mined areas is became compulsory after the supreme court orders. The technologies such as Miyawaki (plantation of 2–4 indigenous trees in 1 km^2 of area), seed casting through drones in unreachable locations, and seed ball methods are being followed by Ministry of Coal, Government of India, for the revegetation of degraded mining area. The sustainable plantation and green belt is successfully developed at Nigahi opencast coal mine (Madhya Pradesh), Jalagam Vengala Rao II coal mine (Telangana), Gevra Coal Mines (Chhattisgarh), and Piparwar coal mines (Jharkhand) by Ministry of Coal, Government of India.

10.5.3 Water Conservation

The prevention of groundwater contamination is one of the important strategies to conserve the freshwater resources. The seepage of contaminants from mine goaf to aquifers through the fractures damages the aquifers and degrades the water quality. The mine goaf is working left after the complete or partial removal of mineral. The position of aquifuge and height of mines control the expansion of mining-induced fractures and failure of aquifers (Chi et al. 2019). The increase in mining height with constant position of aquifuge increases the intensity of water loss resulting in damage of aquifer. The repeated mining of ore seams present at narrow intervals develops the water conductive fractures in the overlying strata leading to loss of surface and groundwater (Ma et al. 2015). The technical measures should be taken before resuming the second mining where ore seams are present adjacent to each other.

The maximum and allowable scale of mining in the ecologically fragile environment should be planned considering water resources carrying capacity (Chi et al. 2021). The water allocation to mining should be rational, cost-effective, and sustainable (Li et al. 2022). The strategic implementation of water conservation policies in the water-constraint mining areas reduces the adverse impacts of mines (Chi et al. 2021). The mine water is successfully being supplied for the domestic and irrigation purposes near coal mines in different parts of the country as per Ministry of Coal, Government of India. The optimization of reasonable mining scale considering the demand of ore and water constraint environmental conditions is sustainable approach of mining and to conserve the ecological environment.

The availability of freshwater is one of major problems faced by mining industries lying near the coastlines. Sea water desalinated via reverse osmosis is used by majority of mining industries, and to reduce operation cost of desalination, many industries share infrastructures in Chile (Alvez et al. 2020). Similarly, the reuse of mine pit water for other purposes would solve the problem of water shortage. In the semiarid region, the reuse of mine pit water for ore beneficiation would reduce the demand of freshwater. The treatment of mine pit lakes is expensive and needs advanced technologies for remediation. The feasibility of mine water reuse during ore beneficiation should be studied considering its chemical composition and influence on the process.

10.5.4 Groundwater Recharge

In the semiarid and arid regions, the vegetation growth depends on the groundwater storage and adverse impacts on the hydrogeological structure due to mines lead to long-term desertification (Shen et al. 2021a). The mining infrastructures cover larger surface area and could be used for the establishments of rainwater harvesting structures. The collected rainwater could be used for storage and recharging the groundwater. The rainwater storage would solve the problem of water shortage for mining activities during the dry period (Gao et al. 2017; Singh et al. 2018). The recharge of aquifers using rainwater in the mining landscapes would slow down the declining rate of groundwater level. However, the recharge rate is less in the hard and rocky terrain of semiarid region due to lack of rainwater infiltration and higher evaporation rate. The artificial recharge of groundwater in hard and rocky terrain of Betwa watershed, Madhya Pradesh, India, a semiarid region, is suggested via construction of check dams and gully plug for sustainable management of water resources (Ahirwar et al. 2020). Similarly, 31% of West Bokaro coalfield mines of Damodar valley, India, are suitable for groundwater recharging sites indicating the feasibility of implementation of artificial recharge technology for sustainable management of water resources (Tiwari et al. 2017).

The wastewater after treatment could be used for the recharging of groundwater in the mining landscapes. The electrocoagulation process is useful for the treatment of wastewater collected from the beneficiation plant and metal (Fe, Cu, Cr, Pb, and Mn) concentrations are observed within the WHO permissible limits (Das and Nandi 2021). The wastewater from beneficiation plants is not suitable for the drinking purposes but for its use for the groundwater recharging is recommended.

10.5.5 Use of advanced Technologies

The proper management of water resources in the mining area and implementation of environment-friendly ore beneficiation technologies prevent the ecosystem degradation and desertification. Mining attracts other economic and developmental activities in the region, which further exacerbate the ecological damage. The proper sorting of ore before beneficiation and thickening of tailings control the water consumption by mining industries (Kinnunen et al. 2021). A huge quantity of tailings generated from pool and heap leaching lead to soil erosion and desertification of the mining area (Xu et al. 2020). The application of *in-situ* leaching technology generates no tailings resulting in less impact on the environment. The disposal of mine waste at least porous and fractured site prevents the contamination of water in the downstream direction (Obiadi et al. 2016). The geological mapping before the waste disposal prevents the pathways of contaminants to water resources.

10.5.6 Information and Awareness

The water security is a major issue in the mining regions for both mines and communities living in the region (Kunz et al. 2020). The community neighboring the mining sites should be informed by the authorities about the possible adverse health impacts and precautions to take during the disaster. It is reported that people, specifically herders, residing near the Sharyn Gol River, Mongolia, are unawareness about its water quality status (McIntyre et al. 2016). The river is contaminated due to coal mines, and proper information is not provided to the community. The local residents and stakeholders of mining authorities should be properly trained about the adverse environmental impacts caused by mines through conferences, seminar, workshops, and training camps.

10.6 CONCLUSIONS

The decline and contamination of water resources in the mining landscapes are one of the main reasons for the desertification. The water resources are overexploited to meet the demand of water for ore beneficiation process. Secondly, the pumping out of in-rushed groundwater from the mine workings results in wastage of water and forms the cone of depression near the mines. In mining landscapes, freshwater contamination is a major concern as local water bodies become contaminated by exposed minerals or ores from excavation and waste disposal sites. The elevated level of contaminants in water hinders the survival of soil microorganism and growth of vegetation. Fertile land becomes barren as a result of the depletion of water resources, eventually turning into a man-made desert after several decades. The water resources management strategies and policies need to be implemented in both semiarid/arid and wet regions of the mines to prevent the desertification.

The implementation of policies for groundwater recharge and water conservation would be beneficial in managing the water scarcity in the mining landscapes. The role of mining authorities and local community is also important in dealing water stress conditions. Proper awareness among the communities about the adverse impacts of mines on water resources forces the mining authorities to adopt technologies that lead to low level of contamination. The demand of water resources for mining can be managed by implemented by advanced technologies (artificial recharge and rainwater harvesting structures) and rational water allocation by government.

The adoption of water conservation strategies to avoid the desertification of the mining landscapes is highly recommended. The low level of environmental impact draws less opposition from community and leads to sustainable mining.

REFERENCES

Ahirwar, S., Malik, M. S., Ahirwar, R., & Shukla, J. P. (2020). Identification of suitable sites and structures for artificial groundwater recharge for sustainable groundwater resource development and management. *Groundwater for Sustainable Development*, 11, 100388.

Albaladejo, J., Martinez-Mena, M., Roldan, A., & Castillo, V. (1998). Soil degradation and desertification induced by vegetation removal in a semiarid environment. *Soil Use and Management*, 14(1), 1–5.

Alun, G., & Li, S. (2017). Actual influence cost estimation of water resources in coal mining and utilization in China. *Energy Procedia*, 142, 2454–2460.

Alvez, A., Aitken, D., Rivera, D., Vergara, M., McIntyre, N., & Concha, F. (2020). At the crossroads: Can desalination be a suitable public policy solution to address water scarcity in Chile's mining zones?. *Journal of Environmental Management*, 258, 110039.

Amin, A. A. (2004). The extent of desertification on Saudi Arabia. *Environmental Geology*, 46, 22–31.

Antunes, I. M. H. R., Gomes, M. E. P., Neiva, A. M. R., Carvalho, P. C. S., & Santos, A. C. T. (2016). Potential risk assessment in stream sediments, soils and waters after remediation in an abandoned W4Sn mine (NE Portugal). *Ecotoxicology and Environmental Safety*, 133, 135–145.

Arkoç, O., Ucar, S., & Ozcan, C. (2016). Assessment of impact of coal mining on ground and surface waters in Tozaklı coal field, Kırklareli, northeast of Thrace, Turkey. *Environmental Earth Sciences*, 75, 1–13.

Awotwi, A., Anornu, G. K., Quaye-Ballard, J. A., & Annor, T. (2018). Monitoring land use and land cover changes due to extensive gold mining, urban expansion, and agriculture in the Pra River Basin of Ghana, 1986–2025. *Land Degradation & Development*, 29(10), 3331–3343.

Awotwi, A., Anornu, G. K., Quaye-Ballard, J. A., Annor, T., Forkuo, E. K., Harris, E., ... & Terlabie, J. L. (2019). Water balance responses to land-use/land-cover changes in the Pra River Basin of Ghana, 1986–2025. *Catena*, 182, 104129.

Baawuah, E., Kelsey, C., Addai-Mensah, J., & Skinner, W. (2020). Economic and socio-environmental benefits of dry beneficiation of magnetite ores. *Minerals*, 10(11), 955.

Barcelos, D. A., Pontes, F. V., da Silva, F. A., Castro, D. C., Dos Anjos, N. O., & Castilhos, Z. C. (2020). Gold mining tailing: Environmental availability of metals and human health risk assessment. *Journal of Hazardous Materials*, 397, 122721.

Beck, K., Mariani, M., Fletcher, M., Schneider, L., Aquino-López, M., Gadd, P., Heijnis, H., Saunders, K. M., & Zawadzki, A. (2020). The impacts of intensive mining on terrestrial and aquatic ecosystems: A case of sediment pollution and calcium decline in cool temperate Tasmania. *Australia Environmental Pollution,* 265, 114695.

Bergkamp, G. (1995). A hierarchical approach for desertification assessment. *Environmental Monitoring and Assessment*, 37, 59–78.

Bokar, H., Traoré, A. Z., Mariko, A., Diallo, T., Traoré, A., Sy, A., Soumaré, O., Dolo, A., Bamba F., Sacko, M., & Toure, O. (2020). Geogenic influence and impact of mining activities on water soil and plants in surrounding areas of Morila Mine, Mali. *Journal of Geochemical Exploration*, 209, 106429.

Calderhead, A. I., Martel, R., Garfias, J., Rivera, A., & Therrien, R. (2012). Pumping dry: An increasing groundwater budget deficit induced by urbanization, industrialization, and climate change in an over-exploited volcanic aquifer. *Environmental Earth Sciences*, 66, 1753–1767.

Cánovas, C. R., Macías, F., Basallote, M. D., Olías, M., Nieto, J. M., & Pérez-López, R. (2021). Metal (loid) release from sulfide-rich wastes to the environment: The case of the Iberian Pyrite Belt (SW Spain). *Current Opinion in Environmental Science & Health*, 20, 100240.

Carvalho, F. P., Oliveira, J. M., & Malta, M. (2016). Preliminary assessment of uranium mining legacy and environmental radioactivity levels in Sabugal region, Portugal. *International Journal of Energy and Environmental Engineering*, 7, 399–408.

Chan, W. S., Routh, J., Luo, C., Dario, M., Miao, Y., Luo, D., & Wei, L. (2021). Metal accumulations in aquatic organisms and health risks in an acid mine-affected site in South China. *Environmental Geochemistry and Health*, 43, 4415–4440.

Chaulya, S. K. (2004). Water resource accounting for a mining area in India. *Environmental Monitoring and Assessment*, 93, 69–89.

Chi, M., Zhang, D., Honglin, L., Hongzhi, W., Yazhou, Z., Shuai, Z., Wei, Y., Shuaishuai, L., & Qiang, Z. (2019). Simulation analysis of water resource damage feature and development degree of mining-induced fracture at ecologically fragile mining area. *Environmental Earth Sciences*, 78, 1–15.

Chi, M., Zhang, D., Zhao, Q., Yu, W., & Liang, S. (2021). Determining the scale of coal mining in an ecologically fragile mining area under the constraint of water resources carrying capacity. *Journal of Environmental Management*, 279, 111621.

Cianciolo, T. R., McLaughlin, D. L., Zipper, C. E., Timpano, A. J., Soucek, D. J., & Schoenholtz, S. H. (2020). Impacts to water quality and biota persist in mining-influenced Appalachian streams. *Science of the Total Environment*, 717, 137216.

Clark, E. V., Zipper, C. E., Daniels, W. L., & Keefe, M. J. (2018). Appalachian coal mine spoil elemental release patterns and depletion. *Applied Geochemistry*, 98, 109–120.

Custodio, E., Albiac, J., Cermerón, M., Hernández, M., Llamas, M. R., & Sahuquillo, A. (2017). Groundwater mining: Benefits, problems and consequences in Spain. *Sustainable Water Resources Management*, 3, 213–226.

Das, D., & Nandi, B. K. (2021). Treatment of iron ore beneficiation plant process water by electrocoagulation. *Arabian Journal of Chemistry*, 14(1), 102902.

de Souza Pereira, W., Kelecom, A., Lopes, J. M., do Carmo, A. S., de Azevedo Py Júnior, D., & da Silva, A. X. (2020). Evaluation of the radiological quality of water released by a uranium mining in Brazil. *Environmental Science and Pollution Research*, 27, 36704–36717.

Dobi, F. B., Kouakou, E. K., Nazoumou, Y., Abdou Boko, B., Edimo, S. N., Maina, F. Z., & Konaté, M. (2021). Aquifer depletion in the arlit mining area (Tim Mersoï Basin, North Niger). *Water*, 13(12), 1685.

Dong, S., Zhou, W., & Wang, H. (2021b). Introduction to special issue on mine water inrushes: Risk assessment, mitigation, and prevention. *Mine Water and the Environment*, 40(2), 321–323.

Dong, S., Zhou, W., Liu, Q., Wang, H., Ji, Y., & Ji, Y. (2021). Water hazards in coal mines and their classifications. In: *Methods and Techniques for Preventing and Mitigating Water Hazards in Mines, Professional Practice in Earth Sciences.* Springer, Cham, pp. 1–27. https://doi.org/10.1007/978-3-030-67059-7_1

Dowarah, J., Deka Boruah, H. P., Gogoi, J., Pathak, N., Saikia, N., & Handique, A. K. (2009). Eco-restoration of a high-sulphur coal mine overburden dumping site in northeast India: A case study. *Journal of Earth System Science*, 118, 597–608.

Dube, T., Dube, T., Dalu, T., Gxokwe, S., & Marambanyika, T. (2024). Assessment of land use and land cover, water nutrient and metal concentration related to illegal mining activities in an Austral semi–arid river system: A remote sensing and multivariate analysis approach. *Science of The Total Environment*, 907, 167919.

Foster, S., Chilton, J., Nijsten, G. J., & Richts, A. (2013). Groundwater: A global focus on the 'local resource'. *Current Opinion in Environmental Sustainability*, 5(6), 685–695.

Franks, G. V., Forbes, E., Oshitani, J., & Batterham, R. J. (2015). Economic, water and energy evaluation of early rejection of gangue from copper ores using a dry sand fluidised bed separator. *International Journal of Mineral Processing*, 137, 43–51.

Gairola, S. U., Bahuguna, R., & Bhatt, S. S. (2023). Native plant species: A tool for restoration of mined lands. *Journal of Soil Science and Plant Nutrition*, 23, 1348–1448.

Gao, L., Bryan, B. A., Liu, J., Li, W., Chen, Y., Liu, R., & Barrett, D. (2017). Managing too little and too much water: Robust mine-water management strategies under variable climate and mine conditions. *Journal of Cleaner Production*, 162, 1009–1020.

Gbedzi, D. D., Ofosu, E. A., Mortey, E. M., Obiri-Yeboah, A., Nyantakyi, E. K., Siabi, E. K., Abdallah, F., Domfeh, M. K., & Amankwah-Minkah, A. (2022). Impact of mining on land use land cover change and water quality in the Asutifi North District of Ghana, West Africa. *Environmental Challenges*, 6, 100441.

González-Merizalde, M. V., Menezes-Filho, J. A., Cruz-Erazo, C. T., Bermeo-Flores, S. A., Sánchez-Castillo, M. O., Hernández-Bonilla, D., & Mora, A. (2016). Manganese and mercury levels in water, sediments, and children living near gold-mining areas of the Nangaritza River basin, Ecuadorian Amazon. *Archives of Environmental Contamination and Toxicology*, 71, 171–182.

Guleria, A., Singh, R., Chakma, S., & Birke, V. (2022). Ecological and human health risk assessment of chromite ore processing residue (COPR) dumpsites in Northern India: A multi–pathways based probabilistic risk approach. *Process Safety and Environmental Protection*, 163, 405–420.

Gyawali, B., Shrestha, S., Bhatta, A., Pokhrel, B., Cristan, R., Antonious, G., Banerjee, S., & Paudel, K. P. (2022). Assessing the effect of land-Use and land-cover changes on discharge and sediment yield in a rural coal-mine dominated watershed in Kentucky, USA. *Water*, 14(4), 516.

Haldar, S. K. (2013). *Mineral Exploration.* Elsevier, Amsterdam. https://doi.org/10.1016/B978-0-12-416005-7.00011-8

Hebblewhite, B. (2020). Fracturing, caving propagation and influence of mining on groundwater above longwall panels: A review of predictive models. *International Journal of Mining Science and Technology*, 30(1), 49–54.

Helser, J., Vassilieva, E., & Cappuyns, V. (2022). Environmental and human health risk assessment of sulfidic mine waste: Bioaccessibility, leaching and mineralogy. *Journal of Hazardous Materials*, 424, 127313.

Huang, W., Duan, W., & Chen, Y. (2021). Rapidly declining surface and terrestrial water resources in Central Asia driven by socio-economic and climatic changes. *Science of the Total Environment*, 784, 147193.

Huang, Y., Tian, F., Wang, Y., Wang, M., & Hu, Z. (2015). Effect of coal mining on vegetation disturbance and associated carbon loss. *Environmental Earth Sciences*, 73, 2329–2342.

Jitendra. (2019). World water day: How mining is depleting groundwater in Rajasthan's Alwar. *Down to Earth.* https://www.downtoearth.org.in/news/mining/world-water-day-how-mining-is-depleting-groundwater-in-rajasthan-s-alwar-63684 (accessed 8 February 2024).

Jones, H., & Boger, D. V. (2012). Sustainability and waste management in the resource industries. *Industrial & Engineering Chemical Research*, 51, 10057–10065.

Juřička, D., Muchová, M., Elbl, J., Pecina, V., Kynický, J., Brtnický, M., & Rosická, Z. (2016). Construction of remains of small-scale mining activities as a possible innovative way how to prevent desertification. *International Journal of Environmental Science and Technology*, 13, 1405–1418.

Keovilignavong, O. (2019). Mining governance dilemma and impacts: A case of gold mining in Phu-Hae, Lao PDR. *Resources Policy*, 61, 141–150.

Kinnunen, P., Obenaus-Emler, R., Raatikainen, J., Guignot, S., Guimerà, J., Ciroth, A., & Heiskanen, K. (2021). Review of closed water loops with ore sorting and tailings valorisation for a more sustainable mining industry. *Journal of Cleaner Production*, 278, 123237.

Koch, H., Kaltofen, M., Grünewald, U., Messner, F., Karkuschke, M., Zwirner, O., & Schramm, M. (2005). Scenarios of water resources management in the Lower Lusatian mining district, Germany. *Ecological Engineering*, 24(1–2), 49–57.

Kunz, N. C. (2020). Towards a broadened view of water security in mining regions. *Water Security*, 11, 100079.

Kusin, F. M., Rahman, M. S. A., Madzin, Z., Jusop, S., Mohamat-Yusuff, F., & Ariffin, M. (2017). The occurrence and potential ecological risk assessment of bauxite mine-impacted water and sediments in Kuantan, Pahang, Malaysia. *Environmental Science and Pollution Research*, 24, 1306–1321.

Larrabure, G., & Rodríguez-Reyes, J. C. F. (2021). A review on the negative impact of different elements during cyanidation of gold and silver from refractory ores and strategies to optimize the leaching process. *Minerals Engineering*, 173, 107194.

Li, C., Tian, X., Cao, Z., & Wang, Y. (2022). Research on optimal allocation of water resources in the Western mining area of China based on WEAP. *Applied Water Science*, 12(6), 113.

Li, Y., Li, H., & Pan, J. (2023). Long-term desertification process monitoring and driving factors analysis in rare earth mining area. *Restoration Ecology*, 32, e13994.

Li, Y., Li, H., & Xu, F. (2021). Spatiotemporal changes in desertified land in rare earth mining areas under different disturbance conditions. *Environmental Science and Pollution Research*, 28, 30323–30334.

Lin, G., Jiang, D., Fu, J., Dong, D., Sun, W., & Li, X. (2020b). Spatial relationships of water resources with energy consumption at coal mining operations in China. *Mine Water and the Environment*, 39, 407–415.

Lin, S., Liu, R., Wu, M., Hu, Y., Sun, W., Shi, Z., ... & Li, W. (2020a). Minimizing beneficiation wastewater through internal reuse of process water in flotation circuit. *Journal of Cleaner Production*, 245, 118898.

Liu, P., Hoth, N., Drebenstedt, C., Sun, Y., & Xu, Z. (2017). Hydro-geochemical paths of multi-layer groundwater system in coal mining regions: Using multivariate statistics and geochemical modeling approaches. *Science of the Total Environment*, 601, 1–14.

Liu, S., & Li, W. (2019). Zoning and management of phreatic water resource conservation impacted by underground coal mining: A case study in arid and semiarid areas. *Journal of Cleaner Production*, 224, 677–685.

Ma, L., Jin, Z., Liang, J., Sun, H., Zhang, D., & Li, P. (2015). Simulation of water resource loss in short-distance coal seams disturbed by repeated mining. *Environmental Earth Sciences*, 74, 5653–5662.

MChandrasekhara, R. A. O. (2006). Groundwater depletion in Papagani catchment. *Economic and Political Weekly*, 41, 593.

McIntyre, N., Bulovic, N., Cane, I., & McKenna, P. (2016). A multi-disciplinary approach to understanding the impacts of mines on traditional uses of water in Northern Mongolia. *Science of the Total Environment*, 557, 404–414.

Meißner, S. (2021). The impact of metal mining on global water stress and regional carrying capacities: A GIS-based water impact assessment. *Resources*, 10(12), 120.

Mhlongo, S., Mativenga, P. T., & Marnewick, A. (2018). Water quality in a mining and water-stressed region. *Journal of Cleaner Production*, 171, 446–456.

Ministry of Coal, Government of India. *Greening Initiatives in Coal Sector.* https://coal.nic.in/en/sustainable-development-cell/greening-initiatives (accessed 1 February 2024).

Ministry of Coal, Government of India. *Mine Water Utilization: Every Drop Matters.* https://coal.gov.in/en/sustainable-development-cell/mine-water-utilization (accessed 2 February 2024).

Natarajan, K. A., Subramanian, S., & Braun, J. J. (2006). Environmental impact of metal mining: Biotechnological aspects of water pollution and remediation: An Indian experience. *Journal of Geochemical Exploration*, 88(1–3), 45–48.

Naz, A., Chowdhury, A., Mishra, B. K., & Gupta, S. K. (2016). Metal pollution in water environment and the associated human health risk from drinking water: A case study of Sukinda chromite mine, India. *Human and Ecological Risk Assessment: An International Journal*, 22(7), 1433–1455.

Neiva, A. M. R., Antunes, I. M. H. R., Carvalho, P. C. S., & Santos, A. C. T. (2016). Uranium and arsenic contamination in the former Mondego Sul uranium mine area, *Central Portugal. Journal of Geochemical Exploration*, 162, 1–15

Niane, B., Guédron, S., Feder, F., Legros, S., Ngom, P. M., & Moritz, R. (2019). Impact of recent artisanal small-scale gold mining in Senegal: Mercury and methylmercury contamination of terrestrial and aquatic ecosystems. *Science of the Total Environment*, 669, 185–193.

Nieva, N. E., Borgnino, L., & García, M. G. (2018). Long term metal release and acid generation in abandoned mine wastes containing metal-sulphides. *Environmental Pollution*, 242, 264–276.

Northey, S. A., Mudd, G. M., Werner, T. T., Haque, N., & Yellishetty, M. (2019). Sustainable water management and improved corporate reporting in mining. *Water Resources and Industry*, 21, 100104.

Obasi, P. N., & Akudinobi, B. B. (2020). Potential health risk and levels of heavy metals in water resources of lead–zinc mining communities of Abakaliki, southeast Nigeria. *Applied Water Science*, 10(7), 1–23.

Obiadi, I. I., Obiadi, C. M., Akudinobi, B. E. B., Maduewesi, U. V., & Ezim, E. O. (2016). Effects of coal mining on the water resources in the communities hosting the Iva Valley and Okpara Coal Mines in Enugu State, Southeast Nigeria. *Sustainable Water Resources Management*, 2, 207–216.

Owolabi, A. O., Amujo, K., & Olorunfemi, I. E. (2021). Spatiotemporal changes on land surface temperature, land and water resources of host communities due to artisanal mining. *Environmental Science and Pollution Research*, 28, 36375–36398.

Pattnaik, B. K., & Equeenuddin, S. M. (2016). Potentially toxic metal contamination and enzyme activities in soil around chromite mines at Sukinda Ultramafic Complex, India. *Journal of Geochemical Exploration*, 168, 127–136.

Peng, X., Dai, Q., Ding, G., Shi, D., & Li, C. (2020). Impact of vegetation restoration on soil properties in near-surface fissures located in karst rocky desertification regions. *Soil and Tillage Research*, 200, 104620.

Petavratzi, E., Kingman, S., & Lowndes, I. (2005). Particulates from mining operations: A review of sources, effects and regulations. *Minerals Engineering*, 18(12), 1183–1199.

Pradhan, G., Tripathy, B., Ram, D. K., Digal, A. K., & Das, A. P. (2023). Bauxite Mining Waste Pollution and Its Sustainable Management through Bioremediation. *Geomicrobiology Journal*, 1–10. https://doi.org/10.1080/01490451.2023.2235353

Prosper, L. B., & Guan, Q. (2015, June). Analysis of land use and land cover change in Nadowli District, Ghana. In: *2015 23rd International Conference on Geoinformatics. IEEE,* New York, pp. 1–6.

Pudasainee, D., Kurian, V., & Gupta, R. (2020). Coal: Past, present, and future sustainable use. *Future Energy*, 21–48. https://doi.org/10.1016/B978-0-08-102886-5.00002-5

Punia, A., Siddaiah, N. S., & Singh, S. K. (2017). Source and assessment of heavy metal pollution at Khetri copper mine tailings and surrounding soil, Rajasthan, India. *Bulletin of Environmental Contamination and Toxicology*, 99, 633–641.

Punia, A., & Siva Siddaiah, N. (2017). Assessment of heavy metal contamination in groundwater of Khetri copper mine region, India and health risk assessment. *Asian Journal of Water, Environment and Pollution*, 14(4), 9–19.

Punia, A., Joshi, P. K., & Siddaiah, N. S. (2021). Characterizing Khetri copper mine environment using geospatial tools. *SN Applied Sciences*, 3, 1–17.

Qian, T., Bagan, H., Kinoshita, T., & Yamagata, Y. (2014). Spatial–temporal analyses of surface coal mining dominated land degradation in Holingol, Inner Mongolia. *IEEE Journal of Selected Topics in Applied Earth Observations and Remote Sensing*, 7(5), 1675–1687.

Qiao, W., Li, W., Li, T., Chang, J., & Wang, Q. (2017). Effects of coal mining on shallow water resources in semiarid regions: A case study in the Shennan mining area, Shaanxi, China. *Mine Water and the Environment*, 36, 104–113.

Saarikoski, S., Teinilä, K., Timonen, H., Aurela, M., Laaksovirta, T., Reyes, F., Vásques, Y., Oyola, P., Artaxo, P., Pennanen, A. S., Junttila, S., Linnainmaa, M., Salonen R. O., & Hillamo, R. (2018). Particulate matter characteristics, dynamics, and sources in an underground mine. *Aerosol Science and Technology*, 52(1), 114–122.

Saha, P., Shinde, O., & Sarkar, S. (2017). Phytoremediation of industrial mines wastewater using water hyacinth. *International Journal of Phytoremediation*, 19(1), 87–96.

Sahoo, S., & Khaoash, S. (2020). Impact assessment of coal mining on groundwater chemistry and its quality from Brajrajnagar coal mining area using indexing models. *Journal of Geochemical Exploration*, 215, 106559

Salomons, W. (1995). Environmental impact of metals derived from mining activities: Processes, predictions, prevention. *Journal of Geochemical Exploration*, 52(1–2), 5–23.

Scrivani, R., Gonçalves, R. R. V., Romani, L. A., Oliveira, S. R. M., & Assad, E. D. (2014, July). An approach based on satellite image time series mining to identify region susceptible to desertification. In: *2014 IEEE Geoscience and Remote Sensing Symposium*. IEEE, New York, pp. 847–850.

Semenova, Y., Pivina, L., Zhunussov, Y., Zhanaspayev, M., Chirumbolo, S., Muzdubayeva, Z., & Bjørklund, G. (2020). Radiation-related health hazards to uranium miners. *Environmental Science and Pollution Research*, 27, 34808–34822.

Shen, Z., Zhang, Q., Chen, D., & Singh, V. P. (2021b). Varying effects of mining development on ecological conditions and groundwater storage in dry region in Inner Mongolia of China. *Journal of Hydrology*, 597, 125759.

Shen, Z., Zhang, Q., Piao, S., Peñuelas, J., Stenseth, N. C., Chen, D., Xu, C-Y., Singh, V. P., & Liu, T. (2021a). Mining can exacerbate global degradation of dryland. *Geophysical Research Letters*, 48(21), e2021GL094490.

Simonin, M., Rocca, J. D., Gerson, J. R., Moore, E., Brooks, A. C., Czaplicki, L., ... & Bernhardt, E. S. (2021). Consistent declines in aquatic biodiversity across diverse domains of life in rivers impacted by surface coal mining. *Ecological Applications*, 31(6), e02389.

Singh, P. K., Tiwari, A. K., & Verma, P. (2018). Evaluation of Water Level Behavior in Coal-Mining Area, Adjacent Township, and District Areas of Jharkhand State, India. In: Saha, D., Marwaha, S., Mukherjee, A. (Eds.), *Clean and Sustainable Groundwater in India*. Springer, Singapore, pp. 261–278.

Song, W., & Liang, Z. (2021). Theoretical and numerical investigations on mining-induced fault activation and groundwater outburst of coal seam floor. *Bulletin of Engineering Geology and the Environment*, 80, 5757–5768.

Soni, A. K. (2019). Mining of minerals and groundwater in India. In: *Groundwater-Resource Characterisation and Management Aspects*. IntechOpen, p. 71. doi:10.5772/intechopen.85309

Soni, A. K. (2020). Estimation of mine water quantity: Development of guidelines for Indian mines. *Mine Water and the Environment*, 39(2), 397–406.

Sonkar, A. K., Varshney, R., Ahmed, S. I., Vishwakarma, A. K., & Jamal, A. (2023). Impact of mining industries on the groundwater level fluctuation in Singrauli coalfield area by using remote sensing and GIS, India. *Environmental Quality Management*, 33(1), 311–325.

Stefanowicz, A. M., Stanek, M., Woch, M. W., & Kapusta, P. (2016). The accumulation of elements in plants growing spontaneously on small heaps left by the historical Zn-Pb ore mining. *Environmental Science and Pollution Research*, 23, 6524–6534.

Stoltenborg, D., & Boelens, R. (2016). Disputes over land and water rights in gold mining: The case of Cerro de San Pedro, Mexico. *Water International*, 41(3), 447–467.

Sun, Z., Hu, Y., & Cheng, H. (2020). Public health risk of toxic metal (loid) pollution to the population living near an abandoned small-scale polymetallic mine. *Science of The Total Environment*, 718, 137434.

Tao, W. (2014). Aeolian desertification and its control in Northern China. *International Soil and Water Conservation Research*, 2(4), 34–41.

Tepanosyan, G., Sahakyan, L., Belyaeva, O., Asmaryan, S., & Saghatelyan, A. (2018). Continuous impact of mining activities on soil heavy metals levels and human health. *Science of the Total Environment*, 639, 900–909.

Thakur, T. K., Dutta, J., Bijalwan, A., & Swamy, S. L. (2022b). Evaluation of decadal land degradation dynamics in old coal mine areas of Central India. *Land Degradation & Development*, 33(16), 3209–3230.

Thakur, T. K., Dutta, J., Upadhyay, P., Patel, D. K., Thakur, A., Kumar, M., & Kumar, A. (2022a). Assessment of land degradation and restoration in coal mines of central India: A time series analysis. *Ecological Engineering*, 175, 106493.

Tiwari, A. K., Lavy, M., Amanzio, G., De Maio, M., Singh, P. K., & Mahato, M. K. (2017). Identification of artificial groundwater recharging zone using a GIS-based fuzzy logic approach: A case study in a coal mine area of the Damodar Valley, India. *Applied Water Science*, 7, 4513–4524.

Tiwary, R. K., Kumari, B., & Singh, D. B. (2017). Water quality assessment and correlation study of physico-chemical parameters of Sukinda chromite mining area, Odisha, India. In: *Environmental Pollution: Select Proceedings of ICWEES-2016*. Springer, Singapore, pp. 357–370.

Vendrell-Puigmitja, L., Llenas, L., Proia, L., Ponsa, S., Espinosa, C., Morin, S., & Abril, M. (2021). Effects of an hypersaline effluent from an abandoned potash mine on freshwater biofilm and diatom communities. *Aquatic Toxicology*, 230, 105707.

Wang, T., Xue, X., Zhou, L., & Guo, J. (2015). Combating aeolian desertification in northern China. *Land Degradation & Development*, 26(2), 118–132.

Wright, I. A., Paciuszkiewicz, K., & Belmer, N. (2018). Increased water pollution after closure of australia's longest operating underground coal mine: A 13-month study of mine drainage, water chemistry and river ecology. *Water, Air, & Soil Pollution,* 229, 55

Wu, Y., Zhang, J., & Guo, X. (2017). An indigenous soil bacterium facilitates the mitigation of rocky desertification in carbonate mining areas. *Land Degradation & Development*, 28(7), 2222–2233.

Wu, Y., Zhou, X. Y., Lei, M., Yang, J., Ma, J., Qiao, P. W., & Chen, T. B. (2017). Migration and transformation of arsenic: Contamination control and remediation in realgar mining areas. *Applied Geochemistry*, 77, 44–51.

Xu, F., Li, H., & Li, Y. (2021). Ecological environment quality evaluation and evolution analysis of a rare earth mining area under different disturbance conditions. *Environmental Geochemistry and Health*, 43, 2243–2256.

Yang, Y., Guo, T., & Jiao, W. (2018). Destruction processes of mining on water environment in the mining area combining isotopic and hydrochemical tracer. *Environmental Pollution*, 237, 356–365.

Yang, Y., Guo, T., & Jiao, W. (2018). Destruction processes of mining on water environment in the mining area combining isotopic and hydrochemical tracer. *Environmental Pollution*, 237, 356–365.

Yin, S., Han, Y., Zhang, Y., & Zhang, J. (2016). Depletion control and analysis for groundwater protection and sustainability in the Xingtai region of China. *Environmental Earth Sciences,* 75, 1246

Zhang, F., Huang, G., Hou, Q., Liu, C., Zhang, Y., & Zhang, Q. (2019). Groundwater quality in the Pearl River Delta after the rapid expansion of industrialization and urbanization: Distributions, main impact indicators, and driving forces. *Journal of Hydrology*, 577, 124004.

Zhang, K., Li, H., Han, J., Jiang, B., & Gao, J. (2021). Understanding of mineral change mechanisms in coal mine groundwater reservoir and their influences on effluent water quality: A experimental study. *International Journal of Coal Science & Technology*, 8, 154–167.

Zhang, Y., Li, J., Tian, Y., Deng, Y., & Xie, K. (2021). Virtual water flow associated with interprovincial coal transfer in China: Impacts and suggestions for mitigation. *Journal of Cleaner Production*, 289, 125800.

Zhou, H., Liu, G., Zhang, L., & Zhou, C. (2021). Mineralogical and morphological factors affecting the separation of copper and arsenic in flash copper smelting slag flotation beneficiation process. *Journal of Hazardous Materials*, 401, 123293.

Zhu, G., Xiao, H., Guo, Q., Song, B., Zheng, G., Zhang, Z., Zhao, J., & Okoli, C. P. (2018). Heavy metal contents and enrichment characteristics of dominant plants in wasteland of the downstream of a lead-zinc mining area in Guangxi, Southwest China. *Ecotoxicology and Environmental Safety*, 151, 266–271.

11 Solute Acquisitional Environment of Ib Valley Coalfield

Emphasis on Hydrochemistry and Water Quality Assessment

Abhishek Pandey Bharat and Abhay Kumar Singh

11.1 INTRODUCTION

Mining operations produce a significant amount of minerals and fuels, together with simultaneous discharge of freshwater resources (Cesar Minga et al. 2023; Guo et al. 2019). Extracted minerals and fuels are used in a variety of sectors, but mine water is released in the surrounding areas resulting in environmental degradation globally (Carvalho 2017; Yuan and Guo 2023). Mining, in a coal basin, may have a variety of effects on groundwater quality depending on the hydrogeology, coal seam geochemistry and nearby aquifer geochemistry. Significant quantities of toxic and hazardous trace elements and ionic concentrations are released into the surrounding environment during coal mining and processing, which could negatively impact the ecosystem and human health. Water pollution from mining can contaminate possible drinking water sources by adding an appropriate amount of harmful ions. According to MoEF, India's most coal mining districts have been identified as critically polluted areas (CSE 2012). Pyrite (FeS_2), often associated with coal deposits, may tend to get oxidized when the water table is lowered and comes in contact with these sulphide minerals and further produces ferrous sulphate ($FeSO_4$) and sulphuric acid (H_2SO_4) in solution (Evangelou and Zhang 1995). The water amount discharged from the mines may be sometimes characteristically highly acidic (acid mine drainage) and have high concentration of iron, aluminum and sulphate (Galhardi and Bonotto 2016). Coal mine drainage (CMD) is affected by physical, chemical and biological factors, and its chemistry evolves over time as a result of interactions between rocks and water, weathering and dissolution of mineral species, interactions with the return flow of water from surface activities, ion exchange reactions, recharge of groundwater from atmospheric precipitation and human activities (Zhang et al. 2021). A high concentration of dissolved heavy metals also accounted in the category

DOI: 10.1201/9781003442554-13

of by-products of the mining industry that are resistant to degradation (Singh and Santal 2015). Such highly mineralized discharging water often does not perform well in the context with drinking, irrigation and industry purposes and more often crosses the permissible limits of heavy metals, toxic anions and biological contaminants (Bhardwaj et al. 2020).

Groundwater is an important freshwater source not only for living creatures but also in different industrial sectors, including domestic and agriculture. Uncertainty persists regarding the impact of mining operations on groundwater regime. The main factors influencing groundwater and its relationship to the hydrologic balance are the groundwater flow rate, chemical quality and the rocks' ability to store groundwater. The primary causes of groundwater pollution include tailing and waste dump, mine water pumping, wastewater from beneficiation plants and dust suppression (Jain and Das 2017). Assessment of solute acquisition processes is given particular attention when evaluating effects of mine drainage on the nearby subsurface water regimes and developing in-situ water remediation and treatment programs in mining regions. Water quality in the mining and populous areas faces the combined impact of both natural and anthropogenic activities, which causes the contamination of the available water resources (Khatri and Tyagi 2015). Globally, at least 2 billion people use a drinking water source contaminated with faeces. Contaminated water can transmit diseases such as diarrhoea, cholera, dysentery, typhoid and polio (Qadri and Faiq 2020). In India, more than 20 states are experiencing the endemic fluorosis at an emerging and alarming rate that can put the people at a risk by mean of consuming such contaminated water. Ingestion of excess fluoride, most commonly in drinking water, affects the teeth and bones. It results in major health disorders such as dental fluorosis, skeletal fluorosis and non-skeletal fluorosis. Global demand for freshwater resources has been continuously increasing due to a number of reasons such as rapid population and economic growth, urbanization and industrialization, land use change, intensive agricultural practices and environmental degradation (Maiti and Agrawal 2005). Improved water supply and sanitation, and better management of water resources, can boost countries' economic growth and can contribute greatly to poverty reduction.

Water quality potential for the industrial, agriculture and domestic sectors with the help of hydrological aspects involving chemical constituent assessment to identify contamination, geochemical processes in charge of carrying out chemical evolution and water quality potential comes in the domain of scientific investigations (Mahato et al. 2018). Indexing water quality for variable purposes is one of the important ways to utilize the hydrochemical data (Saeedi et al. 2010; Ramakrishnaiah et al. 2009). Drinking water quality index (DWQI) one side is provided to describe the general situation of water by using mathematical tools to change water quality parameters levels into a numerical score, and thus, it could be suggested to what extent the water is suitable for drinking purpose. On the other side, the irrigation parameters such as salinity hazard, alkali hazard and permeability index collectively demarcate the suitability of water for irrigation use. Scaling and corrosion are the two significant indicators of water quality, which are monitored for assessing feasibility of water passing through water distribution systems. Water's propensity to precipitate oversaturated minerals such as $CaCO_3$ onto the inner surface of pipes is known as scaling (Sunardi et al. 2020). On the other hand, corrosion is the ability of water to dissolve various pollutants such as Fe, Pb, Zn, As, Ni, Cd, Cu, Sn, Bi and Cl. The measures of these water properties could prevent water leaks, decrease the cost of replacing pipes, pumps and equipment, increase efficiency of appliances and longevity of industrial plants. Ib Valley coalfield (IBVC) is one of the 43 industrial clusters of India which are considered Critically Polluted Areas. The coalfield discharges around 30.29 million m^3 of water from open cast mines and 8.07 million m^3 of water from underground mines per annum. It makes it quite viable to assess the water characteristics in and around the mining activities for a strategic water management system. In this perspective, the major objectives covered in this study are to unveil the hydrochemistry of mine water and groundwater for contaminant source evaluation, to understand solute acquisition processes and to assess water quality for domestic, agriculture and industrial sectors.

11.2 STUDY AREA

11.2.1 Geography, Geology and Climate Conditions

The IBVC, a sub-coalfield of Mahanadi valley coalfields and having third most coal reserves in India, has been selected in this research work. The IBVC is spread between the latitudinal variance of 21°30′N to 22°14′N and longitudinal variance of 83°32′E to 84°10′E with a total area of 1,460 km^2 in the state of Odisha (Figure 11.1A) (Coal Atlas of India 1993). The area experiences typical warm to hot tropical climatic conditions and receives around 1,200 mm of annual rainfall. The presence of rock formations from Precambrian age to Gondwana age and to recent alluviums prepares a specific geology of the region which contains a variety of economically important minerals and fuel resources (Goswami et al. 2018). IBVC is known for having Gondwana Group of rocks; however, the Gondwana formations are unconformably underlain by Precambrian rocks such as granites, granitic gneisses and migmatites and overlain by recent alluviums and laterites. The coal-bearing formations are mainly from Karharbari, the older one and the Barakar, the younger one, which are up to 90–125 m and 600 m thick, respectively. The Karharbari Formation consists mainly of coarse-grained sandstone along with coal seams and is restricted to the north-western margin around Siarmal and Gopalpur areas, whilst the Barakar Formation is made up of greyish white to pink, medium to coarse grained, micaceous, sub-arkosic sandstones with calcareous and ferruginous cement, grey micaceous shale, carbonaceous shale, alternating bands of shale and sandstone, fireclay and thick coal seams. Barren measures that are devoid of coal seams and made up mostly of grey shales, carbonaceous shales, sandstones and clay ironstone shales, overly the Barakars. Chaudhury (1988) first identified the Kamthi Formation in the southern limb in Rohini area which he describes as equivalent to Raniganj Formation. The youngest formations are the laterites and the recent alluviums which include well-rounded quartzite boulders and pebbles. Figure 11.1B shows the land use land cover map of the study area where the majority of the land is made up of vegetation cover and agriculture land together with 90% followed by settlements (5%), water bodies (4%) and waste land (1%).

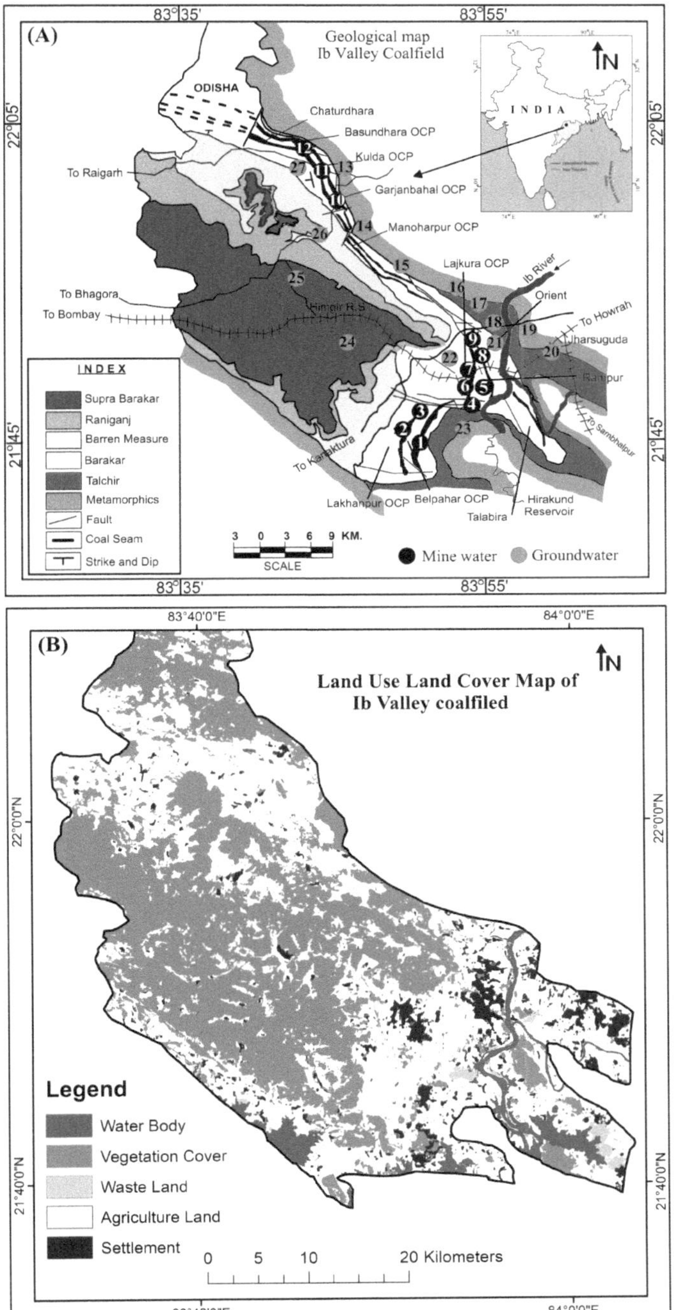

FIGURE 11.1 Study area map covering (A) geology and sampling sites, and (B) Land use land cover details.

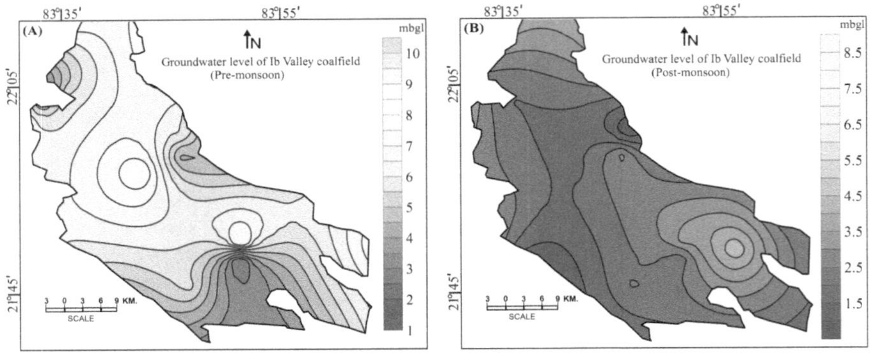

FIGURE 11.2 Groundwater level status of the study area in pre and post-monsoon of the year 2019 (as per CGWB data).

11.2.2 Hydrogeology

Ib river, a tributary of the Mahanadi river, flows through the area and makes the drainage system of the IBVC. It flows from north to south in the area and meets the Hirakund reservoir beyond the mining areas. The nallas such as Chaturdhara, Chelduthi, Lilari, Lambtibahal and Basundhara and the river Bhedan collectively make up the drainage system of the mining area, and the Bhedan river discharges into the Ib valley. Sub-surficial water mainly occurs along the fractures, joints, bedding planes, weathered zones and cavities within the associated rocks of the area. In the pre-monsoon (April 2019) (Figure 11.2A), the water level of the research area ranged from 2.45 to 11.35 meters below ground level (mbgl), while in the post-monsoon (November 2019) (Figure 11.2B), the water level ranged from 0.35 to 11.3 mbgl (CGWB 2020).

11.3 MATERIALS AND METHODOLOGY

11.3.1 Sampling and Analytical Techniques: Quality Control

In this study, a total of 102 water samples covering 65 mine water, 37 groundwater samples were collected from the selected locations during pre-monsoon (2019 and 2022) and post-monsoon (2019 and 2020). The pre-washed highly dense polyethylene bottles of 500 ml volume were used to collect the water samples for the analyses of parameters such as electrical conductivity (EC), total dissolved solid (TDS), pH, total hardness (TH), Cl^-, F^-, NO_3^-, HCO_3^-, SO_4^{2-}, Mg^{2+}, Ca^{2+}, K^+ and Na^+. The physical parameters such as EC, pH and TDS were measured in the field by a portable EC/pH meter, while the remaining parameters were measured in the laboratory after filtering them through 0.45 μm Whatman® nylon filters. The analytical procedures elaborated by the American Public Health Association (APHA) were followed (APHA 2017). The major ions such as SO_4^{2-}, NO_3^- and F^- were measured by UV-VIS Spectrometer at different wavelengths. The respective titrimetric methods were used in analysing the samples for parameters such as HCO_3^- by acid–base titration, Ca^{2+} and Mg^{2+} by EDTA method and Cl^- by argentometric method. The Na^+ and K^+ were measured by Flame photometer. After every 10 samples, the standards of known quantity were performed. For data quality assurance, the calculated values of major ions were converted into meq/L and put in Eq. (11.1). The charge balance error (CBE) was within ±10% for mine water and ±5% for groundwater samples.

$$\text{CBE} = \frac{\sum \text{Cations} - \sum \text{Anions}}{\sum \text{Cations} + \sum \text{Anions}} \times 100\% \quad (11.1)$$

11.3.2 Overview of Water Quality Indexing

Table 11.1 describes water quality indices for drinking, irrigation and industrial purpose. By using these indices, a better water quality assessment can be made.

11.4 RESULT AND DISCUSSION

11.4.1 pH and Total Dissolved Solid

The statistical summary of the studied mine water and groundwater samples has been provided in Table 11.2. The wide variance of pH ranging from 3.26 to 8.18 in pre-monsoon and from 3.38 to 7.93 in post-monsoon demarcates acidic to alkaline nature of the studied mine waters (Table 11.2). Around 29% samples in pre-monsoon and 41% in post-monsoon mainly from mines such as Lilari, Samleswari, Lajkura and Belpahar expressed pH less than 6.5, elucidating slightly acidic to acidic nature. Other than these samples, all mine waters are well within the drinking water limit of pH being 6.5–8.5 (BIS 2012). pH of the groundwater samples covered the range of 5.88–8.47 (in pre-monsoon) and 5.23–8.52 (in post-monsoon) for the coalfield, and average pH values were observed to be 7.0 in pre-monsoon and 6.57 in post-monsoon respectively. Around 51% groundwater samples were found to show pH < 6.5 showing slightly acidic nature of these waters. Dissolved carbon dioxide could play a key role in controlling the pH of the groundwater. The TDS, which is a measure of dissolved inorganic and organic contents, has been observed to show minimum and maximum values of 190–2,095 mg/L in pre-monsoon and 63–1,681 mg/L in post-monsoon respectively for mine waters. TDS ranged from 37 to 594 mg/L (average 238 mg/L) in pre-monsoon and from 24 to 572 mg/L (average 206 mg/L) in post-monsoon for groundwater samples. Groundwater is the main source of safe drinking water globally and is also crucial for agriculture and industry. Based on TDS values, the water can be categorized in three different classes depending upon the intension of use: drinking use (TDS <1,000 mg/L), agriculture use (1,000 <TDS >3,000 mg/L) and unsuitable for both drinking and agriculture (TDS >3,000 mg/L) (Davis and De Weist 1966). Based on the TDS values, 80% of the mine water and all groundwater samples can be identified as freshwater to be a subject of further analyses for drinking purpose.

TABLE 11.1
Calculations of Water Quality Indices for Drinking, Irrigation and Industrial Uses

Purpose	Indicator	Equation		Water Class	Recommended Value	References
Drinking use	Drinking Water Quality Index (DWQI)	$\text{DWQI} = \sum_{i=1}^{11}[W_i \times (C_i/S_i) \times 100]$ Where, W_i is relative weight of ith parameter and can be calculated as $W_i = w_i / \sum_{i=1}^{11} w_i$ C_i and S_i are concentration and desirable limit of ith parameter.	(11.2)	Excellent Good Poor Very Poor Unsuitable	<50 50–100 100–200 200–300 >300	Uddin et al. (2021)
Irrigation use	Sodium Adsorption Ratio (SAR)	$\text{SAR} = \text{Na}^+ / \sqrt{\frac{\text{Ca}^{2+} + \text{Mg}^{2+}}{2}}$ Where ionic concentration is in meq/L	(11.3)	Low Medium High Very High	<10 10–18 18–26 >26	Richards (1954)
	Permeability Index (PI)	$\text{PI} = \frac{\left(\text{Na}^+ + \sqrt{\text{HCO}_3^-}\right)}{\text{Ca}^{2+} + \text{Mg}^{2+} + \text{Na}^+} \times 100$	(11.4)	Good Suitable Unsuitable	>75% 25–75% <25%	Doneen (1964)
Industry use	Langelier Index (LI)	$\text{LI} = \text{pH} - \text{pHs}$ Where $\text{pHs} = 9.3 + A + B - \text{C} - \text{D}$ $A = [\text{Log(TDS)} - 1]/10$ $B = -13.12 \times \text{Log}(°\text{C} + 273) + 34.55$ $C = \text{Log(Ca hardness)} - 0.4$ $D = \text{Log(Alkalinity)}$	(11.5)	Scaling No scaling Corrosion	LI >0 LI = 0 LI < 0	Langelier (1936) and Sunardi et al. (2020)
	Ryznar Stability Index (RSI)	$\text{RSI} = 2\text{pHs} - \text{pH}$	(11.6)	Heavy Scaling Scaling No corrosion no scaling Corrosion High corrosion	<5.5 5.5–6.2 6.2–6.8 6.8–8.5 >8.5	Ryznar (1944)

TABLE 11.2
Statistical Summary of the Analysed Parameters and Comparison with Drinking Water Standards

			pH	EC	TDS	F^-	Cl^-	HCO_3^-	SO_4^{2-}	NO_3^-	Ca^{2+}	Mg^{2+}	Na^+	K^+	TH
				µS/cm	mg/L										
Mine water (Nos. 65)	Pre-M (2019, 2022)	Avg.	6.54	995	722	1.13	26.0	95.8	393.1	7.3	90.8	57.7	26.0	23.8	465
		Min.	3.26	271	190	0.28	8.0	0.0	2.9	0.0	5.0	2.7	7.3	6.4	32
		Max.	8.18	2,989	2,095	3.33	80.0	292.7	1516.4	54.0	286.0	244.6	73.5	57.5	1720
		Stdev.	1.55	811	589	0.78	13.9	82.1	472.7	13.9	87.1	65.0	16.0	16.4	476
	Post-M (2019, 2020)	Avg.	6.34	742	560	0.83	17.8	81.8	312.8	5.3	65.1	46.9	15.7	13.5	356
		Min.	3.38	83	63	0.17	5.0	0.0	2.5	0.0	6.7	1.7	3.7	1.4	28
		Max.	7.93	1957	1681	2.20	35.0	195.0	1214.5	21.9	218.7	165.6	47.5	44.0	1200
		Stdev.	1.47	595	462	0.47	7.3	71.5	391.9	7.2	57.8	51.5	10.5	9.7	346
Groundwater (Nos. 37)	Pre-M (2022)	Avg.	7.00	308	238	0.74	24.5	113.4	24.2	8.9	23.8	6.3	32.3	4.0	85
		Min.	5.88	52	37	0.01	2.0	16.0	6.2	0.0	4.8	1.5	1.6	0.2	16
		Max.	8.48	851	594	3.87	124.0	216.8	107.4	83.7	80.7	13.2	111.5	12.8	256
		Stdev.	0.82	234	155	1.09	38.6	74.9	30.9	24.9	21.8	4.4	32.3	4.4	70
	Post-M (2019, 2020)	Avg.	6.57	281	206	0.71	29.8	94.2	12.5	13.0	21.5	7.0	24.0	3.6	83
		Min.	5.23	32	24	0.03	4.0	8.2	0.4	0.0	3.4	0.8	0.8	0.5	16
		Max.	8.52	856	572	4.55	140.0	214.2	60.7	126.2	82.4	16.0	88.2	13.0	256
		Stdev.	0.90	223	148	1.15	35.8	64.5	17.2	32.8	19.8	4.2	25.1	3.7	60
BIS (2012)	Acceptable Limit (AL)		6.5–8.5	NR	500	1.0	250	NR	200	45	75	30	NR	NR	200
	Permissible Limit (PL)		NR	NR	2,000	1.5	1,000	NR	400	NR	200	100	NR	NR	600
	Samples exceeding AL		41%	–	25%	14%	0	–	3%	7%	19%	21%	–	–	17%
	Samples exceeding PL		–	–	4%	14%	0	–	23%	–	8%	8%	–	–	21%

Note: Pre-M and Post-M represent pre-monsoon and post-monsoon, respectively.

11.4.2 Total Hardness

TH is measured by the equivalent quantity of calcium carbonate ($CaCO_3$) in milligrams per litre. There are two types of hardness: transient hardness, which may be addressed by boiling the water, and permanent hardness, which requires special treatment procedures to handle the sulphate and chlorate elements in the water. The main cations that cause hardness are magnesium (Mg^{2+}) and calcium (Ca^{2+}), which are found in water in significant amounts and bonded to bicarbonate, carbonate, sulphate or chloride. However, additional heavy metals such as Mn, Fe, Al, Sr, As, Ba and Zn also contribute to the hardness of natural water (WHO 2011). In pre-monsoon, TH level of the mine water ranged from 32 to 1,720 mg/L (average 465 mg/L), while in post-monsoon, it varied from 28 to 1,200 mg/L (average 355.6 mg/L) (Table 11.2). Approximately 45%, 32% and 23% of mine waters fall under the TH <200, 200 <TH >600 and TH >600 ranges, respectively. On the other side, TH in groundwater varied from 16 to 256 mg/L with an average of 85 mg/L in pre-monsoon and from 16 to 256 mg/L with an average of 83 mg/L in post-monsoon. Sawyer and Mc. Carty (1967) described four groups of water according to the TH-concentration-based classification system: soft (TH <75 mg/L), moderately hard (75 <TH >150 mg/L), hard (150 <TH >300 mg/L) and very hard (TH >300 mg/L). The studied groundwater samples are spread in the zone of soft hardness (46%), moderately hard (43%) and hard nature (11%).

Bivariate plot shows that mine water possesses mainly hard-fresh and hard-brackish water nature, while groundwater has prominently soft-fresh water nature at most of the locations (Figure 11.3). Elevated concentrations of calcium and other ion concentrations in water may result in scale accumulation on plumbing fixtures and pipelines, while soft water has the potential to erode pipe interiors, making heavy metals such as zinc, copper, lead, cadmium and iron more soluble in water.

11.4.3 Major Ion Chemistry

10.4.3.1 Anion Chemistry of Mine Water

The average major anion composition, compiling all mine water together, depicts the dominance of SO_4^{2-} with a proportion of 77% followed by HCO_3^- (15%) >Cl^- (7%) >NO_3^- (1%) >F^- (<1%) to the total anion (TZ^-). The SO_4^{2-} varied from 2.9 at Orient 2 to 1516.4 mg/L at Lilari with an average concentration of 389.7 mg/L in pre-monsoon, and from 2.5 at Lakhanpur to 1214.5 mg/L at Lilari with an average of 312.8 mg/L in post-monsoon. The chemical weathering of sulphide bearing minerals is the geogenic source of SO_4^{2-} ions in mine water environment. Anthropogenic inputs are also responsible for adding SO_4^{2-} in water. HCO_3^-, accounting for 15% to the TZ^-, showed the average concentrations of 95.8 mg/L in pre-monsoon and 81.8 mg/L in post-monsoon. The maximum HCO_3^- concentrations were 292.7 and 195 mg/L in the respective seasons. More than 52% mine water expressed higher HCO_3^- than SO_4^{2-} inferring carbonation process as the major governing phenomena in those mine water. HCO_3^- may be introduced into water environment due to uptake of CO_2 either from soil zone gases or direct inputs from atmosphere (Langmuir 1971). The CO_2 is dissolved in water and makes carbonic acid which further breaks

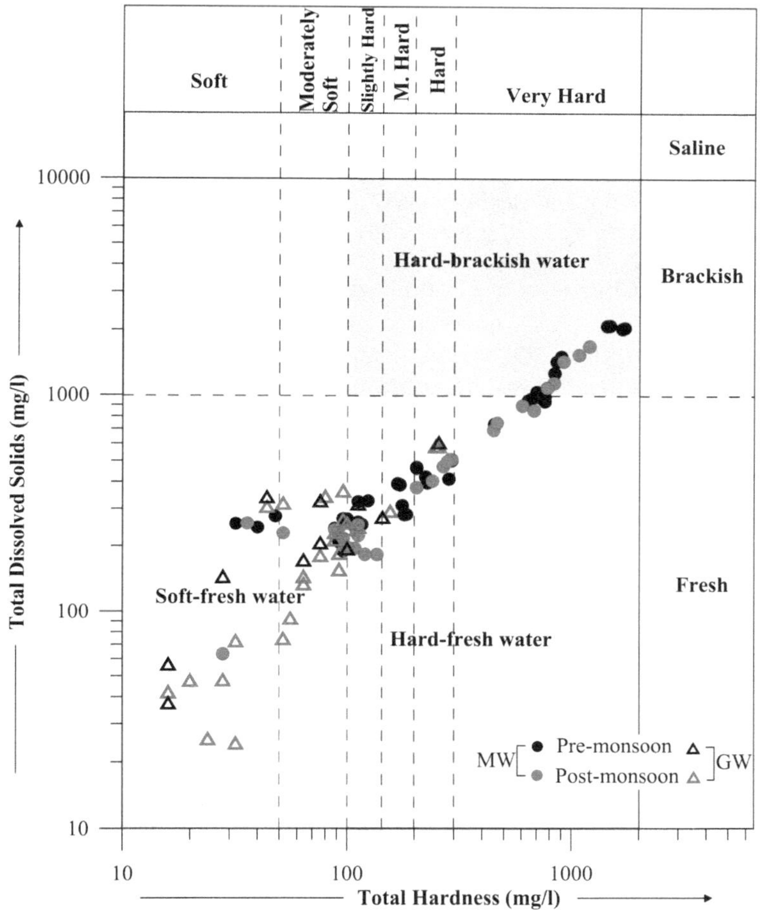

FIGURE 11.3 Total dissolved solid and total hardness–based classification of mine water (MW) and groundwater (GW).

into H^+ and HCO_3^- ions. The acidity generated in this process weathers carbonaceous rock materials associated with the lithology of the study area. Cl^- ranged from 8 to 80 mg/L in pre-monsoon and 5–35 mg/L in post-monsoon. Cl^- in water comes from both natural and anthropogenic sources. In nature element chlorine is associated with different mineral groups comprising halides, oxides, carbonates, borates, silicates, sulphates, arsenates and phosphates, and chemical weathering of these minerals can add Cl^- in water (Yudovich and Ketris 2006). Anthropogenic constraints such as domestic wastes, agricultural return-flow, sewage effluent and septic tank spillages could also be the chief contributors of Cl^- in water (Srinivasamoorthy et al. 2008). The maximum concentrations of NO_3^- were observed to be 54 mg/L in pre-monsoon and 21.9 mg/L in post-monsoon. F^- varied from 0.28 mg/L at Lakhanpur to 3.33 mg/L at Kulda with an average of 1.13 mg/L in pre-monsoon, while from 0.17 mg/L at Lakhanpur to 2.2 mg/L at Garjanbahal with an average of 0.83 mg/L in post-monsoon. Around 65% of the total studied mine water is well below the drinking water limit of F^- (1 mg/L), and around 18% and 17% of samples expressed F^- concentrations higher than the acceptable limit (1 mg/L) and permissible limit (1.5 mg/L), respectively (BIS 2012). F^- may be derived into water from the weathering of fluoride-bearing minerals such as fluorite, chlorite, biotite, apatite, muscovite and sericite. Additionally, the fluoride-containing emissions generated from coal-fired power plants deposit directly into water or deposit firstly on soil with subsequent runoff into water (Chen et al. 2013).

11.4.3.2 Anion Chemistry of Groundwater

The groundwater from IBVC showed a trend of anion compositions in the order of HCO_3^- (55%) >Cl^- (27%) >SO_4^{2-} (11%) >NO_3^- (6%) >F^- (<1%). HCO_3^-, the dominant anion, ranged from 16 to 216.8 mg/L and from 8.2 to 214.2 mg/L in pre-monsoon and post-monsoon, respectively (Table 11.2), while the average HCO_3^- concentrations were observed to be 113.3 and 94.2 mg/L in the respective seasons. The spatial distribution of HCO_3^- in groundwater is shown through contour graphs (Figure 11.4A–C). Carbonation is the most abundant process in increasing the HCO_3^- in water environment. Simultaneously, chemical weathering of silicate minerals could also be one of the probable sources of bicarbonates in water. Cl^-, the second most dominated ion

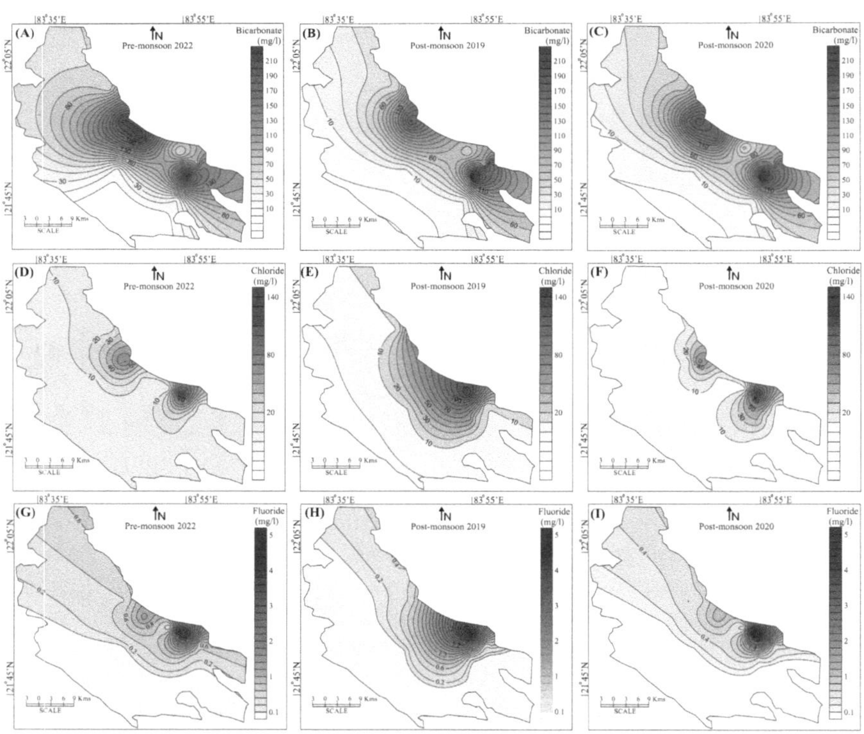

FIGURE 11.4 Spatial and seasonal distribution of HCO_3^-, Cl^- and F^- concentrations in the sub-surficial water environment of Ib Valley coalfield.

in anion chemistry, ranged from 2 to 124 mg/L (average 24.5 mg/L) in pre-monsoon and from 4 to 140 mg/L (average 29.8 mg/L) in post-monsoon (Table 11.2). Cl^- may be sourced from geogenic, atmospheric and anthropogenic constraints depending upon the activities more favourable in the region. The Cl^- distribution in groundwater of the study area is provided in Figure 11.4D–F.

SO_4^{2-} ranged from 6.2 to 107.4 mg/L and from 0.4 to 60.7 mg/L respectively in pre-monsoon and post-monsoon for groundwater. SO_4^{2-} is a good indicator of pollution in mining regions. Groundwater contains sulphate from a wide range of anthropogenic and natural sources, including soils, seawater, sulphate mineral dissolution, oxidation of sulphide minerals, chemical fertilizers, ordinary superphosphate, synthetic detergents and precipitation affected by human activity. Recharge water has also the potential to transfer anthropogenic, pedospheric and atmospheric sulphates into aquifers. Additionally, organic sulphonates, solid or liquid, and other organic forms of reduced S are known to be oxidized step by step by microorganisms to SO_4^{2-}. NO_3^-, contributed 6% to the total anion (TZ^-) of the studied groundwater. It varied from <0.01 to 83.7 mg/L (average 8.9 mg/L) in pre-monsoon and from <0.01 to 126.2 mg/L (average 13 mg/L). Man-made activities are more favourable conditions to increase NO_3^- in groundwater. Municipal wastes, landfill, sewage discharge, industrial effluent, fertilizers, mining wastes, etc., are some important sources of NO_3^-. Around 11% groundwater samples showed NO_3^- concentrations more than 45 mg/L, desirable drinking water limit (BIS 2012).

Fluoride (F^-) is an important water quality parameter, varying from 0.01 to 3.87 mg/L with an average of 0.74 mg/L pre-monsoon and that for post-monsoon, F^- ranged from 0.03 to 4.55 mg/L with an average of 0.71 mg/L (Figure 11.4G–I). Around 14% groundwater samples expressed F^- >1 mg/L, desirable limit for drinking use (BIS 2012). Groundwater fluoride-contaminated geographic areas are primarily identified by the presence of crystalline basement rocks, semi-arid and arid climates, groundwater of the Na–HCO_3 type that is deficient in Ca^{2+}, long groundwater residence times and distance from the recharge area (Handa 1975; Banerjee 2015). Chemical weathering of fluoride-bearing minerals can add F^- in water. F^- even at low concentrations makes the water unsuitable for drinking use. It is due to is higher impact on human health in the form of some critical diseases such as skeletal and dental fluorosis.

11.4.3.3 Cation Chemistry of Mine Water

The distribution of positively charged ions known as cations such as Mn, Fe, Al, Cu, Zn, Ca, Mg, Sr, Na and K in natural water depends upon their availability in the surrounding matrix, atmospheric precipitation, anthropogenic inputs and upon the mineral dissolution characteristics. Out of these cations Ca, Mg, Na and K are dominant in water chemistry. In mine water chemistry, the cations can vary in composition as well as in constituent. Mine water depicted the higher presence of alkaline earths ($Ca^{2+}+Mg^{2+}$) over alkalis (Na^++K^+) by representing the average abundance order of Mg^{2+} (44%) >Ca^{2+} (41%) >Na^+ (10%) >K^+ (5%) to the TZ^+. Mg^{2+} was identified as the dominating cation, and it varied from 2.7 to 244.6 mg/L (average 57.7 mg/L) in pre-monsoon and 1.7–165.6 mg/L (average 46.9 mg/L) in post-monsoon. A good and negative correlation of Mg^{2+} and pH depicts that acidic conditions are more favourable for the availability of alkaline earths such as Mg^{2+} and Ca^{2+} in mine water. Ca^{2+} the second most abundant cation after Mg^{2+} in mine water ranged from 5 to 286 mg/L in pre-monsoon and from 6.7 to 218.7 mg/L in post-monsoon. The average Ca^{2+} concentrations were observed to be 90.8 mg/L (pre-monsoon) and 65.1 mg/L (post-monsoon). Chemical weathering of minerals such as calcite, gypsum, anhydrite, feldspars, pyroxenes and amphiboles could be natural reasons to release alkaline ions into water environment (Vinnarasi et al. 2021; Neogi et al. 2017). Similarly, ion exchange of minerals in rocks and soils also contributes to increased Ca^{2+} and Mg^{2+} in water.

The alkali metals such as sodium (Na^+) and potassium (K^+) together contributed a proportion of 15% to the TZ^+ for the mine water. The Na^+ ranged from 7.3 to 73.5 mg/L (average 25.9 mg/L) in pre-monsoon 3.7–47.5 mg/L (average 15.7 mg/L) in post-monsoon. Some of the major sources of Na^+ in water are atmospheric precipitation, halite dissolution and the decomposition of silicate minerals. Concentrations of K^+ in mine water were found to be 6.4–57.5 mg/L in pre-monsoon and from 1.4 to 44 mg/L in post-monsoon. The potassium-bearing minerals such as orthoclase or biotite found associated with granitic, granite-gneissic rocks in the terrain could be the probable source of K^+ in mine water chemistry (Singh et al. 2012). Coal mines including Lilari, Samleswari and Lajkura where lower pH values were observed demonstrated higher K^+ than Na^+. This could be related to the acidic nature of the water combined with the organic environment, which can also promote the chemical weathering of potassium feldspar in the rock formations (McMahon et al. 1995). When both the cations and anions are added together, they compose a major portion of TDS, and the dominance of any ion in the whole composition is a subject of geochemical mechanism working to that particular location. Figure 11.5A–D present the seasonal and location-wise variance in major ion composition against TDS which could help to assess the overview of the mine water chemistry.

11.4.3.4 Cation Chemistry of Groundwater

The cation composition of groundwater followed the abundance order of Na^+ (40%) >Ca^{2+} (38%) >Mg^{2+} (19%) >K^+ (3%). Na^+, the dominant cation to the total cation (TZ^+), showed its minimum and maximum concentrations of 1.6 and 111.5 mg/L in pre-monsoon and 0.8 and 88.2 mg/L in post-monsoon, respectively. The spatial and temporal variations in Na^+ concentration are shown in Figure 11.6A–C. According to Berner and Berner (1987), the primary sources of Na^+ in the aquatic system are silicate weathering, evaporite dissolution and atmospheric deposition. The average concentrations of Na^+ were 32.3 and 23.9 mg/L in the respective seasons. Ca^{2+}, the second most dominating cation, varied from 4.8 to 80.7 mg/L (average 23.8 mg/L) in pre-monsoon and from 3.36 to 82.4 mg/L (average 21.5 mg/L) in post-monsoon (Figure 11.6D–F). Similarly, Mg^{2+} varied from 1.5 to 13.2 mg/L in pre-monsoon and from 0.8 to 16 mg/L in post-monsoon. The most frequent sources of calcium and magnesium in water are weathering and dissolution of calk silicates (plagioclase, amphiboles, pyroxenes, olivine and biotite) and calcium carbonate (calcite and dolomite) (Wang et al. 2023). K^+, the least dominating cation, showed minimum and maximum concentrations of 0.2–12.8 mg/L in pre-monsoon and 0.5–13 mg/L in post-monsoon for groundwater. K^+ usually exists at low concentrations in groundwater because of weak mobility (Wu et al. 2017b).

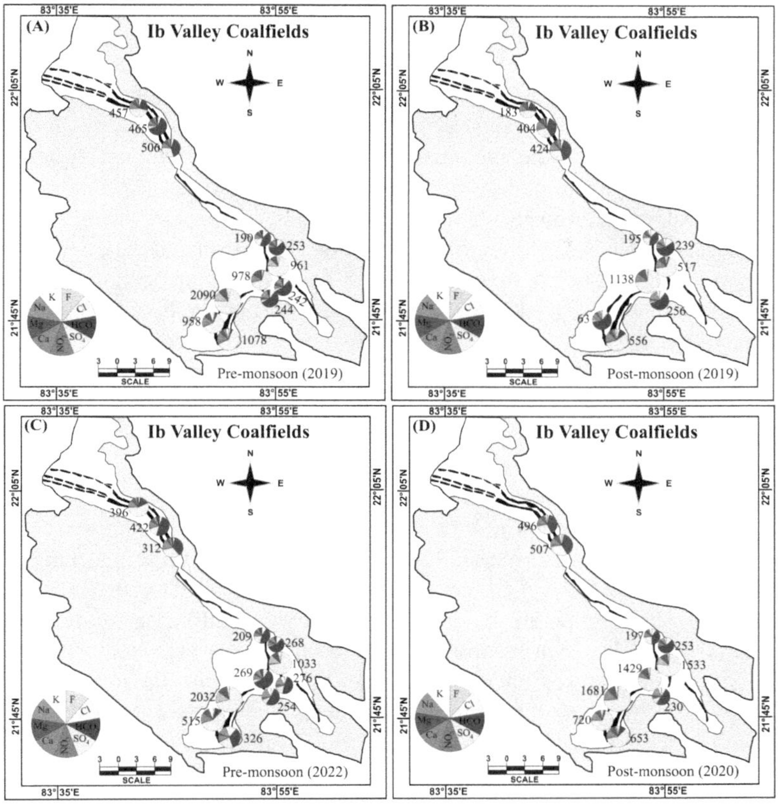

FIGURE 11.5 Proportions of major ionic constituents against total dissolved solids (mg/L) of mine waters.

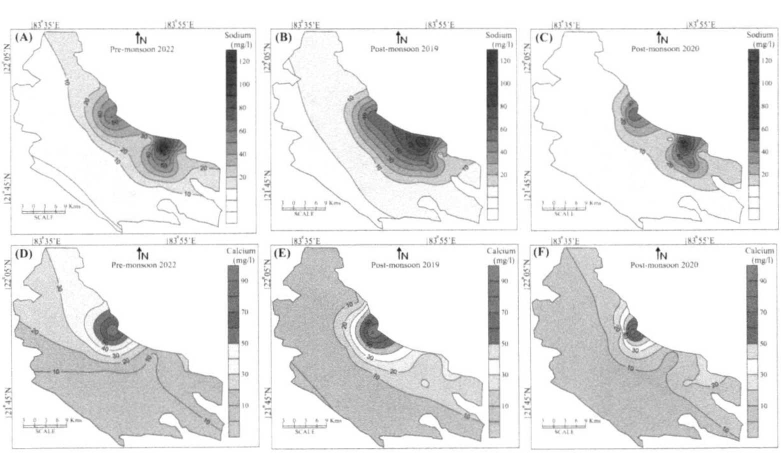

FIGURE 11.6 Spatial and seasonal distribution of Na^+ and Ca^{2+} concentrations in the sub-surficial water environment of Ib Valley coalfield.

11.4.4 Hydrochemical Facies

The diagnostic chemical nature of water solutions in hydrological systems is shown by hydrogeochemical facies (Piper 1944). The majority of graphical techniques are intended to show both the relative proportions of specific major ionic species and the TDS concentrations at the same time. Piper trilinear diagram is frequently used to ascertain the relationships between various dissolved constituents and the classification of water based on its chemical characteristics (Sikakwe et al. 2024). The plot also describes the water chemistry along with its origin and mixing of different water types in one frame. A diamond-shaped field intervenes to indicate the composition of water with regard to both cations and anions, and two triangles at the lower left and lower right describe the relative composition of cations and anions (Figure 11.7A and B). The diamond-shaped field is further subdivided into ten zones as shown in Figure 11.7A. In order to ascertain the geochemical facies of mine water and groundwater, the percentage of equivalent concentrations for each cation and anion (measured in milliequivalent/litre or meq/L) was computed in relation to the total concentration of cations and anion respectively.

Around 91% mine water (in zone 1) dictated influence of alkaline earth metals ($Ca^{2+}+Mg^{2+}$) over alkali metals ($Na^{+}+K^{+}$), while 66% mine water (in zone 4) inferred strong acids ($SO_4^{2-}+Cl^{-}$) exceeding weak acids ($HCO_3^{-}+CO_3^{2-}$). Around 78% groundwater samples are having dominance of alkaline earths ($Ca^{2+}+Mg^{2+}$) over alkalis ($Na^{+}+K^{+}$), while around 57% groundwater represents strong acids ($SO_4^{2-}+Cl^{-}$) exceeding weak acids ($HCO_3^{-}+CO_3^{2-}$). Overall, the mine waters from IBVC show the four major hydrochemical facies such as Ca–Mg–SO_4 (zone 6), Ca–Mg–SO_4–Cl (zone 9), Ca–Mg–HCO_3 (zone 5) and Na–K–HCO_3–Cl (zone 10).

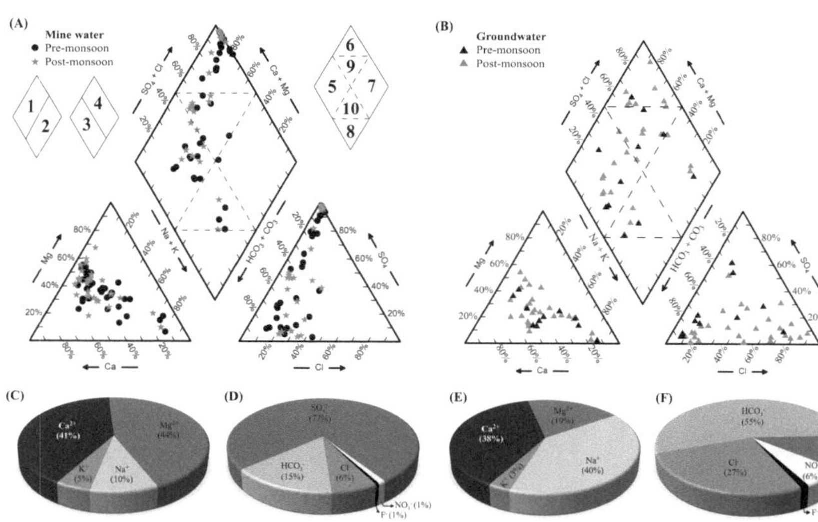

FIGURE 11.7 Piper plots describing hydrochemical facies of (A) mine water, and (B) ground water. Cationic and anionic compositions of mine water (C and D) and groundwater (E and F) respectively.

Majority of the groundwater comprises the hydrochemical facies such as Ca–Mg–HCO_3 (zone 5), Ca–Mg–SO_4 (zone 6) and Ca–Mg–SO_4–Cl (zone 9). Some of the groundwater also obtained hydrochemical facies such as Na–K–HCO_3–Cl (zone 10) and Na–Cl (zone 7). The water samples part of zone 5 might have experienced carbonation phenomena in association with silicate and carbonate mineral weathering due to which these waters have pursued such hydrochemical facies. The water samples covered in zone 6 implicate their origin through sulphate mineral decomposition. However, zones 9 and 10 represent mixed environment of geochemical processes resulting in no one cation–anion pair exceeding 50% of the water composition.

11.4.5 Solute Acquisition Processes

11.4.5.1 Mine Water Environment

The range of $Na^+/(Na^+ + Ca^{2+})$ being 0.03–0.9 and that of $Cl^-/(Cl^- + HCO_3^-)$ being 0.07–1.0 in collaboration with TDS concentrations confirm dominance of rock–water interaction rather than evaporation or precipitation phenomena governing mine water chemistry (Gibbs 1970). The sulphides such as pyrite, melnikovite, marcasite, chalcopyrite and sphalerite are found as mineral matter either in veins or cleat infillings in the coal formations and coal-bearing mudstones in the study area (Senapaty and Behera 2015). After blasting and removal of the overburden material, these minerals get in touch with oxygen and water and get oxidized resulting in low pH and high TDS of the water drainage. Proportions of SO_4^{2-} and HCO_3^- reflect relative dominance of two major sources of protons, i.e. sulphide oxidation and carbonation. High SO_4^{2-} and low $HCO_3^-/(HCO_3^- + SO_4^{2-})$ ratios (<0.5) (Figure 11.8A) suggest

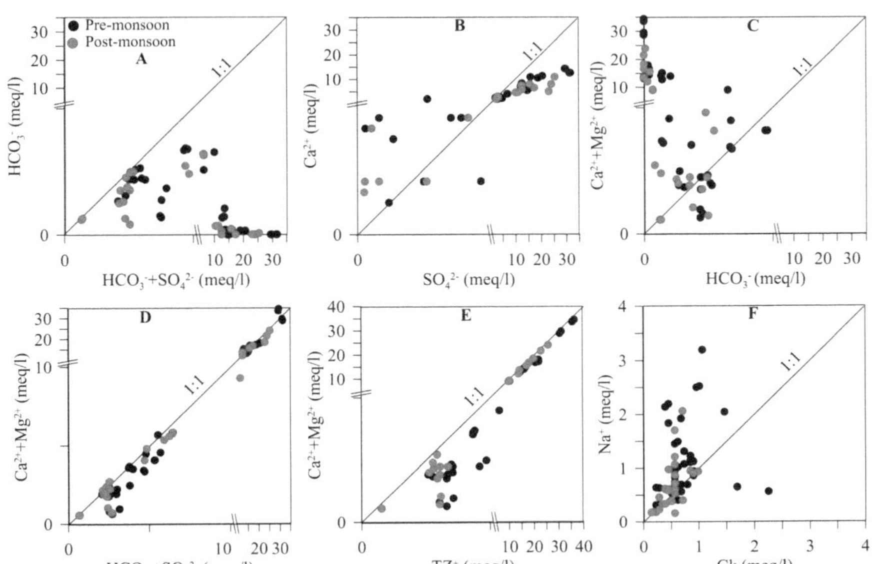

FIGURE 11.8 Scatter plots describing solute acquisition processes for mine water environment.

that either sulphide oxidation or coupled reactions involve both sulphide oxidation and carbonic acid weathering control mine water chemistry (Singh et al. 2011). Low Ca^{2+}/SO_4^{2-} ratio (<1.0) limits input from gypsum dissolution and suggest pyrite oxidation as major source of sulphate ions (Figure 11.8B). Poor correlation between ($Ca^{2+}+Mg^{2+}$) and HCO_3^- and high ($Ca^{2+}+Mg^{2+}$)/HCO_3^- ratio (average 17.9) suggest that pure carbonate dissolution would not be the major source of Mg^{2+} and Ca^{2+} (Figure 11.8C). Better correlation of ($Ca^{2+}+Mg^{2+}$) with ($HCO_3^-+SO_4^{2-}$) suggests that sulphide weathering plays a significant role in controlling chemical composition of mine water (Figure 11.8D) (Singh et al. 2016). The strong correlation of TDS with SO_4^{2-} (0.9), Ca^{2+} (0.9), Mg^{2+} (0.9) and pH (−0.8) together indicates higher dissolved load at low pH environment. Plotted points along the 1:1 equiline in ($Ca^{2+}+Mg^{2+}$) vs total cations (TZ^+) plot (Figure 11.8E), signifying major contribution of $Ca^{2+}+Mg^{2+}$ to TZ^+. High Na^+/Cl^- ratio (average 1.6) (Figure 11.8F) and K^+/Cl^- (average >0.2) (Gupta et al. 2016) unveil that silicate weathering is an important contributor of Na^+ and K^+, with a limited contribution of alkalis from halite dissolution, atmospheric precipitation or anthropogenic inputs.

11.4.5.2 Groundwater Environment

Gibbs (1970) presented ratios of Na/(Na + Ca) and Cl/(Cl + HCO_3) to comprehend the natural processes that govern the chemistry of groundwater, such as precipitation, evaporation, rainfall and rock weathering. The values of these ratios ranged from 0.16 to 0.88 and from 0.03 to 0.73, respectively. When the values of these ratios were plotted against TDS of the samples, it was observed that most of the studied groundwater samples in the present work expressed rock–water interaction phenomena to control water chemistry. The rock–water interaction is a complex phenomenon which involves several processes in its domain, out of them weathering and dissolution of minerals, ion exchange, reverse ion exchange, dissolution and precipitation of mineral phases interventions are important. The molar ratios of Ca^{2+}/Na^+, HCO_3^-/Na^+ and Mg^{2+}/Na^+ are able to identify specifically which weathering process is occurring in the water environment (Gaillardet et al. 1999). The values of Ca^{2+}/Na^+, Mg^{2+}/Na^+ and HCO_3^-/Na^+ for the groundwater varied from 0.07 to 2.95, from 0.01 to 4.25 and from 0.19 to 9.74, respectively. Geochemical data of the studied groundwater samples explain that carbonate and silicate weathering could be one of the important geochemical processes in groundwater environment.

According to Grasby et al. (2000) and Ganyaglo et al. (2011), the groundwater system in many mining regions is complicated and comprising several aquifers with varying geochemical properties. Most of the studied groundwater is rich in HCO_3^- to TZ^-, and the average values of $HCO_3^-/(HCO_3^-+SO_4^{2-})$ being 0.84 denote that carbonation process is the most dominant weathering phenomenon in the region (Figure 11.9A). Since Cl^- is the second most dominating anion after HCO_3^-, it could also play key role in controlling chemical composition of groundwater. The average value of $HCO_3^-/(HCO_3^-+Cl^-)$ for the groundwater is found to be 0.7, implying that along with carbonation process there exist other geochemical processes responsible for adding chloride into the groundwater and increasing overall total dissolved solid loading of groundwater. Majority of the groundwater samples are on or near the

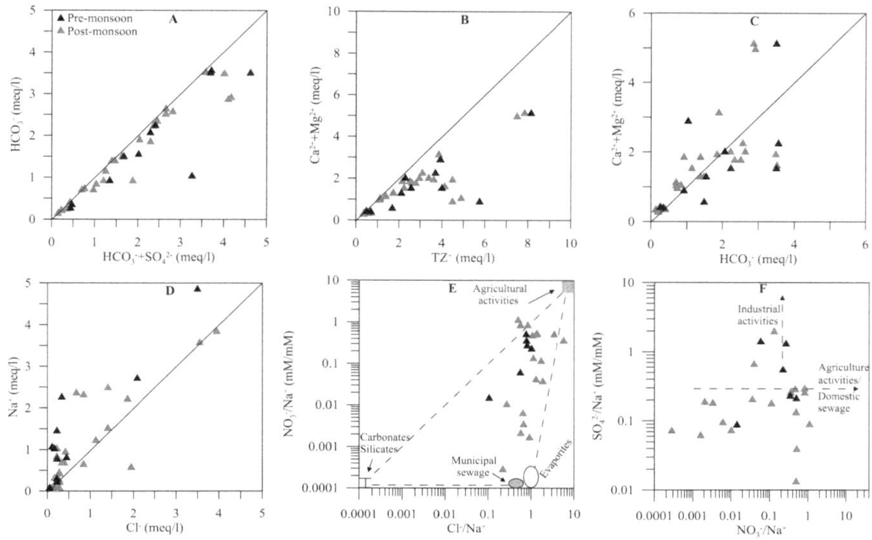

FIGURE 11.9 Scatter plots describing solute acquisition processes for groundwater environment.

equiline (1:1) in Figure 11.9B, signifying that alkaline ions are more responsible to control water cationic composition at the concerned sites. However, few groundwater samples are deviated from 1:1 line towards x-axis (Figure 11.9B), inferring that these waters need some other cations such as Na^+ and K^+ to balance the total cations equivalent load at these sites. In Figure 11.9C, groundwater samples are spread on equiline as well as deviated both sides from equiline. The access HCO_3^- in the samples (deviated towards x-axis) can be equivalently balanced if concentration of alkalis is added on y-axis. This means these waters could have faced weathering of sodium bearing silicate minerals which released both Na^+ and HCO_3^- into water environment. Figure 11.9D shows that some groundwater samples are on the equiline, whereas most of the samples are spread towards y-axis. Halite dissolution demands equal equivalent concentrations of Na^+ and Cl^- which means when plotting these two ions on bivariate plot, the samples should be spread over equiline (1:1). In the majority of the examined groundwater, an excess of Na^+ over Cl^- indicates a non-atmospheric or halite dissolution source, but rather implies weathering of Na silicates accounting for a significant portion of the water composition (Tiwari et al. 2016). Molar ratios of NO_3^-/Na^+ and Cl^-/Na^+ as depicted in Figure 11.9E denote that agricultural activities along with municipal sewage discharge could be responsible for increasing NO_3^- and Cl^- levels in groundwater. Similarly, molar ratio of SO_4^{2-}/Na^+ against NO_3^-/Na^+ shown in Figure 11.9F favours combined impacts of industrial, agricultural and domestic sewages to contribute SO_4^{2-} and NO_3^- into groundwater environment. Overall, natural and anthropogenic activities together control water chemistry of the groundwater system of the area.

11.4.6 Water Quality Assessment and Management

11.4.6.1 Drinking Use

DWQI on one side is provided to describe the general situation of water by using mathematical tools to change water quality parameters levels into a numerical score, and thus it, could be suggested to what extent the water is suitable for drinking purpose (Ramakrishnaiah et al. 2009). DWQI can be calculated using Eq. 11.2 (Table 11.1). wi is the weight attributed to the element according to its relative apperceptive effects on human health and significance on drinking purposes (Yidana and Yidana 2010). The highest weight of 5 was assigned to parameters such as TDS, F and NO_3, 3 was assigned to Cl, SO_4, TH, Fe and Mn, and 2 was assigned to Ca, Mg and Zn. The computed DWQI ranged from 12.3 to 1882.6 for mine water and from 7.9 to 137.8 for groundwater (Figure 11.10). Overall, 71% samples are representing an excellent to good water quality for drinking use and could be suitable for human consumption by mean of the water parameters used in DWQI. Mine water quality is majorly being affected majorly by increased concentrations of Mn, Fe and TDS. Whereas, groundwater quality is being influenced by Mn, F^- and Fe for drinking use. These parameters can be given further attention when treatment methods are concerned.

11.4.6.2 Irrigation Water Quality Assessment

The evaluation of water quality for irrigation use could be done based on its salinity, alkali hazards and permeability characteristics, and three important irrigation parameters, i.e. salinity hazard, sodium adsorption ratio (SAR) and permeability index (PI), can be focused for this purpose. Salinity hazard can be measured based on the (EC values. It can be categorized into four different categories, i.e. C1 (low, EC<250 μS/cm), C2 (medium, 250 <EC >750 μS/cm), C3 (high, 750 <EC >2,250 μS/cm) and C4 (very high, EC >2,250 μS/cm) (USSL 1954). EC, for mine water, ranged from 271 to 2,989 μS/cm with an average of 995 μS/cm in pre-monsoon and

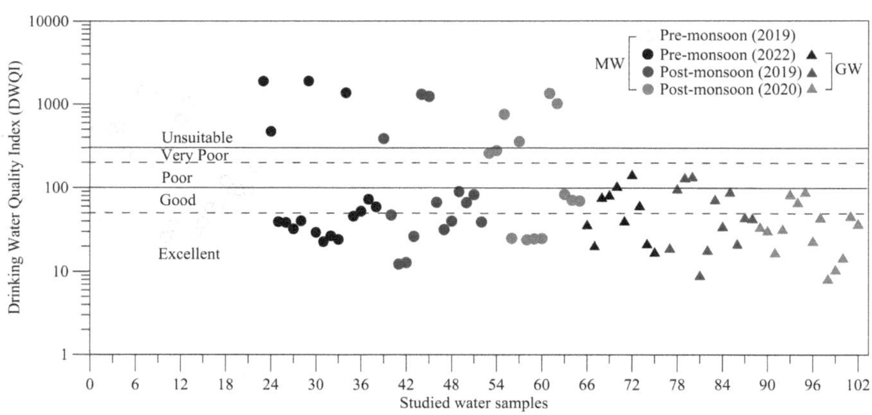

FIGURE 11.10 Drinking water quality index for Ib Valley coalfield.

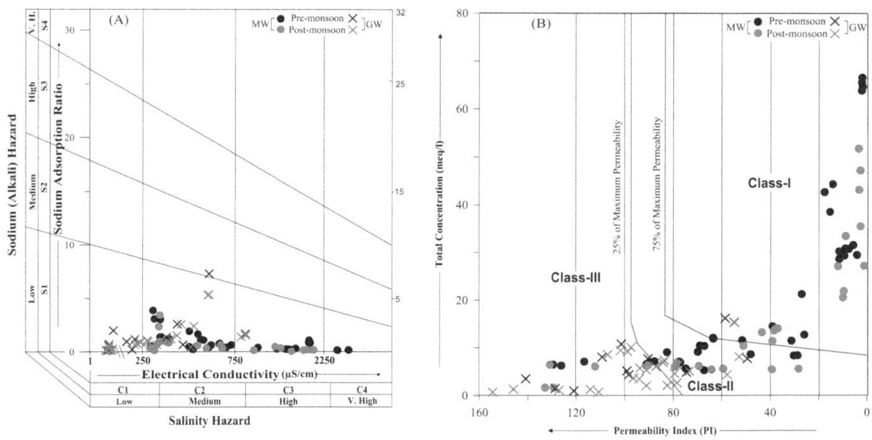

FIGURE 11.11 Classification of irrigation water (A) US salinity diagram (USSL) (Richard 1954) (B) Permeability index (PI) vs Total concentration (Doneen 1964).

from 83 to 1,957 with an average of 742 µS/cm in post-monsoon (Figure 11.11A). For groundwater, EC ranged from 52 to 851 µS/cm for pre-monsoon and from 32 to 856 µS/cm for post-monsoon (Figure 11.11A). Around 73% of the total studied samples showed low to medium salinity hazard, while 23 and 4% of samples expressed high and very high salinity hazard, respectively (Richards 1954). High salinity levels in irrigation water can have direct harmful consequences by closing the gap between the internal and exterior plant water potentials. Furthermore, there is a possibility that the medium's osmotic potential would decrease, which could have a negative impact on crop yields, plant development and other factors.

The Na or alkali hazard on the other side is the presentation of the relative and absolute proportions of the cations such as Na^+, Ca^{2+} and Mg^{2+} in the form of SAR as formulated in Eq. 11.3 (Table 11.1) (Mukhopadhyay et al. 2022; Richards 1954). Based on the SAR values, most of the mine water and groundwater samples showed low alkali hazard (SAR <10) (Figure 11.11A), favouring suitability for irrigation use. When water has higher proportion of Na^+ ion in comparison to other cations, it could probably impact or deteriorate soil physical properties resulting in the dispersion of soil clay, causing the soil to become hard and compact when dry and, hence, decreasing the permeability of the soil (Sunitha and Reddy 2019). The permeability index (PI) is a water movement capacity of soil, and for water entity, it depends on the dissolved concentrations mainly of Na^+, Ca^{2+}, Mg^{2+} and HCO_3^- ions (Eq. 11.4). The classification of water based on PI values has been provided in Table 11.1 (Doneen 1964). The PI values varied from 1.16 to 154.4 covering both mine water and groundwater samples. Around 67% of the total water samples are classified into class-I and class-II (Figure 11.11B), which show suitability of these water samples for irrigation use. Remaining 33% samples are covered in class-III with higher PI values suggesting that water with higher PI and Na^+ and lower TDS could decrease soil permeability (25% of the maximum permeability) if such water is used in irrigation. The PI value below 25% of the maximum permeability expresses unsuitability of the concerned water for irrigation use. The main reason behind it is related to the high concentration of sodium (Na^+) ions in water (Chen et al. 2019). Further, a part of carbonate (CO_3^{2-}) and bicarbonate (HCO_3^-) may also precipitate as $MgCO_3$ and $CaCO_3$ by removing the Ca^{2+} and Mg^{2+} ions from the irrigated water and increasing the proportion of Na^+ ions in the concerned water.

11.4.6.3 Scaling and Corrosion Characterization for Industries

Corrosion and scaling are the two important water characters that permit the water to corrode the material flowing within and adding precipitates onto the surface of pipelines in the inner part, respectively (Awasthi et al. 2024; Tang et al. 2006). Corrosion effect in such a way can release Pb, As, Fe and other elements into the transporting water resulting into an unfit supply in the distributing systems for the concern use. On the other side, scaling can produce an extra layer of white crust within the pipes and boilers. This results in reduction in water flow within the pipes, clogging and ultimately bearing costly maintenance. The saturation status of minerals, temperature conditions, dissolved gases (CO_2, O_2), alkalinity, hardness, pH and total dissolved solid (TDS) together determine the final nature of water (Uddin et al. 2020). Langelier index (LI) and Ryznar stability index (RSI) are the two important indices that can be used to determine scaling and corrosion properties of water to assess the solubility or deposition criteria of ions (Table 11.1). The calculated LI (Eq. 11.5) for mine water ranged from −6.22 to 0.5, out of which 88 and 12% samples are respectively expressing LI <0 and LI >0 (Figure 11.12A). Around 37% mine water, having −1 <LI >1, possesses equilibrium state with respect to $CaCO_3$ implying that neither scaling nor corrosion is affective. For groundwater samples, LI varied from −4.77 to −0.23, and all groundwaters show negative LI (Figure 11.12B).

Similarly, RSI was calculated using Eq. 11.6, and it varied from 7.18 to 15.82 for mine water and from 8.13 to 14.77 for groundwater. Water showing negative LI values or RSI >6.8 could corrode the inner crust material of the pipes through which the water is passing (Sarin et al. 2004). This process may include dissolution of solid $CaCO_3$ and hazardous elements such as Pb, As and Fe into flowing water. Strong

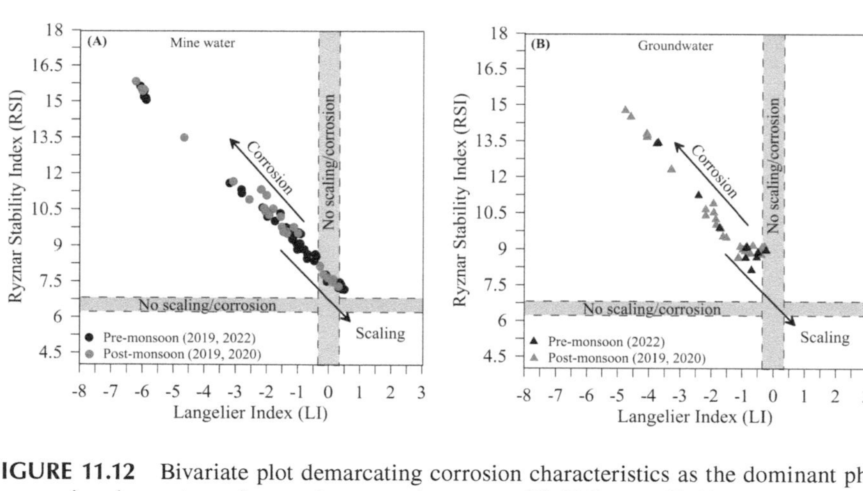

FIGURE 11.12 Bivariate plot demarcating corrosion characteristics as the dominant phenomena in mine water and groundwater environment of Ib Valley coalfield.

correlation of LI with pH (0.98), HCO_3^- (0.76), SO_4^{2-} (−0.79) and Mg^{2+} (−0.78) and that of RSI with pH (−0.96), HCO_3^- (−0.77), SO_4^{2-} (0.75) and Mg^{2+} (0.74) imply that these parameters mostly control scaling or corrosion properties of mine water.

11.5 CONCLUSION

This study covered both the mine water and the surrounding groundwater environment to explain their hydrochemical interventions by involving a discussion on major ion chemistry, similarity in hydrochemical facies, solute acquisition processes and water quality assessment. The wide range of pH from 3.26 to 8.52 makes the water to have acidic to alkaline nature. Around 25 and 4% studied water showed TDS more than acceptable and permissible limits for drinking use. Overall, 71% samples represent an excellent to good water quality for drinking use and could be suitable for human consumption by mean of the water parameters used in DWQI. Around 73% water expressed low to medium salinity hazard, while sodium adsorption ratio was found within desired limit at each site for agriculture use. The overall hydrochemistry influenced by sulphide oxidation and carbonation as the dominant processes provides corrosion characteristics to these waters. Based on the water quality assessment made in the recent work, further strategies towards water management system could be addressed.

ACKNOWLEDGEMENT

The authors are thankful to the Director of CSIR-Central Institute of Mining and Fuel Research, Dhanbad, and Academy of Scientific and Innovative Research (AcSIR). The first author also expresses the gratitude towards Department of Science and Technology (DST), Govt. of India, for granting INSPIRE Fellowship (DST/INSPIRE/03/2017/001845).

CONFLICT OF INTEREST

No potential conflict of interest was reported by the author(s).

AUTHOR'S CONTRIBUTION

The first author carried out the literature survey, water sampling, analysis and compilation of the analytical results followed by interpretation of the data and writing the chapter. Second author significantly reviewed the draft, interpreted the results and finalized the chapter.

REFERENCES

APHA (American Public Health Association) (2017) *Standard Methods for the Examination of Water and Wastewater*, 23rd edn. American Public Health Association, Washington DC.

Awasthi A, Rishi MS, Panjgotra S (2024) Groundwater quality assessment for drinking and industrial purposes in transboundary aquifers of Gurdaspur district, Punjab, India.

International Journal of Environmental Analytical Chemistry, 104(2), 389–403. doi.org/10.1080/03067319.2021.2020766

Banerjee A (2015) Groundwater fluoride contamination: A reappraisal. *Geoscience Frontiers*, 6(2), 277–284.

Berner EK, Berner RA (1987) *The Global Water Cycle: Geochemistry and Environment*. Prentice-Hall, Englewood Cliffs.

Bhardwaj S, Soni R, Gupta SK, Shukla DP (2020) Mercury, arsenic, lead and cadmium in waters of the Singrauli coal mining and power plants industrial zone, Central East India. *Environmental Monitoring and Assessment*, 192(4), 251.

BIS (Bureau of Indian Standards) (2012) *Standards for Drinking Water*, IS:10500. BIS, New Delhi.

Carvalho FP (2017) Mining industry and sustainable development: Time for change. *Food and Energy Security* 6(2), 61–77. https://doi.org/10.1002/fes3.109

Cesar Minga J, Elorza FJ, Rodriguez R, Iglesias A, Esenarro D (2023) Assessment of water resources pollution associated with mining activities in the parac subbasin of the rimac river. *Water,* 15(5), 965. https://doi.org/10.3390/w15050965

CGWB (2020) *Aquifer Mapping and Management in Jharsuguda District, South Eastern Region*. Central Ground Water Board, Bhubaneswar, Odisha, India.

Chaudhury PN (1988) Report on the Geology and the Coal Resources of Gopalpur Area, Hingir Valley, Ib-River Coalfield. Report, Geological Survey of India.

Chen, J, Huang Q, Lin Y, Fang Y, Qian H, Liu R, Ma H (2019) Hydrogeochemical characteristics and quality assessment of groundwater in an irrigated region, Northwest China. *Water*, 11(1), 96. doi.org/10.3390/w11010096

Chen J, Liu G, Kang Y, Wu B, Sun R, Zhou C, Wu D (2013) Atmospheric emissions of F, As, Se, Hg, and Sb from coal-fired power and heat generation in China. *Chemosphere,* 90(6), 1925–1932. https://doi.org/10.1016/j.chemosphere.2012.10.032

Coal Atlas of India (1993) *Coal Atlas of India*. Central Mine Planning and Design Institute, Calcutta.

CSE (2012) *Coal Mining*. Centre for Science and Environment, https://www.cseindia.org/userfiles/fsheet2.pdf (accessed 5 April 2015)

Davis SN, Weist RJM (1966) *Hydrogeology*. Wiley, New York.

Doneen LD (1964) Notes on water quality in agriculture. Water science and engineering, paper 4001. Department of Water Sciences and Engineering, University of California, California

Evangelou VP, Zhang YL (1995) A review: Pyrite oxidation mechanisms and acid mine drainage prevention. *Critical Reviews in Environmental Science and Technology*, 25(2), pp.141–199. https://doi.org/10.1080/10643389509388477

Gaillardet JDBL, Dupré B, Louvat P, Allegre CJ (1999) Global silicate weathering and CO2 consumption rates deduced from the chemistry of large rivers. *Chemical Geology*, 159(1–4), 3–30. https://doi.org/10.1016/S0009-2541(99)00031-5

Galhardi JA, Bonotto DM (2016) Hydrogeochemical features of surface water and groundwater contaminated with acid mine drainage (AMD) in coal mining areas: A case study in southern Brazil. *Environmental Science and Pollution Research*, 23, 18911–18927. https://doi.org/10.1007/s11356-016-7077-3

Ganyaglo SY, Banoeng-Yakubo B, Osae S, Dampare SB, Fianko JR (2011) Water quality assessment of groundwater in some rock types in parts of the eastern region of Ghana. *Environmental Earth Sciences,* 62(5), 1055–1069. doi.org/10.1007/s12665-010-0594-3

Gibbs RJ (1970) Mechanisms controlling world water chemistry. *Science,* 170(3962, 1088–1090. doi:10.1126/science.170.3962.1088

Goswami S, Saxena A, Singh KJ, Chandra S, Cleal CJ (2018) An appraisal of the Permian palaeobiodiversity and geology of the Ib-River Basin, eastern coastal area, India. *Journal of Asian Earth Sciences,* 157, 283–301. https://doi.org/10.1016/j.jseaes.2017.09.006

Grasby SE, Hutcheon I, Krouse HR (2000) The influence of waterrock interaction on the chemistry of thermal springs in western Canada. *Applied Geochemistry,* 15(4), 439–454. https://doi.org/10.1016/S0883-2927(99)00066-9

Guo Q, Yang Y, Han Y, Li J, Wang X (2019) Assessment of surface–groundwater interactions using hydrochemical and isotopic techniques in a coalmine watershed, NW China. *Environmental Earth Sciences,* 78, 1–11. https://doi.org/10.1007/s12665-019-8053-2

Gupta S, Nayek S, Chakraborty D (2016) Hydrochemical evaluation of Rangit river, Sikkim, India: Using Water Quality Index and multivariate statistics. *Environmental Earth Sciences,* 75, 1–14. doi:10.1007/s12665-015-5223-8

Handa BK (1975) Geochemistry and genesis of Fluoride-Containing ground waters in India. *Groundwater*, 13(3), 275–281. doi.org/10.1111/j.1745-6584.1975.tb03086.x

Jain MK, Das A (2017) Impact of mine waste leachates on aquatic environment: A review. *Current Pollution Reports*, 3, 31–37. https://doi.org/10.1007/s40726-017-0050-z

Khatri N, Tyagi S (2015) Influences of natural and anthropogenic factors on surface and groundwater quality in rural and urban areas. *Frontiers in Life Science*, 8(1), 23–39. doi.org/10.1080/21553769.2014.933716

Langelier WF (1936) The analytical control of anti-corrosion water treatment. *Journal American Water Works Association*, 28(10), 1500–1521. https://www.jstor.org/stable/41226418

Langmuir D (1971) The geochemistry of some carbonate ground waters in central Pennsylvania. *Geochimica et Cosmochimica Acta*, 35(10), 1023–1045. https://doi.org/10.1016/0016-7037(71)90019-6

Mahato MK, Singh PK, Singh AK, Tiwari AK (2018) Assessment of hydrogeochemical processes and mine water suitability for domestic, irrigation, and industrial purposes in East Bokaro Coalfield, India. *Mine Water and the Environment,* 37(3), 493–504. doi:10.1007/s10230-017-0508-7

Maiti S, Agrawal PK (2005) Environmental degradation in the context of growing urbanization: A focus on the metropolitan cities of India. *Journal of Human Ecology*, 17(4), 277–287. doi.org/10.1080/09709274.2005.11905793

McMahon PB, Vroblesky DA, Bradley PM, Chapelle FH, Gullett CD (1995) Evidence for enhanced mineral dissolution in organic acid-rich shallow ground water. *Groundwater*, 33(2), 207–216. doi.org/10.1111/j.1745-6584.1995.tb00275.x

Mukhopadhyay BP, Chakraborty A, Bera A, Saha R (2022) Suitability assessment of groundwater quality for irrigational use in Sagardighi block, Murshidabad district, West Bengal. *Applied Water Science*, 12(3), p.38. doi.org/10.1007/s13201-021-01565-4

Neogi B, Singh AK, Pathak DD, Chaturvedi A (2017) Hydrogeochemistry of coal mine water of North Karanpura coalfields, India: Implication for solute acquisition processes, dissolved fluxes and water quality assessment. *Environmental Earth Sciences*, 76, 1–17. doi.org/10.1007/s12665-017-6813-4

Piper AM (1944) A graphic procedure in the geochemical interpretation of water-analyses. *Eos, Transactions American Geophysical Union,* 25(6), 914–928. https://doi.org/10.1029/TR025i006p00914

Qadri R, Faiq MA (2000) Freshwater pollution: Effects on aquatic life and human health. In *Fresh Water Pollution Dynamics and Remediation*, pp. 15–26. doi.org/10.1007/978-981-13-8277-2_2

Ramakrishnaiah CR, Sadashivaiah C, Ranganna G (2009) Assessment of water quality index for the groundwater in Tumkur Taluk, Karnataka State, India. *E-Journal of Chemistry*, 6(2), 523–530. doi.org/10.1155/2009/757424

Richards LA (1954) *Diagnosis and Improvement of Saline and Alkali Soils*. US Dept Agriculture Handbook #60, Washington DC, 166 pp.

Ryznar JW (1944) A new index for determining amount of calcium carbonate scale formed by a water. *Journal-American Water Works Association,* 36(4), 472–483. https://doi.org/10.1002/j.1551-8833.1944.tb20016.x

Saeedi M, Abessi O, Sharifi F, Meraji H (2010) Development of groundwater quality index. *Environmental Monitoring and Assessment*, 163, 327–335. doi 10.1007/s10661-009-0837-5

Sarin P, Snoeyink VL, Bebee J, Jim KK, Beckett MA, Kriven WM, Clement JA (2004) Iron release from corroded iron pipes in drinking water distribution systems: Effect of dissolved oxygen. *Water Research,* 38(5), 1259–1269. https://doi.org/10.1016/j.watres.2003.11.022

Sawyer CN, Mc Carty PL (1967) Chemistry for sanitary engineers, and classification of naturally soft and naturally hard waters to sources and hardness of their water supplies. *Journal of Hydrology*, 518.

Senapaty A, Behera P (2015) Stratigraphic control of petrography and chemical composition of the lower Gondwana coals, Ib-valley coalfield, Odisha, India. *Journal of Geoscience and Environment Protection,* 3(4), 56. doi:10.4236/gep.2015.34007

Sikakwe GU, Eyong GA, Ojo SA (2024) Geochemical modeling and hydrochemical analysis for water quality determination around mine drainage areas. *Water Environment Research*, 96(1), e10937. doi.org/10.1002/wer.10937

Singh AK, Mahato MK, Neogi B, Mondal GC, Singh TB (2011) Hydrogeochemistry, elemental flux, and quality assessment of mine water in the Pootkee-Balihari mining area, Jharia coalfield, India. *Mine Water and the Environment*, 30, 197–207. doi.org/10.1007/s10230-011-0143-7

Singh AK, Mahato MK, Neogi B, Tewary BK, Sinha A (2012) Environmental geochemistry and quality assessment of mine water of Jharia coalfield, India. *Environmental Earth Sciences*, 65, 49–65. doi.org/10.1007/s12665-011-1064-2

Singh AK, Varma NP, Mondal GC (2016) Hydrogeochemical investigation and quality assessment of mine water resources in the Korba coalfield, India. *Arabian Journal of Geosciences*, 9, 1–20. doi.org/10.1007/s12517-015-2298-1

Singh NP, Santal AR (2015) Phytoremediation of heavy metals: The use of green approaches to clean the environment. *Phytoremediation: Management of Environmental Contaminants*, 2, 115–129. doi.org/10.1007/978-3-319-10969-5_10

Srinivasamoorthy K, Chidambaram S, Prasanna MV, Vasanthavihar M, Peter J, Anandhan P (2008) Identification of major sources controlling groundwater chemistry from a hard rock terrain: A case study from Mettur taluk, Salem district, Tamil Nadu, India. *Journal of Earth System Science,* 117, 49–58. https://doi.org/10.1007/s12040-008-0012-3

Sunardi S, Ariyani M, Agustian M, Withaningsih S, Parikesit P, Juahir H, Ismail A, Abdoellah OS (2020) Water corrosivity of polluted reservoir and hydropower sustainability. *Scientific Reports,* 10(1), 11110. https://doi.org/10.1038/s41598-020-68026-x

Sunitha V, Reddy YS (2019) Hydrogeochemical evaluation of groundwater in and around Lakkireddipalli and Ramapuram, YSR District, Andhra Pradesh, India. *HydroResearch*, 2, 85–96. https://doi.org/10.1016/j.hydres.2019.11.008

Tang Z, Hong S, Xiao W, Taylor J (2006) Characteristics of iron corrosion scales established under blending of ground, surface, and saline waters and their impacts on iron release in the pipe distribution system. *Corrosion Science*, 48(2), 322–342. https://doi.org/10.1016/j.corsci.2005.02.005

Tiwari AK, De Maio M, Singh PK, Singh AK (2016) Hydrogeochemical characterization and groundwater quality assessment in a coal mining area, India. *Arabian Journal of Geosciences*, 9, 1–17. doi.org/10.1007/s12517-015-2209-5

Uddin MG, Nash S, Olbert AI, (2021) A review of water quality index models and their use for assessing surface water quality. *Ecological Indicators*, 122, 107218.

Uddin MR, Hossain MM, Akter S, Ali ME, Ahsan MA, (2020) Assessment of some physicochemical parameters and determining the corrosive characteristics of the Karnaphuli estuarine water, Chittagong, Bangladesh. *Water Science,* 34(1), 164–180. https://doi.org/10.1080/11104929.2020.1803662

USSL (US Salinity Laboratory) (1954) Diagnosis and improvement of saline and alkali soils. In *US Department of Agriculture Hand Book No 60*. USSL.

Vinnarasi F, Srinivasamoorthy K, Saravanan K, Gopinath S, Prakash R, Ponnumani G, Babu C (2021) Chemical weathering and atmospheric carbon dioxide (CO_2) consumption in Shanmuganadhi, South India: Evidences from groundwater geochemistry. *Environmental Geochemistry and Health*, 43, 771–790. doi.org/10.1007/s10653-020-00540-3

Wang S, Chen J, Jiang W, Zhang S, Jing R, Yang S (2023) Identifying the geochemical evolution and controlling factors of the shallow groundwater in a high fluoride area, Feng County, China. *Environmental Science and Pollution Research,* 30(8), 20277–20296. https://doi.org/10.1007/s11356-022-23516-5

WHO (2011) *World Health Organisation, Guidelines for Drinking-Water Quality*, 4th edn. World Health Organisation, Geneva, Switzerland.

Yidana, SM, Yidana A (2010) Assessing water quality using water quality index and multivariate analysis. *Environmental Earth Sciences*, 59, 1461–1473. doi.org/10.1007/s12665-009-0132-3

Yuan R, Li Z, Guo S (2023) Health risks of shallow groundwater in the five basins of Shanxi, China: Geographical, geological and human activity roles. *Environmental Pollution,* 316:120524. https://doi.org/10.1016/j.envpol.2022.120524

Yudovich YE, Ketris MP (2006) Chlorine in coal: A review. *International Journal of Coal Geology,* 67(1–2), 127–144.

Zhang J, Chen L, Hou X, Lin M, Ren X, Li J, Zhang M, Zheng X (2021) Multi-isotopes and hydrochemistry combined to reveal the major factors affecting Carboniferous groundwater evolution in the Huaibei coalfield, North China. *Science of the Total Environment,* 791:148420. https://doi.org/10.1016/j.scitotenv.2021.148420

12 Groundwater Level Behaviour Assessment Using Sensor-Based Water Level Recorder and GIS in Umaria Coalfield, Madhya Pradesh, India

Suraj Kumar, Saurabh Kumar Singh, and Ashwani Kumar Tiwari

12.1 INTRODUCTION

Clean and safe water access is fundamental to any community's well-being and sustainable development. In India, like many other regions worldwide, groundwater serves as a vital freshwater source for various purposes, including domestic use and agricultural activities. The availability of clean drinking water continues to be a pressing need, considering that approximately 30% of the urban and 90% of the rural population in India rely entirely on untreated surface or groundwater sources (Kumar et al., 2005). The declining groundwater levels and deterioration of its quality are due to several factors, such as population growth and freshwater demands, agriculture, climate change, urbanisation, erratic monsoon, mining and other industrialisation (Neves and Matias, 2008; Cidu et al., 2009; Bhuiyan et al., 2010; Chandra et al., 2015; Kong et al., 2016; Tiwari et al., 2016; Tiwari et al., 2017; Sidhu et al., 2021). Excessive depletion of groundwater resources affects major regions of South and Central Asia, North Africa, North China, Australia, Middle East and local areas across the globe (Konikow and Kendy, 2005)

In India, agricultural activities depend on groundwater resources and irrigate around 50% of the total irrigation area (Central Water Commission, 2000). Moreover, Shah et al. (2000) reported that groundwater wells fulfil the demands of around 60% of irrigated food production. Overexploitation of groundwater resources for different purposes is causing groundwater level depletion and reduction in supply in several regions of the country (Singh and Singh, 2002). Declining groundwater level (1–2 m/year) is an alarming and challenging issue in several parts of India (Singh and Singh, 2002). Moreover, overexploitation of groundwater resources also causes several other

 DOI: 10.1201/9781003442554-14

problems, such as reducing river base flow, saline water intrusion, diminishing wetlands, degrading water quality, reducing soil moisture and local subsidence (Singh and Singh, 2002, Custodio, 2002). Sidhu et al. (2021) worked on an assessment of groundwater depletion in the Punjab state of India and reported that the groundwater level is rapidly declining in the state due to the overpumping of groundwater resources for the irrigation of rice and wheat cropping systems. Moreover, they have suggested critical remedial measures such as soil mulching, micro-irrigation, deficit irrigation, conservation tillage, cultivation of drought-resistant crop varieties and crop diversification to maintain the groundwater level in Punjab.

Groundwater level monitoring is helpful in assessing changes in the quality and quantity of groundwater, long-term water level trends and variations understanding, groundwater sustainable development, which ensures availability for present and future generations and reduces the mitigating risks to aquifers and others. Hence, regular groundwater level monitoring studies are essential for sustainable groundwater resource management in any area. Therefore, this study aims to assess groundwater level behaviour in Umaria Coalfield, Madhya Pradesh, India.

12.2 STUDY AREA

Umaria is a district of Madhya Pradesh, India, situated in the eastern region of Madhya Pradesh (Figure 12.1). It shares borders with Satna district to the north, Dindori district to the south, Shahdol to the east and Katni district to the west (CGWB, 2013). The district has four seasons, i.e., summer (March to mid-June), cold (December–February), monsoon (mid-June to September) and transition period

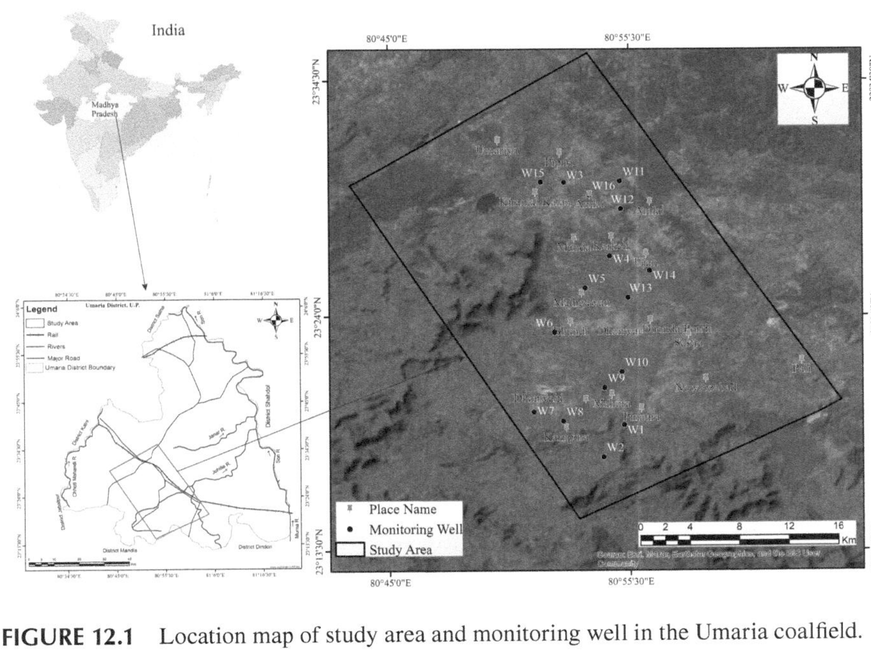

FIGURE 12.1 Location map of study area and monitoring well in the Umaria coalfield.

(October and November) (CGWB, 2013). The average total annual rainfall in the district is around 1248.8 mm. The month of August receives the highest amount of rainfall, accounting for approximately 30% of the total, while from October to May, only 13% of the total rainfall occurs (CGWB, 2013). In the Umaria district, the Johilla River holds the utmost importance as a tributary of the Son River. The Johilla River originates from the Amarkantak plateau in the Maikhal ranges, situated in the neighbouring Anuppur district. The district exhibits a diverse landform, which includes hills, plains and plateaus, characterised by rugged topography (CGWB, 2013). The study area lies in the south Rewa Gondwana basin and mainly consists of the Supra Barakar and Barkar formations, Talchir and Metamorphics formations (Srivastava and Anand-Prakash, 1984).

12.3 METHODOLOGY

In the Umaria coalfield, a total of 16 wells were selected to monitor the groundwater level during the post-monsoon and pre-monsoon seasons. The post-monsoon groundwater level data were collected during the month of November in the year 2022, while the pre-monsoon groundwater level data were collected in the month of June in the year 2023, respectively (Figure 12.1). A sensor-based water level recorder was used to monitor the groundwater level information during both seasons. Geographic information system (GIS) software was used to create water-level contours and the water level fluctuation (WLF) map (Figure 12.2).

12.4 RESULTS AND DISCUSSION

Depth to groundwater level monitoring plays a vital role in understanding the water's availability and aquifers' health, forecasting droughts or contamination issues and managing sustainable water resources. Moreover, groundwater level monitoring studies can be critical for assessing the long-term effects of mining on groundwater quality and quantity. In this study, 16 wells were monitored to determine groundwater depth during post- and pre-monsoon seasons and the trend of water level fluctuation in the area. The statistical analysis of groundwater level monitoring data during both seasons is provided in Table 12.1.

12.4.1 Groundwater Level in Pre-monsoon

During the pre-monsoon season, depth to water level varies from a minimum of 1.9 m below ground level (mbgl) to a maximum of 11.8 mbgl in the study area. The groundwater level of the study area has an average value of 6.7 mg during pre-monsoon (Table 12.1). Only one well (i.e. W3 from Pipria village) out of a total of 16 wells has a water table below 2 mbgl in the study area (Figure 12.3). However, only two wells (i.e. W2 and W8) from Kranpura village have a water level range from 2.1 to 4.0 mbgl (Figure 12.3). Furthermore, five wells (i.e. W1, W4, W6, W7 and W9) from Pinura, Munda Karkeli, Bhudhi, Dhanwahi and Mahura villages have water levels ranging from 4.1 to 6.0 mbgl and two wells (i.e. W5 and W6) from Majhgawa and Amha villages have water levels between 6.1 and 8.0 mbgl in the study area,

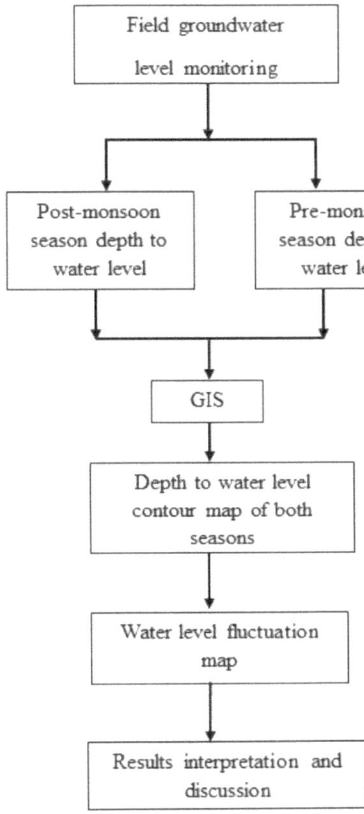

FIGURE 12.2 Methodology adopted for carrying out the research in the Umaria Coalfield.

TABLE 12.1
Statistical Analysis of Groundwater Level Monitoring Data During Both Seasons

Statistical Analysis	Pre-monsoon Water Level (mbgl) ($n=16$)	Post-monsoon Water Level (mbgl) ($n=16$)
Minimum	1.9	1.3
Maximum	11.8	9.7
Average	6.7	4.5
Standard deviation	3.2	2.7

respectively (Figure 12.3). However, six wells (i.e. W10, W11, W12, W13, W14 and W15) from the villages of Mahura, Amlai, Ujan and Kirantal Kalan have water levels greater than 8 mgbl in the study area (Figure 12.3). In the study area, well 13 has the maximum water level, and well 03 has the minimum water level during the pre-monsoon season.

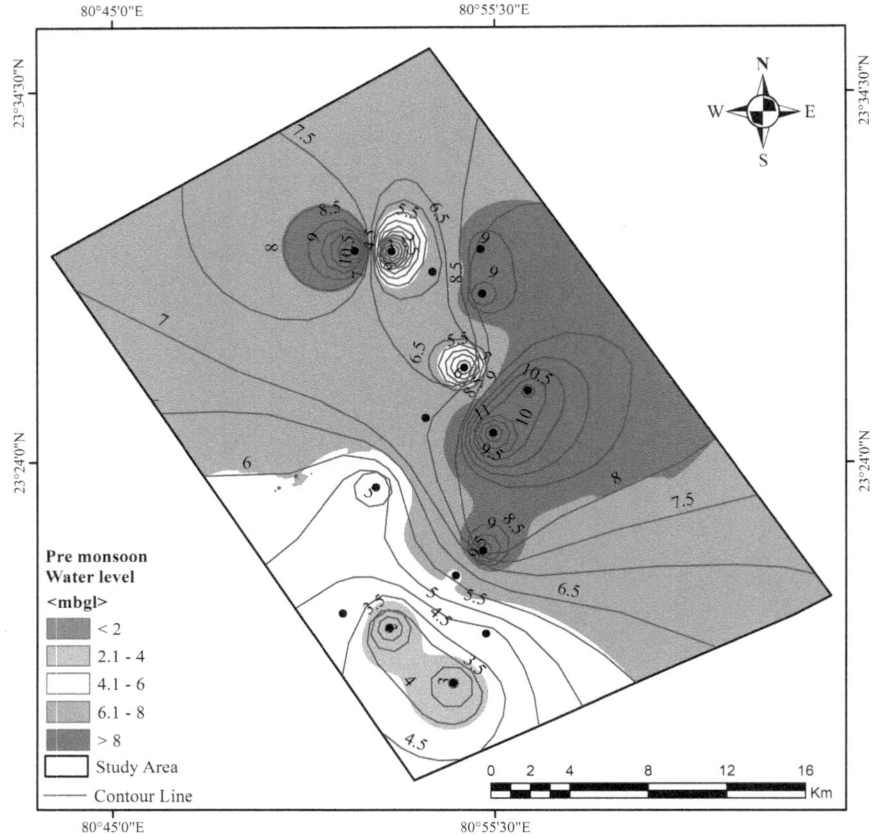

FIGURE 12.3 Depth to water level map in pre-monsoon in the Umaria Coalfield

12.4.2 Groundwater Level in Post-monsoon

During the post-monsoon season, depth to water level varies from a minimum of 1.3 m below ground level (mbgl) to a maximum of 9.7 mbgl in the study area. The groundwater level of the study area has an average value of 4.5 mgbl during post-monsoon (Table 12.1). During the post-monsoon, only four wells (i.e. W3, W4, W2 and W8) out of a total of 16 wells have a water table below 2 mbgl in the study area (Figure 12.4). These four wells are from the Pipria, Mund Kareli and Karnpura villages. However, five wells (i.e. W1, W6, W7, W9 and W16) have a water level range from 2.1 to 4.0 mbgl, and these wells are from the villages of Pinaura, Bhundhi, Dhanwahi and Mahura in the study area (Figure 12.4). Furthermore, two wells (i.e. W5 and W150) from Majhgawa and Kirantal Kalan villages have water levels ranging from 4.1 to 6.0 mbgl, and three wells (i.e. W11, W12 and W13) from close to Amlai and Ujan villages have water levels between 6.1 and 8.0 mbgl in the study area, respectively. However, two wells (i.e. W10 and W14) from the villages of Mahura and Ujan have water levels greater than 8 mgbl in the study area (Figure 12.4). In the study area,

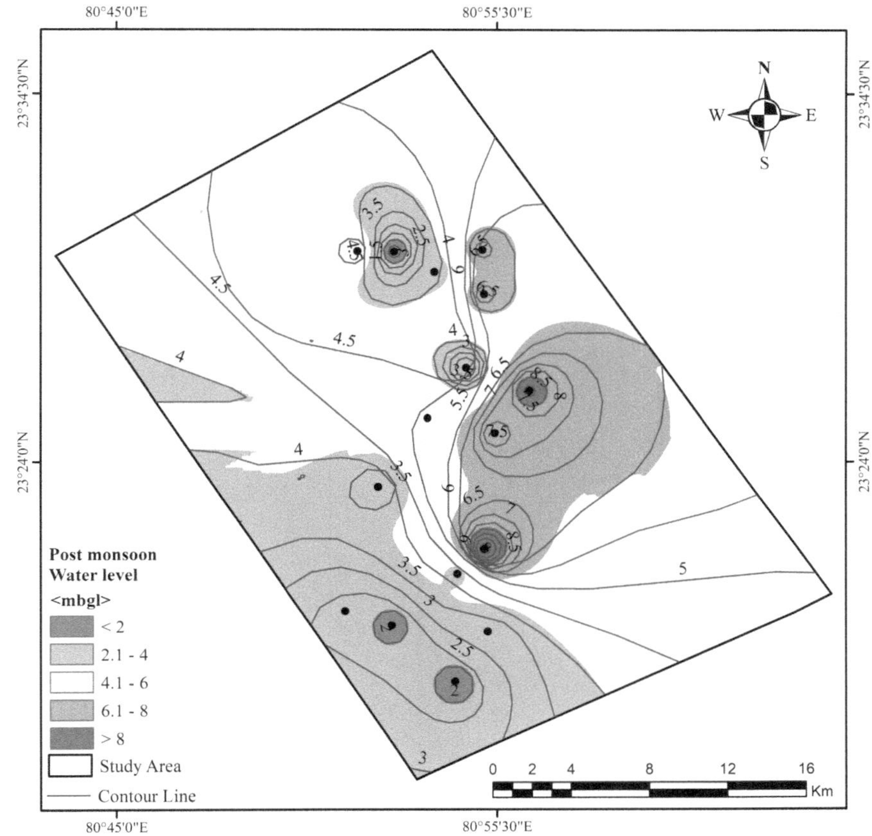

FIGURE 12.4 Depth to water level map in post-monsoon in the Umaria Coalfield.

well 10 has the maximum water level, and well 03 has the minimum water level during the post-monsoon season.

12.4.3 Water Level Fluctuation

The WLF in the study area varies from a minimum of 0.1 mbgl to a maximum of 6.3 mbgl with an average value of 2.2 mbgl in the study area (Figure 12.5). We have found a very low WLF (< 1 mbgl) to low WLF (1.1–2 mgbl) in wells located in the Dhanwahi, Karnpura, Mahura and Pinaura villages in the southern region of the study area. Moreover, one well (W3) of the Pipria has very low WLF in the northern part of the study area (Figure 12.5). However, we have found that one well (W13) from the Ujan village area and one (W15) from the Kirantal Kalan village area have a very high WLF in the study area, respectively (Figure 12.5). Moreover, the rest of the wells have medium WLF in the study area (Figure 12.5). The WLF is dependent on various factors. Subhash et al. (2015) suggested that the elevation, geological structures and soil types of Dhanbad district of Jharkhand were the most significant

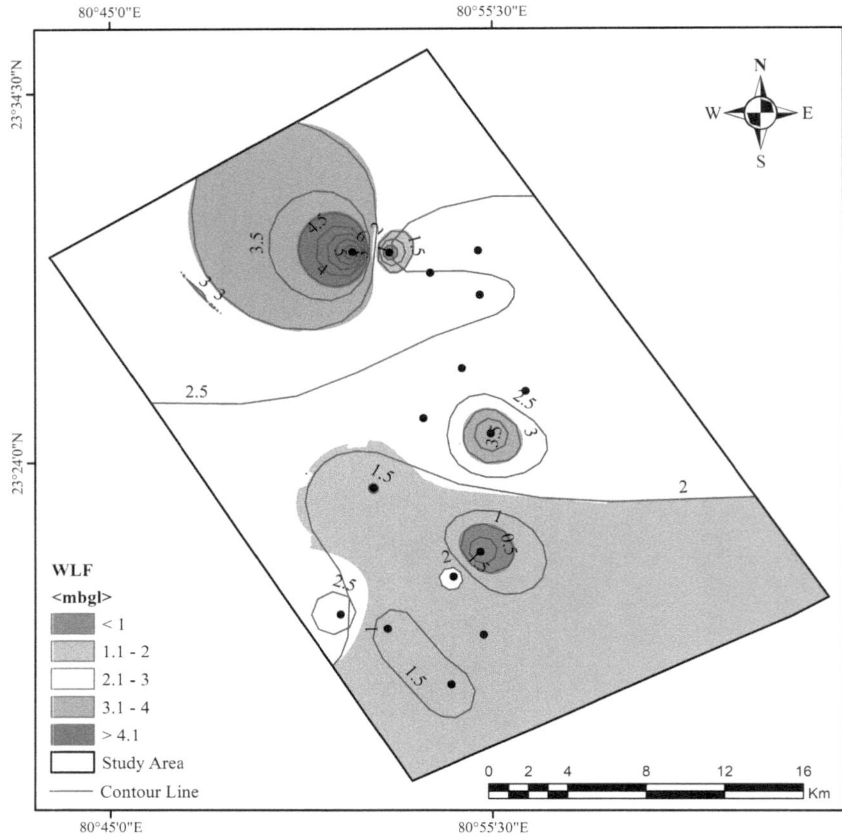

FIGURE 12.5 Groundwater level fluctuation map in the Umaria Coalfield.

factors that affect the water level fluctuation in the district. Similar factors mainly influenced the WLF in the West Bokaro Coalfield area (Tiwari et al., 2016). observation has been reported by Tiwari et al. (2016). Moreover, Tiwari et al. (2016) reported that the mining activities and related industries were also affecting the groundwater level of the study. Furthermore, Singh et al. (2018) reported that both natural and mining and associated activities affected the WLF in the East Bokaro, West Bokaro and Jharia coalfields.

12.5 CONCLUSION

Monitoring of groundwater level depth plays a vital role in understanding the water's availability and aquifers' health, forecasting droughts or contamination issues and managing sustainable water resources for the area. Furthermore, it can be important to assess the long-term effects of mining on groundwater quality and quantity. The depth to groundwater level ranges from 1.9 to 11.8 mbgl with an average value of 6.7 mbgl in the pre-monsoon season while varies from 1.3 to 9.7 mbgl with an average value of 4.5 mbgl during post-monsoon season in the study area. The WLF in the study area varies from 0.1 to 6.3 mbgl, with an average value of 2.2 mbgl. Wells from the Dhanwahi, Karnpura, Mahura and Pinaura villages in the southern region of the study area have low WLF. In contrast, wells from the Ujan and Kirantal Kalan village areas have a very high WLF in the study. GIS-based maps could be helpful for policymakers in making decisions about groundwater resource management in the mining region. Groundwater level fluctuation may affect natural and anthropogenic activities in the study. This study may be helpful in the management of water resources in the mining areas.

ACKNOWLEDGEMENTS

The authors thank the Vice-Chancellor of JNU, New Delhi, for the research facilities. AKT thanks the SERB for the financial support through the SRG Project (SRG/2022/000355).

REFERENCES

Bhuiyan, C. (2010). Hydrogeological factors: their association and relationship with seasonal water-table fluctuation in the composite hardrock Aravalli terrain, India. *Environmental Earth Sciences*, 60, 733–748.

Central Water Commission (2000) *Water and Related Statistics*. New Delhi: CWC.

CGWB (2013) *District Ground Water Information Booklet, Umaria District, Madhya Pradesh, Published by Ministry of Water Resources*. Bhopal: Central Ground Water Board, North Central Region, Government of India.

Chandra, S., Singh, P. K., Tiwari, A. K., Panigrahy, B. P., & Kumar, A. (2015). Evaluation of hydrogeological factors and their relationship with seasonal water table fluctuation in Dhanbad district, Jharkhand, India. *ISH Journal of Hydraulic Engineering*, 21(2), 193–206.

Cidu, R., Biddau, R., & Fanfani, L. (2009). Impact of past mining activity on the quality of groundwater in SW Sardinia (Italy). *Journal of Geochemical Exploration*, 100(2–3), 125–132.

Custodio, E., 2002. Aquifer overexploitation: what does it mean? *Hydrogeology Journal*, 10(2), 254–277.

Kong, X., Zhang, X., Lal, R., Zhang, F., Chen, X., Niu, Z., ... & Song, W. (2016). Groundwater depletion by agricultural intensification in China's HHH plains, since 1980s. *Advances in Agronomy*, 135, 59–106.

Konikow, L. F., & Kendy, E. (2005). Groundwater depletion: A global problem. *Hydrogeology journal*, 13, 317–320.

Neves, O., & Matias, M. J. (2008). Assessment of groundwater quality and contamination problems ascribed to an abandoned uranium mine (Cunha Baixa region, *Central Portugal). Environmental Geology*, 53, 1799–1810.

Shah, T., Molden, D., Sakthivadivel, R. & Seckler, D. (2000) *The Global Groundwater Situation: Overview of Opportunity and Challenges*. Colombo: International Water Management Institute.

Sidhu, B. S., Sharda, R., & Singh, S. (2021). Spatio-temporal assessment of groundwater depletion in Punjab, India. *Groundwater for Sustainable Development*, 12, 100498.

Siebert, S., Burke, J., Faures, J. M., Frenken, K., Hoogeveen, J., Döll, P., & Portmann, F. T. (2010). Groundwater use for irrigation–a global inventory. *Hydrology and Earth System Sciences*, 14(10), 1863–1880.

Singh, D. K., & Singh, A. K. (2002). Groundwater situation in India: Problems and perspective. *International Journal of Water Resources Development*, 18(4), 563–580.

Singh, P. K., Tiwari, A. K., & Verma, P. (2018). Evaluation of water level behavior in coal-mining area, adjacent township, and district areas of Jharkhand State, India. In: Aha, D., Marwaha, S., Mukherjee, A., (Eds), *Clean and Sustainable Groundwater in India*, Springer, pp. 261–278.

Srivastava, S. C., & Prakash, A. (1984). Palynological succession of the lower Gondwana sediments in Umaria coalfield, Madhya Pradesh, India. *Journal of Palaeosciences*, 32(1–3), 26–34.

Tiwari, A. K., Nota, N., Marchionatti, F., & De Maio, M. (2017). Groundwater-level risk assessment by using statistical and geographic information system (GIS) techniques: a case study in the Aosta Valley region, Italy. *Geomatics, Natural Hazards and Risk*, 8(2), 1396–1406.

Tiwari, A. K., Singh, P. K., Chandra, S., & Ghosh, A. (2016). Assessment of groundwater level fluctuation by using remote sensing and GIS in West Bokaro coalfield, Jharkhand, India. *ISH Journal of Hydraulic Engineering*, 22(1), 59–67.

Index

Note: **Bold** page numbers refer to tables and *italic* page numbers refer to figures.

FSC
www.fsc.org
MIX
Papier | Fördert gute Waldnutzung
FSC® C083411

Kolibri 360 GmbH
Ferdinand-Jühlke-Straße 7
99095 Erfurt, Deutschland
produktsicherheit@kolibri360.de